Technical Indicators and Safety Design of Freeway in High Altitude Area

Jianbei Liu

Technical Indicators and Safety Design of Freeway in High Altitude Area

SSTP

Springer

Jianbei Liu
CCCC First Highway Consultants Co., Ltd.
Xi'an, China

ISBN 978-981-99-0619-2 ISBN 978-981-99-0620-8 (eBook)
https://doi.org/10.1007/978-981-99-0620-8

Jointly published with Shanghai Scientific and Technical Publishers, Shanghai, China
The print edition is not for sale in China (Mainland). Customers from China (Mainland) please order the print book from: Shanghai Scientific and Technical Publishers.
ISBN of the Co-Publisher's edition: 978-754-78-6179-0

This Springer imprint is published by the registered company Springer Nature Singapore Pte Ltd.
The registered company address is: 152 Beach Road, #21-01/04 Gateway East, Singapore 189721, Singapore

Foreword

Permafrost is a "roadblock" to road construction in High Altitude Area. In the past 60 years since the Qinghai-Tibet Highway was opened to traffic in 1954, along with the occurrence and development of different forms of permafrost engineering distresses, Chinese scientists have gradually gained a deeper understanding of the physical and mechanical properties of permafrost and a more systematic knowledge of the complexity of permafrost engineering. Since the Qinghai-Tibet Railway was completed and opened to traffic in 2006, global warming and permafrost degradation have caused engineering distresses such as railway subgrade subsidence and cracking. Over the past decades, the practice of many key national permafrost projects has fully proved that the scientific and technological progress in the field of permafrost engineering will be a long-term process of spiral development.

On the road to explore the construction technology of road engineering in permafrost regions, Chinese scientific and technological workers have never stopped. From the 1970s to the end of the 1990s, they formed the permafrost engineering research methods and testing techniques based on the successive renovation and reconstruction of the Qinghai-Tibet Highway, gradually laid the foundation of China's permafrost engineering research, and established the distress mechanism analysis, distress treatment technology and theoretical system of highway permafrost engineering in China. At the beginning of the twenty-first century, through the engineering practice and system integration of the Qinghai-Tibet Railway, the theoretical exploration and technical design of "cooling subgrade" were further integrated into the research on permafrost engineering, and a large number of research results with international advanced level were achieved. In 2011, in order to start the reconstruction work of Yushu after the earthquake, China decided to build the Gonghe to Yushu Freeway in Qinghai Province, which once again set off the climax of permafrost engineering research.

Compared to Qinghai-Tibet Railway and Qinghai-Tibet Highway of Class II level, the construction of large scale, high standard, and heavy load freeway on permafrost foundation faces various problems such as engineering scale effect, closed heat storage effect of thick pavement structure in large section and strong heat absorption effect of black pavement, which may lead to greater engineering risks. Theoretical innovation and technical breakthroughs must be made in the construction of freeways in permafrost areas.

It is gratifying to see that the series of books on "Key Technologies of Freeway Construction in High Altitude and Cold Regions" has shown us a series of important achievements obtained by Chinese scientific researchers, which include subgrade, pavement, bridges, tunnels, environmental protection, monitoring and early warning, and other professional contents, and has created the series represents the latest achievements in road permafrost engineering research in China and the world. The editor-in-chief of the series has more than 40 years of experience in scientific research and design of road engineering in permafrost regions, and has a high-end research and development platform of the State Key Laboratory of Road Engineering Safety and Health in High Altitude Areas. The editorial team includes design masters and renowned experts in the field of highway permafrost engineering in China, as well as young talents who have been conducting special research for a long time. Their profound technical accumulation, theoretical background, and rich practical experience have played an important role in guaranteeing the academic and technical level of the series.

Since September 2013, when General Secretary Xi Jinping first proposed the initiative of jointly building the Silk Road Economic Belt, the Belt and Road Initiative has become a national strategy to deepen China's reform and opening up, to realize the Chinese dream, to achieve common development in the world and to build a community of human destiny, and the basis for realizing these great strategic ideas is transported. The "Land Silk Road Economic Belt" is the core corridor for the interconnection of Asia, Europe, and Africa, spanning from east to west across the high altitude and areas including Qinghai-Tibet Plateau, the Karakorum Mountains, the Pamir Plateau, Siberia, and other high latitude and cold regions in the northern hemisphere, and involving 12,000km of principal arterial roads. I believe the publication of the series will play an important role in ensuring the construction of large-scale road projects through High Altitude Areas, supporting the transportation industry to seize the development opportunities of "One Belt, One Road", and helping China's "standards and technologies to go abroad".

February 2019

Jianlong Zheng
Academician of the Chinese Academy
of Engineering
Xi'an, China

Preface

Tibet is the gateway to the southwest of our great motherland and a strategic place for national defense. After 40 years of hard work in reform and opening up, Tibet's transportation has undergone earth-shaking changes, which has played a major role in regional economic development, social progress, improvement of people's living standards, and consolidation of the southwest frontier of the motherland. However, as Tibet is increasingly connected with other regions and constrained by the objective natural conditions of the region and the related construction technology bottlenecks, the existing highway transportation system can no longer meet people's needs for rapid transportation and material transportation. By the end of 2015, China's highway mileage was 142,600 km, ranking the first in the world, and Tibet is the only province in China that has not been connected with the external highways, it has become an "island" in the national highway network. How to promote the modernization of transportation in Tibet and the prosperity and stability of regional economic development has always been a concern for the Party, the country, and even the people of the whole country.

Influenced by the special geographical location, low-pressure, and oxygen-deficient environment, harsh natural conditions, complex geological conditions, and fragile ecological environment of the Qinghai-Tibet Plateau, the construction of freeways in high-altitude areas in Tibet will face a series of difficulties and challenges. From a worldwide perspective, there are few mature experiences to learn from, making it far more difficult to build freeways on the Qinghai-Tibet Plateau than in plain areas. While solving the worldwide problem of permafrost, we are also facing a series of unsolved technical problems: the average altitude along the Qinghai-Tibet Highway is more than 4,000 m, and the Tanggula Mountain Pass even reaches 5,231 m. In the low-pressure and oxygen-deficient environment, the attenuation of vehicle comprehensive performance makes the value of longitudinal slope in high-altitude areas necessarily different from that in plain areas; high-altitude oxygen-deficient environment will cause altitude sickness and slow driving behavior, and long-distance continuous driving is likely to cause fatigued driving. During the arrangement process of service facilities, factors such as altitude sickness and fatigued driving should be fully considered; long-term snow freezing has seriously affected the driving safety of freeways, such as the sharp decline of driving stability, which must be considered in terms of traffic safety facilities. In order to overcome the above technical problems and ensure highway traffic safety in high-altitude areas, it is urgent to carry out forward-looking research and technical breakthrough.

According to the series of research projects of the national science and technology support plan project "Highway construction technology in high-altitude areas", this book mainly focuses on the changes in the comprehensive performance of typical vehicles entering Tibet under the conditions of low-pressure and oxygen-deficient environment, as well as the changes of driving psychology and physiology and driving behavior on highways in plateau areas Research on traffic safety assurance technology, integration of key technologies for highway safety design under the special plateau environment, and development of project demonstration and application, to provide basic test, research and project demonstration research results for ensuring highway traffic safety in plateau areas and realizing rapid development of highways in Tibet, which have been summarized, collated and compiled.

This book is divided into seven chapters. Chapter 1 introduces the main technical problems faced by the selection of technical indicators and geometric design of freeway construction in the Qinghai-Tibet Plateau areas. From the perspective of ensuring the safety of highway system operation, it studies the safety-related contents of vehicles entering Tibet, drivers and passengers, highway conditions, and environment; Chap. 2 introduces the traffic operation characteristics in high-altitude areas; Chap. 3 studies the dynamic characteristics of typical vehicles in the low-pressure and oxygen-deficient environment at high-altitude; Chap. 4 studies the characteristics of drivers' psychological and physiological changes in the low-pressure and oxygen-deficient environment at high-altitude; Chap. 5 studies the freeway operating speed model in high-altitude areas; Chap. 6 studies the highway route safety design technology in high-altitude areas; Chap. 7 introduces the application and demonstration of relevant technologies.

The book was presided over and written by Liu Jianbei. The specific compilation division is as follows: Chaps. 1 and 2 were written by Liu Jianbei, Wang Shuangjie, and He Yulong; Chap. 3 was written by Liu Jianbei and Guo Zhongyin; Chap. 4 was written by Zhang Zhiwei and Ma Xiaolong; Chap. 5 was written by Liu Jianbei and Gao Jinsheng; Chap. 6 was written by Liu Jianbei, He Yulong and Liu Benmin; Chap. 7 was written by Liu Jianbei and Shi Heng. Thank Deng Hanyue, Zhu Lingjie, Zhang Yanning, Feng Bingbing, Zhang Yuan, Li Dapeng, Zhang Meimei, Zhang Lufei, Hou Yangyang, etc., for their research and support for this book.

Limited to the level, it is inevitable that there are inappropriate points in the book. You are welcome to criticize and correct.

Xi'an, China Jianbei Liu
February 2019

Summary

Based on the special geographical location and low pressure environmental conditions in the high-altitude area of the Qinghai Tibet Plateau, taking full account of the attenuation of vehicle comprehensive performance under the low pressure and oxygen deficient environment, the sharp decline of driving stability in the situation with lasting snowing and freezing, as well as the high possibility to cause the drivers' altitude sickness and fatigued driving, this book comprehensively and systematically introduces the prediction model of freeway operating speed in the high altitude regions under the low pressure and oxygen deficient environment, the key technologies for speed control and safety design of freeway, the key technical indicators and value selection standards for geometric design of freeway, the allocation standards for service facilities and other research results. For demonstration and application, it takes the Huashixia-Dawu Highway as an example, and provides the improvement proposals on the highway alignment design and speed control facilities.

The main readers of this book are engineering technicians, scientific researchers, equipment developers, and technical managers in the fields of highway design, transportation, traffic safety management and other professional fields, as well as teachers and students in colleges and universities majored in the fields.

Contents

Chapter 1
Introduction

1.1 Background

Tibet Autonomous Region, regarded as "the last pure land on the earth", is the dream heaven of people around the world because of its unique geographic location and climate, pristine ecological environment as well as its mysterious Tibetan religion and culture. Meanwhile, as the country's southwestern portal, its development, prosperity and steadiness are highly concerned by the Party, government and people from all nationalities. While highway traffic system is always the key element of regional economy development. But due to its nature conditions and related techniques' bottleneck, the development of Tibet's highway traffic system is restricted. China owns 142,600 km freeway by 2015, which ranks No. 1 worldwide. Even so, the freeway traffic system of Tibet grows so slow that there is not one qualified freeway that capable of withstanding nature disaster, offering all weather service and keeping close connection with other places. To build the Qinghai—Tibet Freeway, a part of the national road G6 (Beijing to Lasha Freeway), is included in the *Plan of National Highway* issued in June 2013. Along with the moving forward of the Belt and Road Initiative, President Xi states precisely on the Sixth Central Symposium on Tibet Work that we need to fasten the construction of Tibet's comprehensive transport system and make Tibet the important channel of China to South Asia.

By the end of 2016, the Beijing Golmud section of Beijing-Tibet Freeway had been completed, but Golmud Lhasa section had not yet started. The Golmud Lhasa Freeway connects Qinghai Province and the Tibet Autonomous Region. Its construction will change the situation of no freeway to outside in Tibet's history, shorten the distance between Tibet and other regions, arrange close connection between Tibet and other cities, and will prompt the economy development and social welfare improvement in Tibet and Qinhai province at the same time. The Golmud Lhasa section of the Qinghai—Tibet Highway is about 1138 km long, of which 528 km passes through the plateau permafrost region with extremely harsh environment and complex and changeable geological conditions. Since the construction of the Qinghai—Tibet Highway in the permafrost region of the Qinghai—Tibet Plateau in

J. Liu, *Technical Indicators and Safety Design of Freeway in High Altitude Area*,
https://doi.org/10.1007/978-981-99-0620-8_1

1953, Chinese researchers and technicians in permafrost engineering have continuously promoted the development and progress of road construction technology in permafrost regions through long-term research and exploration for more than 60 years, ensuring the healthy operation of the Qinghai—Tibet Highway, and laying a solid technical foundation for the construction of the Qinghai—Tibet Freeway.

However, after determining the corridor for Qinghai-Tibet Freeway construction, the first important technical problem to overcome is the selection of technical standards. The freeway is designed for high-speed running of vehicles. Affected by the special terrain, geology, environment and other factors in the Qinghai—Tibet Plateau, the construction of the freeway in the Qinghai—Tibet Plateau will face a series of difficulties and challenges. While solving the worldwide problem of permafrost, it is necessary to carry out forward-looking research and technical breakthrough on how to overcome the key problems that seriously affect the driving safety of freeway, such as the attenuation of vehicle's comprehensive performance in the low pressure and oxygen deficient environment, the altitude sickness caused by high altitude, the slow driving behavior, and the sharp decline of driving stability caused by long-term snow freezing, so as to ensure the traffic safety of freeway in high altitude areas. At present, there is still a blank in the research of related fields all over the world.

The Research Project of the National Science and Technology support program of "Technologies of Freeway Construction in High Altitude and Cold Areas" specially established the research program of "Research on Freeway Safety Design Technology under the Special Environment of the Qinghai-Tibet Plateau". The aim was to reveal the high-altitude comprehensive performance changes of large typical freight vehicles, summarize the psychophysiological changes and driving behavior changes of highway driving in the plateau area, and research and develop new safety protection facilities suitable for ice and snow environment. Accordingly, researches, such as the parameters of freeway geometry indicators in high-altitude areas, principles and methods for selecting technical standards, and traffic safety assurance technologies., will be carried out to integrate key technologies for freeway safety design in plateau special environment, and work on project demonstration and application, in order to provide the basic experiments, research and project demonstration support to ensure traffic safety of freeways in plateau areas, as well as realize the rapid development of freeways in Tibet.

Due to the special geographical environment characteristics of the Qinghai—Tibet Plateau, the freeway operation in this region will face huge traffic safety challenges in the future. The construction of Qinghai—Tibet Plateau Freeway mainly faces the following technical problems of traffic safety:

(1) **There are Few Research Results of Highway Design Technology in High-altitude Areas All Over the World**

There are many research contents on highway alignment design indicators worldwide, but few studies on highway technical standards, geometric alignment indicators and parameters are carried out according to the characteristics of plateau low pressure and oxygen deficient environment, especially for freeways. Some Chinese scholars have made some research achievements on the Qinghai—Tibet Highway. However, as the Qinghai—Tibet Highway is a Class II highway, whether the basis for selecting technical standards, determining design indicators and parameters is suitable for the freeway needs further research. In the aspect of longitudinal slope design, although certain achievements have been made in the research of highway profile in plateau areas in China, many problems still remain in the experimental stage, and there is no systematic and in-depth research on the maximum grade, the relationship between grade reduction and driving behavior in plateau high-altitude areas. In terms of interval arrangement of service areas, qualitative methods are mainly used at present, focusing on the physiological needs of passengers and vehicle operation. In view of the low pressure and oxygen deficient environment on the plateau, it is also necessary to consider driving fatigue, human physiological needs, safety requirements for driving and medical assistance, and further study the reasonable spacing of service facilities on the Qinghai—Tibet Plateau Freeway in combination with the geological conditions along the line, traffic flow and domestic and foreign experience in the layout of service facilities. In terms of superelevation value, different countries have certain differences, and the selection of the maximum superelevation value should depend on the actual situation.

Therefore, the technical indicators of highway design in high altitude areas must be further systematically studied in combination with the actual environmental characteristics.

(2) **The Special Geographical Environment of the Plateau and High Altitude has A Huge Impact on the Safety of People and Vehicles**

The low oxygen content in the air on the Qinghai-Tibet Plateau will have a great impact on both vehicles and personnel.

(1) Impact on Vehicles

At high altitude, the oxygen content in the air is only half of that in the plains. The thin air will lead to low oxygen content in the air intake of the vehicle engine and affect the performance of the vehicle. The working process will be: the engine fuel mixture combustion is not complete, in cold state lubricating oil viscosity increases, engine running resistance increases, power loss increases while output reduces, and the coolant boiling point is lower than that in the plains area; thus, the vehicle appears "open pot", i.e. the engine get high temperature, and the vehicle can not work normally. In spring, autumn and winter, when vehicles run in the flat road and downhill road, the engine temperature drops fast, which affecting the enhancement of engine power.

Altitude increasing and vehicle performance reducing also affect the safety of vehicle operation. Especially for trucks with full load, the climbing ability decrease significantly in the area, which will lead to the reduction of vehicle running speed, enlarge running speed difference between passenger cars and trucks and then cause traffic accidents.

(2) Impact on Personnel

In the high altitude environment, the oxygen content is significantly reduced. Hypoxia can cause headache, chest tightness, shortness of breath, palpitations, nausea and vomiting, cyanosis of lips, insomnia, dreaminess, and blood pressure may also increase. A few people may gradually become more severe due to fatigue, cold, upper respiratory tract infection and other reasons, and develop into high altitude pulmonary edema or high altitude brain edema. High altitude pulmonary edema and brain edema have a rapid onset and high mortality. After people in the plain area enter the plateau, the general incidence rate is 30~40%. The proportion of brain edema and pulmonary edema in the population with altitude reaction is 10~20%. At an altitude of 5100 m, 90% of drivers will experience altitude reaction. In general, when reaching the altitude above 3000 m, due to the reduction of oxygen content, there will be a certain degree of altitude sickness, and with the rise of altitude, the risk of altitude sickness increases rapidly.

Affected by the oxygen deficient environment in plateau, drivers are more likely to experience driving fatigue and improper operation behavior. It mainly shows that the attention span, attention transfer, short-term memory and complex thinking judgment are reduced, which leads to the decline of the driver's driving ability and seriously affects the driving safety. With the increase of altitude, the driver's reaction time increases, and the driving agility decreases. Fatigue, misoperation and other behaviors have a great impact on driving safety.

Therefore, it is a necessary guiding factor for safety design to study and master the influence of high altitude areas on vehicle dynamics and braking performance, and understand the driver's psychological and physiological behavior.

(3) **Facing the Unique Natural Climate Conditions of the Qinghai - Tibet Plateau, It Is Necessary to Update and Improve the Design Methods and Technologies of Freeways**

China is a vast country with significant regional differences. There are some differences in the implementation and use of current technical standards and specifications such as *Technical Standards for Highway Engineering* and *Specifications for Design of Highway Routes* nationwide. Influenced by the special geographical location, harsh natural environment and other factors in Tibet, the design methods and key technologies of freeways in high altitude areas need to be updated and expanded.

In view of the reduction of vehicle performance in high altitude areas and the easy overheating of the engine due to the reduction of heat dissipation capacity, when the altitude exceeds 3000 m, the longitudinal slope shall be reduced in the highway geometric design. At present, vehicles in high altitude areas are generally equipped with turbocharging systems. The proportion of heavy trucks in the traffic along the line and the external characteristics of their engines are quite different from those in inland China; The minimum radius of horizontal curve, curve length and other indicators will change greatly considering the weakening condition of pavement friction capacity for the highway in the plateau region which is cold and easy to freeze; In the layout of facilities in freeway service areas in China, there is a lack of quantitative analysis and scale feasibility analysis. In general, it focuses on the physiological needs of passengers and the limit of vehicle operation, while ignoring the impact of factors such as the environment along the freeway, the two-way effect of traffic flow, and the fatigued driving characteristics of drivers. In view of the low pressure and oxygen deficient environment in the plateau, there are few studies on the vehicle refueling demand, human physiological needs, safe driving requirements, combined with the geological conditions along the line, traffic flow, domestic and foreign service facilities layout and reasonable spacing.

In addition, the terrain on the plateau fluctuates gently, with good intervisibility. Highway alignment indicators in Qinghai—Tibet region are generally high, and there is basically no lateral interference of the highway in permafrost regions. Drivers usually expect to pass through the cold and oxygen deficient areas at a high speed. At the same time, most of the roads to Tibet are unmanned areas, and the highways in plateau areas have obvious long-distance operation characteristics, which makes some sections prone to overspeed. It is necessary to develop the design and control technology of highway alignment and speed adaptation matching based on the distribution of operating speed.

Therefore, the construction of Qinghai—Tibet Freeway needs to combine the characteristics of regional natural climate, from the perspective of safety, to meet the driver's comfort and vehicle performance stability as the goal, study and determine the main technical indicators and parameters of Qinghai—Tibet Freeway using threshold, and establish a method system suitable for highway geometric design in plateau areas.

(4) **Frequent and Diverse Adverse Meteorological Conditions Put Forward Higher Requirements for the Freeway's Ability to Actively Resist Disasters and Maintain Traffic**

Rain, snow, ice and freezing disasters occur frequently in high altitude areas. Compared with the inland areas in China, the probability of accidents increases significantly. After the completion of the freeway, the phenomenon of braking sideslip on the ice and snow roads is also very prominent. After the sideslip of vehicles, it often leads to traffic accidents. Ice and snow will become one of the important factors affecting traffic safety. According to the analysis of relevant data, the duration of continuous freezing along the Qinghai—Tibet Highway in Tibet is 20–30 days every year. In high altitude areas and areas with frequent cold air activities, road icing starts early, ends late, and lasts a long time. At the same time, according to the statistics of geological disaster data in the past 50 years, the frequency of earthquakes in Tanggula Mountain Pass is the highest, and the frequency of snow disasters in the section from Kunlun Mountain to Tanggula Mountain is the highest, with an average altitude of more than 4500 m.

In addition, rubber products such as leather pipes, brake master cylinders, leather cups of slave cylinders and tires are easy to crack and break at low temperatures. If hydraulic braking device is adopted, the cylinder cup will shrink excessively, sometimes causing sudden failure of braking. In case of snow or ice on the windshield, it will hinder the driver's vision, which is not conducive to driving safety. How to ensure the traffic safety of regional freeways and improve the ability of highways to actively prevent and resist disasters are all urgent problems to be solved before the construction of freeways on the Qinghai—Tibet Plateau.

Along the Qinghai—Tibet Highway, it is not only be low pressure, cold and oxygen deficient, but also happen natural disasters frequently, mainly as snow disasters and earthquakes. After the completion of the Qinghai—Tibet Freeway, the total mileage is about 1110 km. Along the line network, it is single and sparsely populated. The active disaster resistance and traffic protection of the freeway will face huge challenges. Therefore, in view of the major technical requirements of how to ensure the driving safety under special environment for freeways in the Qinghai—Tibet Plateau, and in combination with the special geographical and climatic characteristics of the plateau, it is necessary to study and master the characteristics and laws of vehicles, drivers and highway environment in the plateau before the construction of the Qinghai—Tibet Plateau Freeway; Determine the key indicators and parameter thresholds of the Qinghai—Tibet Freeway route that meet the environmental requirements and meet the intrinsic safety, and propose the technical indicator system and methods suitable for the Qinghai—Tibet Plateau environment; In the stage of highway construction design, research and construct the safety guarantee plan, and propose the freeway driving safety guarantee technology under special meteorological conditions; It is of great significance to "nip in the bud" from the perspective of improving intrinsic safety and strengthening system active intervention.

1.2 Introduction to Research Status Worldwide

1.2.1 Vehicle Dynamic Characteristics at High Altitude

With the gradual increase of altitude, the Qinghai—Tibet Plateau is characterized by a decrease in atmospheric pressure, air density, oxygen content, ambient temperature, large temperature difference between day and night, long annual low temperature period, dry climate, less precipitation, and large evaporation. The changes of environmental parameters are as follows: the atmospheric pressure drops by about 9%, the air density drops by 6~10%, and the oxygen content drops by about 10% every 1000 m above sea level; The annual average temperature decreases by 5~7 °C for every 1000 m elevation increase. The area with an altitude of more than 4000 m is a fixed cold area, with an annual average temperature of less than −4 °C and a cold period of more than 5 months. In such areas, vehicle engines, especially diesel engines of trucks, will face the following major risks and problems:

i. Supercharger overtemperature and overspeed. With the increase of altitude, the intake air volume of naturally aspirated diesel engine decreases due to the reduction of intake air pressure, and the performance of diesel engine will deteriorate accordingly, resulting in its inability to meet the operating requirements in plateau areas. After the supercharging technology is adopted, this change can be effectively compensated. However, the compensation ability of the supercharging system is mainly limited by the temperature at the turbocharger turbine inlet and the speed of the supercharger. Therefore, it is necessary to prevent the supercharger from overheating and overspeed.
ii. Acceleration of diesel engine speed. As the comprehensive efficiency of the supercharger will decline with the increase of altitude, the lower the speed of the supercharger, the greater the decline of its comprehensive efficiency. This is reflected in the total power characteristics of the diesel engine: the lower the speed of the diesel engine, the greater the reduction of the torque value of the diesel engine, so that the speed of the maximum torque point of the diesel engine increases with the increase of altitude. Generally speaking, when the rotating speed of plateau diesel engine is 800 ~ 1000r/min, the torque value of total power characteristics will decrease by 7 ~ 10% for every 1000 m increase in altitude.
iii. Working heat load of diesel engine. Due to the low air density in the plateau area, the boiling point of water will decrease with the increase of altitude, and the working heat load of diesel engine will increase. Poor cooling effect will bring a series of problems to the engine.

In order to study and evaluate the environmental adaptability of vehicles and power equipment, the United States, Britain, Germany, Japan and other countries began to invest heavily in developing environmental simulation equipment and environmental laboratories as early as the 1950s. For example, the vehicle environment laboratory of the Royal Army Research Institute, the static and dynamic environment test equipment for locomotives and carriages of the Vienna International Vehicle Research and Test Center, and the weapon environment test equipment of the Aberdeen Test Ground in the United States.

In the 1960s, American scholar Fosberry used the "All Climate Laboratory" to conduct high-altitude atmospheric conditions simulation tests on a variety of naturally aspirated diesel engines, mainly to adapt to the commercial needs of internal combustion engine manufacturers and users, and to modify the power of diesel engines. The application of turbocharging technology in vehicle internal combustion engines enables people to find the most effective measures to solve the performance degradation of internal combustion engines at high altitude (low pressure). In 1981, Jebashvili, a Soviet scholar, conducted tests on a turbocharged diesel engine on a simulation test bench. Under full load and rated speed, the atmospheric conditions of 700~3500 m above sea level are simulated. In order to study the possibility of using turbocharging to compensate for the power drop of diesel engines at high altitude, a turbocharging intercooling system is equipped during the test. The general rules obtained from the test further improve the power conversion formula for turbocharged diesel engines recommended by the International Committee on Gas Turbines (CIMAC). Ricardo Company, Cummins Company, Caterpillar Company, etc. have conducted special research on the combustion heat release of diesel engines, conducted special cooling capacity tests on the diesel engine cooling system, converted the thermal energy distribution of modern diesel engines into efficiency distribution, and evaluated the thermal load state of plateau diesel engines. In 1997, Olesiak deduced the brake temperature and wear theoretical equations. In 2004, Choi used the finite element analysis method to study the thermoelastic contact of the disc brake under multiple brake applications, and obtained the pressure field distribution and brake temperature.

Yu Qiang and others in 2000 explored the influence of altitude on engine driving power. With the increase of altitude, the density of air entering the cylinder decreases, and the work consumed in the compression stroke decreases when the engine brakes. When the engine brakes, not only the work consumed in the compression stroke decreases, but also the work consumed in the exhaust process also decreases. In 2001, Li Wenxiang and others carried out vehicle performance tests, road tests on CA150PL2G1 in Golmud Lhasa section of Qinghai—Tibet Highway, and analyzed its power performance and economy at 3900 m above sea level. In 2006, Li Zhiyu pointed out that when driving in high altitude and low air pressure areas, the power and economy of the engine will decline, the lubricating oil is easy to deteriorate, and the braking performance will become worse, and proposed that the overall layout and structural design of automobile manufacturing should be considered. In 2006, Ye Linbao of Shanghai Jiaotong University and others conducted field bench and road tests on turbocharged diesel engines at altitudes of 20, 2200 and 3800 m. The results showed that placing the maximum torque point in the highest efficiency area of the turbocharger compressor, while the efficiency at the rated power point should be about 66%, can improve the comprehensive performance of diesel engines at plateau and reduce the deterioration of engine performance caused by the increase of altitude; Increasing the compression ratio and starter power is an effective way to ensure the plateau starting performance.

In 2009, Zhang Zhiqiang and others summarized the characteristics of plateau environment and comprehensively and deeply analyzed the impact of plateau environment on vehicle diesel engines. Through the plateau adaptability analysis of special vehicles, it is found that the acceleration characteristics and climbing performance are the main factors restricting the plateau adaptability of vehicles. Especially, the acceleration characteristics and torque characteristic curves of diesel engines are studied, and relevant technical measures are proposed, which provides ideas for further improving the adaptability of diesel engines to plateau environment. In 2009, Li Siding and others briefly analyzed the climate and geographical characteristics of plateau and cold region, focusing on the hazards and impacts of plateau and alpine environment on vehicle performance, use, maintenance and management, aiming at reasonable selection, correct use and scientific management of vehicles, and giving full play to their due effectiveness. In 2010, Chen Yuguang and others put forward a control method of diesel engine intake pressure under variable altitude environment, aiming at the problem of deterioration of vehicle diesel engine's external characteristics when operating in plateau areas. This method dynamically adjusts the pressure ratio of the adjustable pressure increasing system of the secondary turbine by adjusting the flow coefficient of the bypass valve of the high-pressure turbocharger. The simulation results show that the method greatly reduces the influence of atmospheric pressure changes on the external characteristics of diesel engines, and significantly improves the power performance and fuel economy.

In 2012, Shen Jiangwei and others analyzed the sharp decline of vehicle driving performance in high altitude and low pressure areas: poor power and economy, low braking performance, reduced reliability and durability, and easy to boil the engine, which seriously reduced the efficiency of the vehicle. The research shows that when the average altitude rises to more than 4500 m, the engine driving force of vehicles and other mechanical equipment is about 60% of that in plain areas. On the basis of this altitude, when the altitude increases by 1000 m, the engine driving force of vehicles and other mechanical equipment decreases by about 10%. Other altitude sicknesss of automobiles are shown in the following aspects: boiling of cooling water and overheating of engine due to low atmospheric pressure; The braking efficiency of vehicles with pneumatic braking devices will be significantly reduced; Because the gasoline is easy to volatilize, the engine is overheated, and the fuel supply pipeline is often blocked. In 2013, Cheng Yuan and others conducted research on the impact of plateau and alpine environment on vehicle braking performance. By analyzing the relationship between the air pressure, the mass concentration of oxygen in the air and the altitude, as well as the performance of the vehicle's braking performance in the plateau and cold environment, it is concluded that the vehicle's braking performance can be improved by improving the vacuum degree of the vacuum booster and changing the position of the vacuum pipe joint.

To sum up, scholars all over the world have conducted research on power reduction, acceleration characteristics and torque characteristic curves of vehicle diesel engines in high altitude areas, and proposed plans and measures for development and improvement of different types of engines to adapt to plateau environment. Some scholars also explored the main factors that affect vehicle dynamic performance in plateau environment from the perspective of the whole vehicle, such as the effect of effective thermal efficiency and air density on resistance. There are many types of vehicle types and engines targeted in the above research, but the above achievements involve earlier vehicle types. In recent years, with the rapid development of economic construction in Tibet, the composition of vehicles entering Tibet has undergone significant changes. The research vehicle types and engines involved in the above achievements are no longer suitable for the dynamic performance, overall dimensions and other conditions of the dominant vehicle types in the actual operation of highways in Qinghai—Tibet, At the same time, the development trend of dominant models of highway operation in Qinghai—Tibet region has not been considered, and the existing research results cannot be used to provide suggestions and basis for highway geometric design in Qinghai—Tibet region. Therefore, it is necessary to investigate the characteristics of the dominant models of highways in Qinghai—Tibet region at the current stage, and combine the development trend of motor vehicles in China and the region to study the variation rules of vehicle power and braking performance of the dominant models of highways in the region at high altitude.

1.2.2 Psycho-physiological Characteristics of Drivers in High Altitude Areas

(1) **The Characteristics of the Driver's Physiological, Psychological, and Behavior Changes**

Since the 1990s, with the development of new theories of road alignment design, the development of psycho-physiological testing equipment, and the application of medicine, psychology, physiology and ergonomics to the field of transportation, more and more scholars have begun to pay attention to the correlations of the leading technical indicators of roads and the driver's psycho-physiological response.

Traffic Psychology started in 1912 when Harvard University psychologist A. Spoger used psychology to study the causes of tram accidents. In 1972, K. S. Rutley and D. G. W. Mace found that the workload caused by different intersections can be reflected by the psychological indicator of heart rate variation. In 1980, Iwagawa, Furuya Shiro, etc. in Japan used heart rate, blood pressure and other indicators to study the relationship between the road alignment indicator in forest areas and the driver's psychological and physiological response when driving. Around 1988, Crundall, Davide, Underwood and Geoffrey in the U.K. studied the difference in ECG between experienced and non-experienced drivers during a 20-min driving process under different cognitive loads generated by different road types. In 2004 Stuart T. Godley, Thomas J. Triggs and Brian N. Fieldes studied the relationship between lane width and driving load and speed and found that narrow lane width increases driving load and reduces running speed.

The application of EEG and ECG methods to study road traffic problems in China started relatively late. In the early 1990s, Ren Futian first put forward the "New Theory of Road Alignment Design", which proposed that the road alignment design should be based on the traffic needs and psycho-physiological response of users. Furthermore, each element of the route should be designed with a dynamic point of view and strive for harmony. In 2003, Zheng Ke studied the relationship between heart rate, respiration rate and freeway alignment using a dynamic electro-cardio graph, an auxiliary skin galvanometer and a respirator. In 2005, Wang Shuling used a dynamic electrocardiograph to study the relationship between heart rate and the longitudinal profile of a two-lane highway in mountainous areas and suggested that the slope of a two-lane highway in mountainous areas should not exceed 7%. In the mountainous area, when the slope is 3, 4, 5, 6, and 7%, the corresponding extreme values of the slope length are 700, 550, 450, 400 and 400 m, respectively.

Other studies of drivers' driving states have been at regular altitude areas such as plains or mountains. For instance, Zhang Kairan analysed the stress characteristics of drivers with less driving experience in two typical situations based on heart rate and heart rate increment tests. Yao Na et al. monitored the changes in heart rate, running speed, and the driver's surrounding environment parameters when the driver passed the black spot to study the relationship between driver stress and traffic accidents.

In recent years, there has been some progress in the research on the driving characteristics of road traffic systems in some particular natural environment areas. However, there is still a lack of research on the relationship between natural environmental factors. Song Changping from Qinghai Occupational Disease Precaution Institute conducted a neuro-behavioural function test for car drivers in a low-concentration carbon monoxide driving environment and studied the changes in drivers' attention, sensitivity, accuracy and reaction time. Other research is mainly focused on diseases and physical fitness of the human body in the plateau environment rather than the driving state of vehicle drivers. By quantitative analysis, Pan Xiaodong et al. of Tongji University used the driving test to study the correlation between road alignment and the driver's psychological stress and physiological burden.

Hao Xiaohong from Inner Mongolia Agricultural University researched highway alignment design in grassland based on drivers' psycho-physiological responses. In the research project "*Research on Driving Behavior and Its Influence in Desert Highway Traffic Environment*", Xinjiang Agricultural University used the driving adaptability testing equipment to analyse the driver's overall characteristics on desert highways. Moreover, their correlation with driver's age, driving age, continuous driving time, temperature and other factors at the testing time. In addition, the variations of the driver's psychological and physiological characteristics in the unique desert environment were also discussed.

CCCC First Highway Consultants Co., Ltd. (hereinafter referred to as CCCC FHCC) and Beijing University of Technology have studied the relationship between driving workload in plateau areas and key road alignment indicators. The study selected the Qinghai-Tibet Highway as the test section in the plateau area. Observed the variation pattern of driver's workload in the plateau area when the altitude is 3200–3300 m, and the altitude is 4300–4400 m. The study obtained the safety threshold of highway driving workload in high-altitude areas. It then provided a basis for evaluating the impact of highway alignment design on driving workload in high-altitude areas.

After analysing the research worldwide, it is found that there are few systematic comparative studies on the influence of plateau environment on vehicle drivers. Foreign research on driving characteristics mainly focuses on the driver's adaptability, based on the accident tendency theory and puts forward the requirements for the driver's physiological and psychological quality. Through the psychological or physiological index test, it is evaluated whether the driver is suitable for driving. However, there are very few domestic studies on drivers' psychological and physiological characteristics in the plateau environment and their changing rules.

(2) **The Operation and Workload of Vehicle Drivers**

In the late 1970s and early 1980s, researchers in Japan began to use multi-channel physiological recorders on heart rate, blood pressure and other physiological indicators to study the relationship between the workload, psychological load and traffic safety of drivers when driving. In 1976, Yajima, Kazuyoshi, Ikeda, and Kenji from Japan collected data on the psychophysiological responses of drivers during 10 h of daytime driving and 24 h of continuous driving tests to study the fatigue characteristics of drivers. Li Qing, Maehara Naoki, etc., used multi-channel physiological recorders to study the driver's heart rate and blood pressure changes while driving. Iwakawa, Furuya Shiro, etc., used heart rate, blood pressure and other indicators to study the road alignment index in forest areas and the driver's psychological and physiological response.

In 2005, Tania Dukic of Chaimers University of Technology in Sweden and L. Hanson, K. Holmqvist, and C. Wartenenberg et al. from Lund University studied the influence of button location on drivers' visual behaviour and sense of safety. Benjamin W. Tatler of Sussex University, Roland J. Baddeley and Iain D. Gilchrist of Bristol University, etc., studied the features of selected fixation points by using pictures of different brightness, contrast, colour and shape. Torbjom Falkmer and Nils Peter Gregersen of Linkoping University in Sweden used eye trackers to measure the eye movement behaviour of 20 experienced and 20 inexperienced drivers in real traffic environments. The analysis verified some of the conclusions of other researchers.

Shinji Miyake integrated three physiological variables of heart rate variability, finger plethysmography amplitude, respiration and six load factors of NASA-TLX into one index, carrying out a multivariate analysis of load from physiological and subjective aspects, and obtained good results. It is an innovation and attempt at the load measurement method.

Wu Rina took the forest highway as the research object and found that the longitudinal slope and the running speed significantly impact the driver's heart rate change rate. When the driver goes up, and down the slope in a straight line with no load, the influence of the longitudinal slope is greater than the speed. However, with a heavy load, the effect of speed is greater than that of the longitudinal slope. Li Xianghong researched speed-limited and non-speed-limited roads on grassland highways. By recording the subjects' heart rate, heart rate growth rate, heart rate variability, and blinking times, he found that questionnaires, fatigue scales, and driving behaviour test values were consistent when drivers were uncomfortable or tired.

Yang Yushu of Shanghai Jiaotong University collected the ECG signals of 16 test subjects during a 90-min simulated driving operation in the laboratory and analysed seven time–frequency indicators of the ECG signals during the 15-min period at the beginning and end of the test. Four ECG time–frequency indicators (S.D., L.F., H.F., LF/HF) were significantly correlated with fatigue.

Cao Xintao studied the theory, law, evaluation methods and indicators of driving workload. Through many actual driving tests, the relationship model between the driver's physiological response to driving on the longitudinal slope, the alignment index of the longitudinal slope and the vehicle speed is established, and the limit gradient is defined.

Xiao Yuanmei, Wang Zhiming, etc. verified the reliability and validity of two mental load evaluation scales, namely the subjective load assessment technique (SWAT) and the NASA task load index (NASA TLX) scale, and concluded that they both have good performance. The reliability and validity, after appropriate revision, can be used as an effective tool for studying mental load in China.

Due to military needs, a large number of researches on plateau physiological diseases and tolerance in the Qinghai-Tibet Plateau have been conducted. Li Suzhi, the former president of the General Hospital of the Tibet Military Region, concluded that high-altitude pulmonary oedema and high-altitude cerebral oedema are the diseases most easily caused after entering the plateau. Furthermore, vehicle drivers had a higher incidence of high-altitude cerebral oedema, which accounts for 30% of the population entering the plateau. In winter, its incidence is significantly higher than that in summer.

Cui Jianhua, Institute of Mountain Diseases, 18th Hospital of the Chinese People's Liberation Army, summed up the three types of common acute altitude disease when people from non-altitude areas enter areas above 3 000 m above sea level, namely acute altitude sickness, high altitude pulmonary oedema and high altitude cerebral oedema. The symptoms are as followed.

i. According to their frequency, the main clinical symptoms of acute altitude sickness are headache, dizziness, shortness of breath, palpitations, nausea, anorexia, vomiting, etc. Common signs are increased heart rate, hyperpnea, mildly abnormal blood pressure, facial or arms and legs oedema, cyanosis, etc.
ii. In the early stage of high altitude pulmonary oedema, most patients have headaches, dizziness, general weakness, loss of appetite, lethargy, trance, oliguria, etc., followed by cough, palpitations, shortness of breath, chest tightness, etc. The most characteristic clinical manifestation is spitting up pink or white foamy sputum. Severely ill patients may have restlessness, confusion and even coma, and some patients may have symptoms such as nausea, vomiting, abdominal pain, diarrhoea, and fever. In patients with high altitude pulmonary oedema, the body temperature is 37–39 °C, the pulse is 81–121 times/min, the respiration is 20–40 times/min, and the blood pressure is mainly within the normal range.
iii. High-altitude cerebral oedema (also known as high-altitude coma) is characterised by loss of consciousness (coma). Patients often have some premonitory symptoms and signs before the coma and enter the coma with further disease development. Most patients breathe shallowly and rapidly but more rapidly if there are complications. About 50% of patients showed increased heart rates, 40% of patients with regular heart rates, and a few patients with slow heart rates. Blood pressure is mainly within the normal range. Some patients have increased blood pressure and pulse pressure, and a few have decreased blood pressure and even shock.

To sum up, the earlier research on drivers' psychological and physiological characteristics and driving workload mainly focus on common altitude areas. Few are about the psychological and physiological characteristics, operating responses, and drivers' fatigue in low pressure and oxygen deficient environment in high-altitude areas. The discussion of the relationship between drivers' psychological and physiological changes in high-speed driving and the geometric design indicators of highways in plateau areas was even less.

In addition, many physiological indicators reflect the driver's psychological and physiological response in the current research, including heart rate, eye movement, blood pressure, EEG, etc., among which heart rate is the most widely used. In the military medical field, many studies have been carried out on human discomfort and diseases that are easy to cause in plateau areas. It has been found that the physiological symptoms of various diseases are usually changes in heart rate. The test environment of various psycho-physiological test methods is a static test, and there is a lack of research on test indicators in actual road operations. Therefore, to overcome the shortness of the above research, the author of this book selects the psycho-physiological response index represented by the heart rate index by comparing the observation results of different psycho-physiological indicators in the actual car and simulated driving control tests. Further research is done on the relationship between the driver's psychological and physiological characteristics, operational response and fatigue state and the leading technical indicators of highways in the plateau area under high-speed driving.

1.2.3 Characteristics of Freeway Traffic Operation in High-Altitude Areas

(1) Characteristics of Traffic Accidents and Traffic Safety Service Levels in High-altitude Areas

Xiao Runmou and others from Chang'an University rely on the Langmusi-Chuanzhusi highway reconstruction project on the G213 line to research vehicle driving safety and identify the main reasons for the frequent traffic accidents and some traffic safety measures on the plateau long straight section. In *Analysis of 1894 Cases of Road Traffic Accidents on the Qinghai-Tibet Plateau*, Wang Qian et al. concluded that: **i.** The sample traffic accidents mainly occurred on the Qinghai-Tibet Highway, accounting for 68.0%, and were primarily found in the areas with higher altitudes and relatively straight roads. **ii.** Vehicle collision and rollover are the main types of accidents, accounting for 51.1%. **iii.** Traffic accidents caused severe injuries, and about 54.0% required hospitalisation; 108 cases died, and the mortality rate was 57%. **iv.** The time for transition is long, and most do not take any measures before treatment. **v.** Most traffic accidents on the Qinghai-Tibet Highway occurred among those who travelled from the plains to the plateau for the first time, accounting for 61.1%. The causes of the accident are as follows: **i.** Unique climate environment at high altitude: People from the plain area often experience hypoxia symptoms when they first arrive at the plateau. **ii.** Fatigued driving: Some drivers rush day and night to reach their destination in time. **iii.** Unfamiliar road conditions: Due to a perennially frozen layer, the Qinghai-Tibet line is prone to road puddles when the cold and hot seasons alternate. There are many temporary access roads due to frequent maintenance or the perennial snow in high-altitude areas, and the road surface is often covered.

Foreign scholars' research on traffic accidents in high-altitude hypoxia areas is rarely involved, and the leading research focuses on the impact of bad weather on road traffic safety.

Anna K. Andersson, in *Winter Road Conditions and Traffic Accidents in Sweden and U.K. (2010)*, studied the relationship between road conditions and traffic accidents in Sweden and the U.K. in winter. Sweden has 720 outdoor observation stations, 200 equipped with cameras, and they observe data every 30 min. The leading observation indicators include road surface temperature, air temperature, relative humidity, precipitation, wind speed and wind direction. Research shows that: in January 2005 and January 2006, the number of accidents caused by road slippage accounted for the vast majority of the total number of accidents. The number of traffic accidents increased significantly when the road surface temperature was lower than −3 °C or snow began to fall, and most traffic accidents occurred within 2 h of snowfall. Kazushi SANO et al., *Analysis of Freeway Traffic Accidents Without a Median in Cold and Snowy Regions* (2009), studied roads with medians in cold and snowy regions of Japan, using five Years of traffic accident data. The study showed that the highest traffic fatalities occurred on icy roads without a median.

Road traffic safety prediction is vital in comprehensive traffic planning and management. Traffic accident forecasting can be divided into macro forecasting and micro forecasting. Macro forecasting refers to the forecasting and analysis of the overall traffic safety level of a country, a city or a region. Macro forecasting mainly analyses and studies the impact of social and economic development, population changes, car ownership, non-motor vehicle ownership, and laws and policies on traffic safety. It provides a theoretical basis for the country or region to promulgate policies and regulations on traffic safety.

The model established by Smeed in 1949 is considered to be the most classic safety prediction model. Smeed believes that traffic hazard (expressed as the number of deaths per vehicle) is inversely proportional to the level of motorisation, and many of the models are based on the work Smeed has done. In addition, there are the Trinca model, Koornstra model, Towill model, etc. These models all reflect the problem from one aspect of the traffic safety system.

Since the 1980s, several researches on macro traffic safety have been done. Among them, the Beijing University of Technology and Beijing Institute of Transportation Engineering Research have more positive results. The representative researches of Beijing University of Technology include the grey evaluation method of traffic safety, the prediction model of traffic accident frequency constructed by the time series method, the generation model of urban road traffic accident established by the method of system dynamics, and a bipolar fuzzy mathematical model for macro evaluation of traffic safety.

Macro forecasting is challenging to predict the accident on a specific road, so it is difficult to propose specific road safety design and guidance suggestions for arranging traffic engineering facilities. Therefore, specific evaluation or prediction of the safety of road facilities requires micro-forecasting analysis. The focus of micro forecasting is on specific road facilities. By analyzing the road conditions, traffic conditions, landscape and other factors that affect traffic safety, the accident of the facility can be estimated, and then the safety level of the facility can be evaluated and predicted to meet the needs of road safety and traffic facility design. The research object of this book is freeway, the accident frequency prediction model to be developed in this book belongs to the category of micro-prediction.

The research of road safety and the establishment of an accident frequency prediction model can be divided into two categories according to the characteristics of road facilities: road sections and nodes. Among them, node security research can be divided into urban, suburban and rural according to the different regional attributes; According to the different types, it can be divided into intersections, roundabouts and interchanges. Intersections can be divided into signalised and non-signalled intersections according to their types. Non-signalled intersections can be further divided according to the number of approaches and the right of way. The safety research of road sections can also be divided into urban, rural, etc. Regarding the different attributes of the area, according to the different grades of roads, it can be divided into freeways, multi-lane roads, two-lane roads, etc.

The research outcomes of the two-lane accident prediction method in the United States have been widely recognised. They have been included in the Interactive Highway Safety Design Model (IHSDM) and the *Highway Safety Manual* (HSM). The latest research results appear as a sample chapter of HSM (first edition). This new method is based on historical accident data prediction, mathematical and statistical models, before and after analysis, and expert experience prediction methods. During the research, the highway data and accident data of Washington state and Minnesota state were collected, and the accident frequency prediction model of the road section and the accident frequency prediction model of three kinds of intersections were developed respectively. These models can be combined to predict the whole road. The accident situation, including accident number, accident severity distribution, accident type distribution, etc., can also predict the accident situation after safety improvement.

At present, there are many studies on freeway safety in China. Most of the literature focuses on the analysis of safety status, safety evaluation, accident law research and preventive measures, and most are aimed at specific freeways. Liu Xiaoming from the Beijing University of Technology has studied the impact of road conditions on traffic accidents. Moreover, he believes that other influencing factors of accidents, such as the driver's age, gender, driving proficiency, and vehicle conditions, have the same impact on the accident rate of each section. It is assumed that the accident rate of a section is mainly related to the road conditions where the section is located. Therefore, the number of lanes, road alignment, vertical and horizontal slopes, radius, and whether there are three-dimensional intersections, bridges, and other structures near the section are the main influencing factors of traffic accidents. Based on that, a quantitative method for freeway accident prediction is established. By which, the accident data of the Beijing section of Jing-shi Freeway is analyzed, and the results show that the effect of interchange and curvature factors on the accident rate is greater than that of bridges and cross-section forms. Based on the analysis of the safety mechanism of road design elements, longitudinal design elements and statistical analysis, Chen Yongsheng from the Beijing University of Technology established an accident frequency prediction model, which provides a theoretical basis for road safety design and black spot identification.

In Chap. 4 "About Traffic Safety Investigation Method" of HSM First Edition, the method of using the safety service level to rank the traffic safety level situation is introduced. Similarly to the widely used road capacity manual, HSM defines the potential safety performance indicators of road traffic facilities, lays a theoretical foundation for road safety, provides the basic knowledge required for road safety evaluation, and establishes the level of traffic safety. The relationship between it and its influencing factors, among which is the theoretical basis of the safety performance model has paved the way for further research on the safety service level. It defines the potential safety performance indicators of road traffic facilities, lays a theoretical foundation for road safety, and provides the basic knowledge required for road safety evaluation. Moreover, it establishes the relationship between traffic safety level and its influencing factors, among which the theoretical basis of the safety performance model has paved the way for further research on the safety service level. Safety service level is deemed to be another new development direction in the field of traffic safety research. At present, there are relatively few studies on systematic research on safety service levels.

In 2003, Jake Kononov and Bryan Allery introduced the concept of safety service level and the safety service level analysis framework based on the safety performance function (SPF). Using the accident rate (the frequency of accidents per mile per year) and the annual average daily traffic volume as the safety service level classification indicators, a generalised linear regression accident prediction model for two-way six-lane freeways in American cities is established. Furthermore, the expected value of the accident prediction model is used as the standard. Using 1.5 times the standard deviation to divide the safety service level into four levels: LOSS I, LOSS II, III, and IV (Fig. 1.1). They emphasised that the safety service level can describe the traffic safety level, but it cannot reflect the traffic safety problem, which requires specific diagnosis and analysis.

Zhong Liande of Beijing University of Technology studied the safety service level of rural sections and urban sections of freeways. On the basis of a metric equation for the relationship between the mean value of freeway accidents, the section length, and the annual average daily traffic volume, he adopted the concept of "confidence interval" from probability theory to grade the security service level. The resulting graded graph is shown in Fig. 1.2.

To sum up, the current research on traffic safety in high-altitude areas mainly focuses on analysing traffic operation and accident characteristics and barely relies on

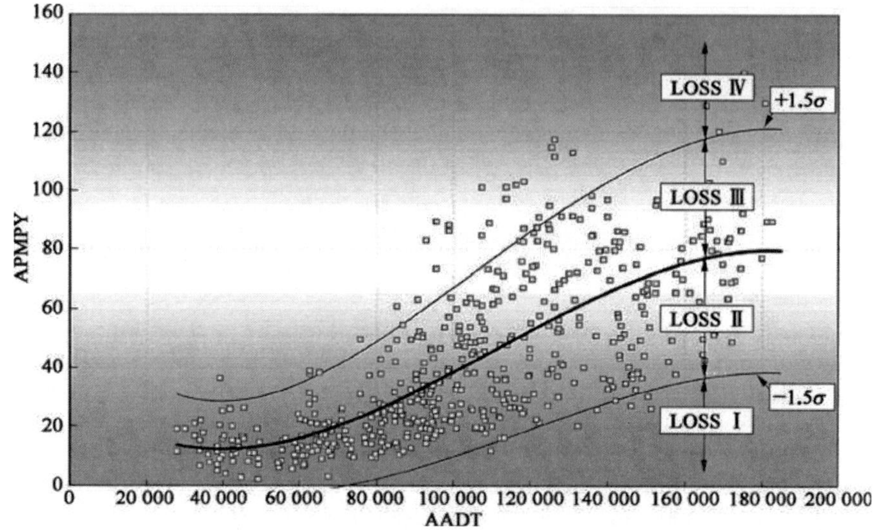

Fig. 1.1 Schematic diagram of the safety service level of two-way six-lane highways in U.S. cities (Total Sccidents)

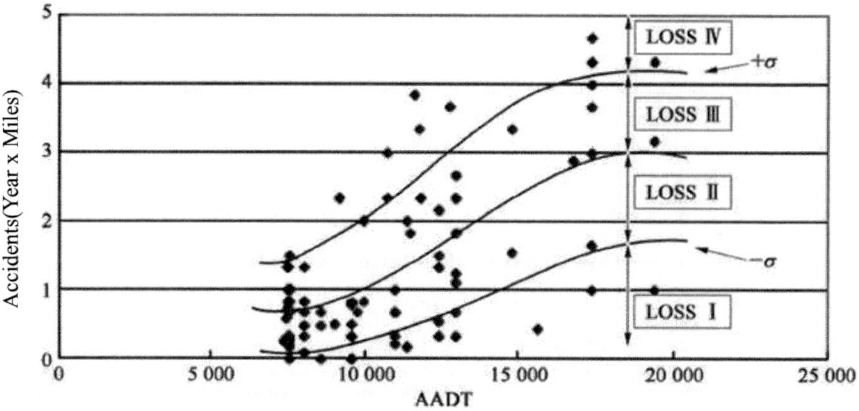

Fig. 1.2 Four-lane highway LOSS in the rural section

statistical knowledge to investigate the correlation between various factors. In recent years, there has been significant progress in the methods and models of accident prediction, the research objects are however mainly focused on plains and mountain roads, and accident prediction research in unique high-altitude environments is rarely involved. Researchers worldwide have done many works on the impact of natural conditions on traffic safety. However, they mainly focus on the impact of bad weather, and less on the impact of oxygen content and altitude on traffic safety.

Therefore, this book introduces the key characteristics (altitude, oxygen content) of the environment in high-altitude areas and studies the relationship between traffic flow, traffic accidents and environmental characteristics through statistical knowledge. It reveals the mechanism of high-altitude accidents, establishes accident models, and analyses the definition, quantification and application of the concept of the safety service level.

(2) **Road Operating Speed at Different Altitudes**

Many operating speed prediction models are mainly divided into two categories according to the research methods: the first type is the statistical regression model, which adopts the actual measured operating speed to establish the relationship between operating speed and linear indicators. The second type is the relative factor restricted model. The research method is not to directly establish the relationship between speed and road alignment but to establish the relationship between various speed limits and road properties and determine the operating speed according to these speeds, such as the famous Brazilian models. Compared with the low-altitude environment, the high-altitude environment is quite different to the unique climate. However, there is no relevant research on the operating speed model for this situation.

Since the 1970s, the U.S. Federal Highway Administration (FHWA) has conducted much research on road safety and developed a well-known road safety evaluation and design software package—IHSDM, the only released systematic road safety design application system worldwide. FHWA established the empirical regression formula of the road operating speed by using the regression method to measure the vehicle speed of more than 200 measuring points in 6 states and constituted the road operating speed prediction model.

The Institute of Highway Science of the Ministry of Transport of the People's Republic of China (hereinafter referred to as the "MOT") has established a speed simulation system SF_Sim5.0 in the national "Ninth Five-Year" scientific and technological research project "Highway Investment Comprehensive Benefit Analysis System" to simulate vehicle speed and fuel consumption. Fan Zhenyu and Zhang Jianfei proposed in the article *Research and Calibration of Highway Operating Speed Calculation Models* a operating speed calculation model for freeways and secondary highways. The models include operating speed calculation for straight lines (including large radius curves) and sharp curve sections.

In the research on the operating speed of freeways, the subject of "*Design Method and Standard of Freeway Operating Speed*" of the Western Transportation Science and Technology Project is the basis of MOT's research, *Technical Standards for Highway Engineering* (JTG B01-2003) and Code for Design of Highway Routes (JTG D20- 2006). According to the driving characteristics, the freeway is divided into straight sections, flat curve sections, longitudinal slope sections, and horizontal and vertical combined sections. Operating speed prediction model under linear conditions. The research results have been incorporated into the *Guidelines for Safety Evaluation of Highway Projects* (JTG/TB05-2004) of the Ministry of Communications. Then, according to the different influencing factors of the road section, the investigation and modelling are carried out respectively, and the prediction model of operating speed under different alignments is established. The research results have been incorporated into the *Guidelines for Safety Evaluation of Highway Projects* (JTG/TB05-2004) of the MOT.

In 2010, CCCC FHCC put forward the classification standard of operating speed based on vehicle wheelbase and power-to-weight ratio in the project "Highway Operating Speed System, Safety Evaluation and Engineering Application Technology Research" in the Western Transportation Construction Science and Technology Project of the MOT. They established the application model of highway operating speed. The research results of this project have been incorporated into the MOT's "Highway Project Safety Evaluation Specification" (JTG B05-2015). The research results show that: the vehicle can reach the desired speed in the condition of no roadside interference on the straight line section; the operating speed of the small radius flat curve section changes drastically; the altitude of 4000 m is the critical point of the influence of the altitude on the operating speed, the influence of the longitudinal slope on operating speed is more significant than that in the low altitude area.

It can be seen that the research on operating speed is relatively developed, and a modelling mechanism of various methods and concepts has been adopted. At the same time, many results have been obtained, and they are widely used in various highway construction and service processes. However, there are few types of research on the freeway operating speed model in the low pressure and oxygen deficient environment in the high-altitude area of the Qinghai-Tibet Plateau. There are several key reasons: First, the research on vehicle performance in the low pressure and oxygen deficient environment started late, and the research on the operating speed prior to the study of vehicle performance is limited; at the same time, the construction of plateau highways relatively slow, and the data that can be used for research and investigation are in lack.

(3) **Dynamic Speed Control (Speed Limit) Technology of Freeway**

Since 1960, the United States has conducted a number of studies on the relationship between speed and security. Solomon studied the main rural roads in the United States in 1964, and selected 28 roads from 35 regions for research. Among them, 3/4 are rural two lane roads with a speed limit of 55 ~ 70 m/h. By comparing the speed of vehicles involved in accidents with that of vehicles not involved in accidents, it is found that there is a U-shaped relationship between accidents and speed. The speed corresponding to the lowest accident rate is slightly higher than the average operating speed of the road, while the speed corresponding to the higher accident rate is on both sides of the average speed. At the same time, the relationship model between speed difference and traffic accident rate is proposed for two-way four lane freeway, as shown in Fig. 1.3.

In 1974, due to the energy crisis, the U.S. government promulgated and implemented the national maximum speed limit, which prohibits driving at speeds higher than 55 miles/h (90 km/h). A large number of studies show that after the implementation of the national maximum speed limit, the number of road traffic accidents and deaths in the United States have decreased significantly.

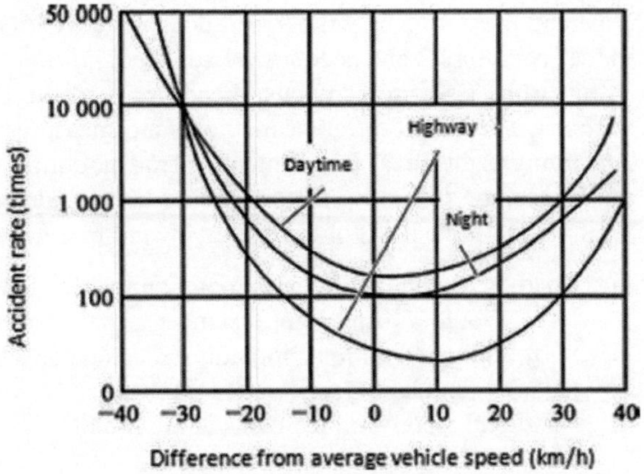

Fig. 1.3 Relationship between accident rate and average speed difference

Farmer et al. collected the data from 1990 to 1997, and used the time series model to analyze the fatalities and mortality in 31 states (24 of which have increased the speed limit, and 7 of which have not). They believed that on the interstate highways in the 24 states where the speed limit has been increased, the fatalities and mortality have increased by 15 and 17% respectively; there were no remarkable changes referring to the characteristics of traffic accidents on the non-interstate highways.

Najjar et al. collected the traffic accident data of Kansas from 1993 to 1998 (except 1996) using the three segment sequence method. It was found that there was no significant change in the number of accidents, deaths and mortality on interstate highways. However, from 1997 to 1998, accidents on rural two lane roads increased significantly. As for whether the increase or decrease of the speed limit value will affect the driver's choice of operating speed, 0ssiander, Jernigan and Lynna, Kay Fitzpatrick and Esterlitz all quantitatively described the relationship between the increase of speed limit and operating speed. The results show that when the speed limit is increased, the operating speed of vehicles will also increase, which will indirectly affect the traffic safety.

Synthesized the studies on the relationship between speed limit and traffic safety by various scholars, it can be seen that with the increase or decrease of speed limit, traffic safety will rise and fall in different situations. But in general, with the reduction of the speed limit, the situation of traffic safety will decline.

When the roads at specific sections or locations cannot meet the statutory speed limit, speed limit shall be adjusted according to local conditions, and speed limit zones shall be considered. According to Speedzone Guild Lines of the Institute of Transportation Engineers (ITE), the speed limit value should be the value close to the 85th percentile speed and divisible by 5, or the upper limit of 10 mile space (10 km/h probability distribution interval). When the difference between the legal speed limit and the operating speed does not exceed 3 mile/h, it is unnecessary to arrange a speed limit zone. If the difference between the speed limit value in the current speed limit zone and the 85th percentile speed does not exceed 3 mile/h, the speed limit value in the speed limit zone need not be adjusted.

Since the right to formulate speed limits in the United States belongs to each state, each state has different methods and measures for formulating speed limits. According to the Texas Department of Transportation's speed limit zone formulation guidelines, the following factors should be considered in the formulation of speed limit zones on the basis of traffic survey and free flow speed analysis.

i. Road factors: including horizontal and vertical alignment, lane form, separator form, shoulder form, sight distance;
ii. Surrounding conditions: including roadside development, surrounding residential areas and land development intensity;
iii. Historical records of accidents along the line.

South Africa has studied the formulation method of speed limit zone as early as the 1980s, and proposed that the formulation of speed limit zone can be divided into the following three steps:

i. Determine the starting point and end point of the speed limit zone.
ii. After considering a variety of operating speed influencing factors, the road section may have multiple theoretically calculated speed limits; select the lowest speed limit. If the lowest speed limit is not very necessary, the next lower speed limit can be selected.
iii. Check whether the length of the speed limit zone meets the specified minimum length, and if not, adjust it. The speed limit zone shall be as long as possible, and the the 85th percentile speed in the speed limit zone shall be consistent as far as possible.

From the data collected wordlwide, on the basis of a large amount of data, foreign countries have carried out a long-term and large amount of research from the aspects of speed and safety, speed and efficiency, speed limit zone arrangement, speed control technology, etc., and established quantitative models of the interaction between speed and various parameters. The formulation methods of speed limit include legal speed limit, the 85th percentile speed method, optimization method, expert system method, etc. The applicability, arrangement method and speed control effect of individual and combined speed control technologies have not been thoroughly and systematically studied.

At present, the research on speed restriction wordlwide mainly focuses on quantitative description of the impact of speed restriction changes on traffic safety and operating efficiency. The methods used are mostly statistical analysis based on long-term data observation, and the data are compared before and after or horizontally. Due to the lack of traffic accident data and traffic flow related reference data in China, it is impossible to compare before and after, so there is no in-depth study on the interaction between speed limit and other factors.

In the book *Highway Speed Limit and Speed Control Technology* of Beijing University of Technology, based on the impact of speed limit on operation efficiency, economy and safety, a decision model of speed limit value is proposed. The

Table 1.1 Minimum length of speed restriction zone in the United States

Speed limit (km/h)	110	100	90	80	70	60	40
Minimum length (km)	10	2.0	0.9	0.8	0.7	0.6	0.4

indoor test method was used to test the attenuation law of the driver's short-term memory. At the same time, the minimum length of the speed limit zone of the speed limit sign in the *Manual on Uniform Traffic Control Devices* (MUTCD) published by the Federal Highway Administration of the United States is determined (Table 1.1). In the study, the speed limit value is selected in steps of 10 km/h, and the minimum length of the restricted area is also considered in steps of 10 km/h. The minimum length of the speed limit zone is composed of two parts: the leading distance of traffic signs and the driver's stable driving distance.

From the data collected wordlwide, there are many achievements in the research of speed limit formulation methods wordlwide, but the speed limit technology under the special geographical and climatic conditions of low pressure and oxygen deficient environment in high-altitude areas has not been involved. The size of speed limit value has a certain relationship with traffic safety. The formulation method of speed limit value in high altitude areas and the determination of specific speed limit value are of great significance for the safe operation of highways in plateau areas. Moreover, due to the lack of control facilities and the relatively low level of traffic safety on plateau roads, it is urgent to conduct in-depth and systematic research on the applicability, arrangement methods and speed control effects of high-altitude highway speed control technology.

1.2.4 Geometric Technical Indicators of Freeway Routes in High Altitude Areas

(1) Adoption of Technical Standards for Freeways in Similar Regions Abroad

At present, although there are many studies on highway alignment in foreign countries, there are few studies on highway alignment parameters in high-altitude areas. The technical standards for freeways of some countries located in cold regions and around the Qinghai–Tibet Plateau can provide some reference.

Canada is located in North America, where the climate is warm and short in summer and cold and long in winter. Canada's highway management system is mainly provincial, and the federal government only plans, coordinates and grants financial subsidies to principal arterials. Canada does not have a set of national technical standards, and basically takes Ontario standards and norms as the nation's. Canada's freeway design is based on reality and does not blindly pursue high standards. The vast majority of freeways use low embankments, and most overpasses do not use approach bridges, but use earth filled approach roads, which not only saves investment, but also is conducive to driving safety, environmental protection and coordination. The horizontal alignment of the highway is basically in harmony with the terrain conditions. When the ground is flat, the route is straight and smooth. When the terrain changes greatly, the horizontal alignment changes with the terrain. The minimum curve radius of highways of the same class with complex terrain is smaller than that of China. The longitudinal plane of the highway varies with the terrain. In urban sections, the longitudinal plane is basically consistent with the terrain, and the vertical elevation is basically flush with the ground line. In mountainous areas, the longitudinal slope of freeway reaches 6% (1,500 ~ 2,000 m above sea level).

Finland is located in Northern Europe, and 1/4 of the country is located in the Arctic Circle, with a cold climate. An important measure to ensure traffic safety in Finland is to limit the maximum running speed. The highway running speed shall be regulated according to the traffic conditions. For example, on major highways, the maximum allowable running speeds are 60, 80 and 100 km/h respectively (120 km/h on principal arterials).

Sweden has the lowest traffic accident mortality rate in the world, with only 6 deaths per 100 thousand vehicles (62 in China). The subgrade width of Sweden freeway is generally 27 m, with two-way four lane cross-section. The width of the carriageway is 3.75 m, the width of the outer shoulder is 2.75 m, and the width of the inner shoulder is narrow, depending on the traffic volume. No curb is arranged for the median, and the width of the median is 4 ~ 20 m, depending on the terrain. In areas close to cities or with less land, it can be less than 4 m and guardrails must be arrange, while no guardrails are arranged for areas larger than 4 m.

Countries and regions around Tibet are mainly India, Nepal and Pakistan, and their traffic development is similar. When designing roads in India, the terrain where the roads are located shall be defined first. The terrain is classified according to the cross slope of the terrain. See Table 1.2 for terrain classification of Indian roads.

After the terrain category is determined, the design standard of the highway must be further determined according to the functional classification of the highway. The types of roads in India include national roads, provincial roads, local roads and rural roads. The national highway and provincial highway of India will generally adopt the freeway construction standard in plain and hilly areas, and the upper limit of design speed is 100 km/h.

Nepal Road Standard 2070 (2010) is a Nepal road design standard jointly issued by the Government of Nepal and the Ministry of Infrastructure and Transport, which is applicable to non urban roads in Nepal. The highway in Nepal is divided into four grades, and the freeway is the first grade standard. Like India, the freeway in

Table 1.2 Topographic classification of indian roads

S/N	Terrain classification	Terrain cross slope (%)
1	Plain	0 ~ 10
2	Hills	10–25
3	Mountain ridges	25–60
4	Escarpment	Greater than 60

Nepal is mainly built in plain and hilly areas, with the maximum design speed of 120 km/h.

(2) **Research Status of Geometric Technical Indicators of Freeway Routes in High Altitude Areas in China**

Cheng Zhan, Liu Jiguo and others analyzed and summarized the main technical problems and principles of overall design for building extra long highway tunnels in this area in combination with the main landform, geology and climate characteristics of the high altitude area. Taking the general proposal design of Galongla Tunnel as an example, the key points that should be paid attention to in the general design of highway tunnels in high altitude area is illustrated. Heilongjiang Provincial Highway Survey and Design Institute and Harbin Institute of Technology undertake a western transportation science and technology project "Research on Horizontal and Vertical Alignment and Cross Section Design Indicators of Roads in Cold Regions". The project aims at designing roads in cold regions, and from the aspects of determination of road design status, test and analysis of pavement adhesion coefficient in cold regions, and safety analysis of roads in cold regions, it studied the design of cross section and vertical section, and gave the suggestion of design indicators of highway alignment in cold regions. Li Songling and Pei Yulong studied the speed limit of the ice snow road in the flat curve section, applied the vehicle kinematics theory, considered the characteristics of drivers driving vehicles in the curve, analyzed the characteristics of vehicles turning on the ice snow road, and established the kinematics model of vehicles driving in the flat curve section. By selecting trucks and cars two types, and according to soft snow pavement, compacted snow pavement and ice pavement three road conditions, they established relationship models between running speed and horizontal curve design indicator respectively. They selected the flat curved line sections with 6%, 8% and 10% superelevation values, used the Matlab simulation technology to solve the vehicle speed limit model, and analyzed the correlation between the speed and the radius of circular curve and superelevation. Based on these, they propose the corresponding safe running speed of vehicles with different radius of circular curve under above three pavement conditions, which provides a theoretical basis for scientifically and reasonably determining the vehicle speed limit value of the ice and snow road section on the horizontal curve.

Fang Jing and Zhou Ronggui from the Research Institute of Highway Ministry of Transport studied the key design parameters of highway routes in plateau areas. Through the observation of the blood oxygen content of drivers at different altitudes in the plateau area, the relationship between the blood oxygen content and the reaction time delay is established, and the reaction time of drivers at high altitudes is calculated based on this, so as to determine the stopping sight distance in the plateau area, and give the value of the vertical curve radius. The minimum radius of circular curve is obtained by taking the maximum superelevation of highway in plateau area as the control standard. Considering the poor vehicle power performance, driving fatigue and tension due to the thin air in high altitude areas, the maximum longitudinal grade reduction and the maximum grade length under different gradients are proposed.

CCCC FHCC, in the western project "Research on Highway Operating Speed System, Safety Evaluation and Engineering Application Technology", combined with the actual survey of Qinghai—Tibet Highway, analyzed and studied the adaptability of alignment standards and specifications in high-altitude areas, the characteristics of vehicle operating speed there and the driver's driving workload, established a prediction model for the operating speed in such areas, and proposed key alignment indicators in plateau areas (Qinghai—Tibet Highway), which provides a basis for the alignment design of the Qinghai—Tibet Plateau.

The *Design Specification for Highway Alignment* puts forward the requirement that the longitudinal slope of highways in plateau areas above 3000 m above sea level should be reduced. With the increase of altitude, the atmospheric pressure, air temperature and density of highways in the plateau area gradually decrease. The reduction of air density will affect the normal operation of the car engine, thus affecting the power performance of the car. The research and test run show that the average power of Jiefang's engine decreases by 11.3% at 1,000 m above sea level; 21.5% at 2,000 m; 33.3% at 3,000 m; 46.7% at 4,000 m; 52% at 4,500 m. According to the test and analysis, when the altitude exceeds 3000 m, the longitudinal slope shall be considered for reduction.

(3) **Reasonable Spacing of Service Facilities**

Freeway service facilities are an important part of freeway construction, and also an important supporting facility for providing traffic safety guarantee and comprehensive services for highway users and vehicles in the freeway operation stage. The reasonable spacing of service facilities is closely related to road network planning, traffic flow and traffic composition, terrain, etc.

The arrangement of freeway service areas in Europe can be roughly divided into three categories: comprehensive service area, parking area and refueling area. The functions of the comprehensive service area are basically the same, mainly including vehicle maintenance, cleaning, refueling/water filling, driver and passenger rest, shopping, catering, etc. The spacing is 30 ~ 50 km. Parking area plays an important role in the arrangement of freeway service areas abroad, and is an important facility to ensure traffic safety and reduce the incidence of accidents. The spacing of parking areas in European countries is different. The spacing of most parking areas is 10 km, so as to ensure that drivers can use parking facilities within about 10 min of driving to facilitate parking at any time. The distance between refueling areas is relatively large, generally about 40 km, because motor vehicles can still drive about 50 km when the fuel is insufficient. The spacing of freeway service areas in some European countries is shown in Table 1.3.

The Federal Highway Administration's *Manual of Geometry Design for American Freeway* arranges the standard spacing between comprehensive service areas as 40 km, and the maximum positioning is 100 km. According to the *Design Essentials for Japanese Freeway*, the standard distance between comprehensive service areas is 50 km, and the maximum distance is 100 km. The standard distance between all rest facilities is 15 km, and the maximum distance is 25 km, which can better meet the

Table 1.3 Spacing of freeway service areas in some European countries

Countries	Facility types	Spacing (km)
UK	Service area	16~17
German	Parking area	5~10
	Service area	50
French	Parking area (Class A)	8~10
	Parking area (Class B)	25~30
	Refueling facility	40~50
	Service area	100
Hungary	Parking area	20~30
Holland	Refueling facility	20~30

needs. In Japan's *Research on the Planning and Design of Rest Facilities on Freeways (II)*, the satisfaction rate of the corresponding spacing arrangement is given. The research shows that the maximum spacing of 25 km for rest facilities can meet 91% of the needs, and the standard spacing of 15 km can meet 98%; the distance between comprehensive service intervals can meet 61% of the needs when the use the maximum distance 100 km, and 89% when use the standard distance 50 km.

Compared with developed countries, China's freeway construction started late, and the construction and management experience of service facilities are also less. The construction of freeway service areas in China is not in harmony with the demand. In recent years, with the acceleration of the pace of freeway construction in China, the requirements for the layout of facilities in freeway service areas are also increasing, and many scholars have also paid attention to this aspect of research. In China, the arrangement of freeway service facilities is mainly based on the *Technical Standards of Highway Engineering* implemented in 2004. According to this standard, the average distance between service areas is 50 km, and the distance between parking areas and comprehensive service areas or parking areas should be 15~25 km.

Jiang Cailiang mentioned in the *Analysis on the Construction Requirements of Freeway Service Areas in Ice and Snow Disaster Areas* that for areas prone to ice and snow or other disasters, the reasonable spacing of freeway service areas can largely avoid the phenomenon of vehicles and passengers staying on the road section. For disaster prone areas, in addition to considering the economic development trend, long-term traffic volume, traffic flow characteristics and landscape coordination along the road in general to determine the appropriate spacing, the spacing of freeway service areas in the disaster prone areas should be appropriately reduced. It is recommended that the service areas should be arrange at 20~30 km intervals.

Jiang Guichuan, Yi Shu, etc. conducted a comprehensive study on the layout of service facilities in freeway under the condition of network. Firstly, they analyzed that the essential difference between freeway network and isolated sections lies in the existence of hub nodes, and then, for various situations such as separate hub interworking, hub interworking adjacent, linear arrangement, circular arrangement, etc., they analyzed the layout of service facilities in corresponding sections in detail, and summarized the principles and methods of reasonable layout of service facilities in various situations.

Zhou Zhitao, Pan Binghong and others divided the service facilities into three categories based on the functional characteristics of the existing service facilities of the freeway. By comprehensively demonstrating the control factors of the distance between various types of service facilities, three key factors, namely, the timely accident rescue distance, the safe driving distance with low fuel volume and the fatigue characteristics based on driver's self-control are selected as the control factors of the distance between the three types of service facilities, and the calculation models of the service facilities spacing are established respectively. Through consulting investigation and regression analysis of 14 typical freeway service areas, the relevant parameters in the models are deeply studied, the reasonable values of the parameters are determined, and the recommended spacing and layout types of the three types of service facilities are proposed.

Based on the research status quo of technical standards and main technical indicators of freeways in cold and high altitude areas wordlwide, the foreign technical standards for freeways basically do not involve the requirements for the use of technical indicators in high altitude areas. The Indian standards have requirements for the design of longitudinal slopes of highways in mountain ranges and steep cliffs with an altitude of more than 3,000 m, but they are only applicable to low-grade highways. In general, the developed countries in cold regions have a high proportion of passenger cars and good dynamic performance of trucks, therefore, the maximum grade specified abroad is not only slightly larger than the domestic standard value, but also can be flexibly used according to the terrain characteristics; the longitudinal slope indicator requirements of countries around the Qinghai—Tibet Plateau are similar to those of China, but the sight distance indicator requirements are higher.

There are many studies on horizontal and vertical alignment indicators of roads in cold regions in China. The study on highway alignment indicators in high-altitude regions is mainly focused on vertical sections with few achievements. The research vehicle type on longitudinal slope reduction belong to an earlier era, which is no longer applicable to the current development status of the automobile industry, and there is no systematic and in-depth study on the relationship between geometric design indicators and driver's psychological and physiological characteristics in high-altitude plateau regions. There are few studies on the reasonable spacing of freeway service areas wordlwide, which mainly focus on qualitative methods, focusing on the physiological needs of passengers and vehicle operation limits, while ignoring the impact of factors such as the environment along the freeway, the two-way effect of traffic flow, and the fatigued driving characteristics of drivers. In view of the low pressure and oxygen deficient environment at high altitude and considering the vehicle refueling demand, human physiological needs, requirements for safe driving, and combined with the geological conditions, traffic flow along the line, and domestic and foreign experience in the layout of service facilities, the reasonable spacing of service facilities of Qinghai—Tibet high altitude freeway must be further studied.

Therefore, the author of this book believes that it is necessary to carry out further systematic study on the key design indicators such as the minimum radius of circular curve, the gradient of the maximum grade, the length of the slope and the grade reduction, the parking sight distance, and the spacing between service facilities in high-altitude areas, so as to provide technical support for the design and reconstruction of highways in plateau areas, thus improving the traffic safety level and traffic guarantee capacity in plateau areas, and providing new basis for the formulation and revision of standards and specifications.

1.3 Main Research Contents

In view of the special geographical environment characteristics of the Qinghai—Tibet Plateau and the main traffic safety technical problems faced by freeway construction in high altitude areas, scientific research is mainly carried out from three aspects. The specific research contents are introduced as follows.

1.3.1 Research on Vehicle Performance and Driving Behavior in Low Pressure and Oxygen Deficient Environment

(1) **Study on the Change of Typical Vehicle Performance and Safety Characteristics**

The research contents of typical vehicle performance and safety characteristics in low pressure and oxygen deficient environment mainly include:

(1) Vehicle Dynamic Performance Theory and Influence Factor Analysis

Starting from the theoretical process of vehicle power generation and transmission, the variety characteristics of vehicle engine starting, acceleration and power transmission in low pressure and oxygen deficient environment are analyzed. The classical relationship model between oxygen content and diesel engine is used to deduce the engine output power of trucks under low pressure and oxygen deficient environment. Select the typical engine basic data to establish the model, use GT-Power software to simulate the environmental parameters, calculate and analyze the output data. Determine the simulated engine performance from the aspects of optimal speed and output power, and finally determine the environmental factors and action forms that affect the vehicle dynamic performance.

(2) Analysis of Vehicle Braking Performance Theory and Influencing Factors

In view of the changes in the braking performance caused by the meteorological environment in Qinghai—Tibet region during the long and great longitudinal slope driving, the classical freight car brake temperature rise model is used to modify the environmental parameters, mainly focusing on the adjustment of atmospheric environmental temperature, atmospheric environmental density and other parameters.

(3) Investigation and Analysis of Highway State in Qinghai—Tibet Area in Low Pressure and Oxygen Deficient Environment

On the spot investigation on the current highways in Qinghai—Tibet region, mainly collect highway design data (especially the geometric design of highway plane and profile, the design of highway cross section, the friction coefficient of pavement, etc.), meteorological environment data (especially oxygen content, the distribution of air pressure along the line, the distribution of pavement temperature, etc.), and highway traffic operation data (basic vehicle types and corresponding distribution, daily traffic volume distribution, traffic accident pattern distribution, etc.). In combination with the data collection of section operating speed, the main form characteristics, maximum running speed, ideal acceleration of vehicles in the low pressure and oxygen deficient environment in Qinghai—Tibet region are basically determined.

(4) Analysis of Changes in Vehicle Safety Features in Qinghai-Tibet Plateau Region

Based on the results of surveys on the road and traffic environment under low pressure and oxygen deficient environment in Qinghai-Tibet Plateau region, the features of typical highway section environment and typical highway freight vehicle features are selected and extracted. In order to obtain vehicle operating feature parameters, field tests were performed in each season at altitudes of 2500, 3000, 3500, 4000, and 4500 m respectively, mainly include vehicle stability test on road curves, climbing capacity comparison test, downhill capability comparison test and brake temperature rise test.

The extracted feature parameters are substituted into the vehicle dynamic performance model and vehicle braking performance model, and to carry out calculations and analysis on the relative performance values of the vehicle, which leads to the analysis of the changes in vehicle safety features, including analysis of its impact of traffic safety on long straight lines, small radius circular curves, and longitudinal slope, etc., and to find out the relations in between.

(2) **Studies on the Features of Psychological and Physiological Changes of Drivers**

From perspectives of hypoxic, unique routes and traffic environment on the plateau, according to psychological and physiological theories both domestic and international, analysis are carried out on the mechanisms of psycho-physiological changes of vehicle driver's in such environments, so as to identify the environmental factors that may affect driver safety.

Based on these environmental factors, to adopt indoor virtual reality simulation and on-site vehicle test after designing the orthogonal test, by using physical and psychological testing equipment, combining drivers' physiological and psychological state scales, drivers' physiological and psychological features data can be acquired, and to propose measures that can significantly and objectively characterize the psycho-physiological state of vehicle drivers, such as eye movements, heart rate variability, etc.

For altitudes at 3500, 4000, and 4500 m, dynamic psycho-physiological parameters and dynamic GPS are used to detect real-time changes in various psycho-physiological parameters and running speed. Main methods used include static measurements, accompanying vehicle records and questionnaires.

At last, combining drivers' psycho-physiological state measures, to establish a correlation between the parameters of natural environment and routes of plateau with the physiological and psychological indicators of drivers, and to analyze the relationship between design of traffic engineering facilities and psycho-physiological measurement indicators, as well as the physiological and psychological patterns of change.

(3) **Studies on Driver's Perception and Operation Ability**

Studies on driver's perception and operation ability in plateau hypoxic environment mainly include:

(1) Characterization of Human Body Changes in Hypoxic Environment

From perspectives of driving operation process, to analyze capabilities, functions and processes required in each action, combining medical research on human body, and to discuss the possible effects of hypoxic environment on human body in terms of awareness, judgment and operating ability. A basic change relationship of the theory can be established after analyzing and discussing the main factors/influencing factors involved in these effects.

(2) Driver's Visual Recognition, Judgment and Operation Test in Hypoxic Environment

First of all, combining the research results of driver's ability assessment and visual recognition evaluation wordlwide, to determine indicators that can be used to evaluate and compare the perception and operating abilities of drivers in high altitude environments, such as response time, maximum information, etc. Based on the research results of features of human physiological changes in the plateau hypoxic environment, preliminary modeling of the relationship between plateau hypoxic environment and driver's perception and operating abilities are established. On the basis of above results, methods of virtual reality simulation and on-site observation are used to carry out tests on drivers in plateau area, and based on the analysis of the test results, further improvement and modification are made to the modeling of the relationship between plateau hypoxic environment and driver's perception and operating abilities, and to finally summarize and determine the changes of driver's perception, reaction and operation ability in the plateau hypoxic environment, and to provide theoretical basis for road safety design in such environment.

(3) Studies on Driving Fatigue Characteristics in Low Pressure and Oxygen Deficient Environment

Driving fatigue is tested by the following three methods, evaluative measurement method, psycho-physiological response parameters test method and simulation measurement method.

The psycho-physiological response parameters test method is carried out at altitudes of 2500, 3000, 3500, 4000 and 4500 m. To make sure drivers experience fatigue, each test driving distance of small car shall not be less than 300 km (continuous driving for 2.5 h at a speed of 120 km/h); that of for medium-sized bus is 250 km (continuous driving for 2.5 h at a speed of 100 km/h); and for medium-sized and trucks, the test driving distances are no less than 200 km (continuous driving for 2.5 h at a speed of 80 km/h).

For simulation measurement method, by designing long distance road models and using driving simulation, and to select driver for long driving simulation test, according to the vehicle trajectory and driver's ECG state changes, calibration and evaluation are made for driver's fatigue status. Meanwhile, self-subjective assessment of driver load status is carried out by means of load scales, combining the two results, a regularity model of driver's fatigue development based on different low pressure and oxygen deficient environments is developed.

(4) **Studies on Freeway Operating Speed Model**

(1) Research and Analysis of Domestic and International Technical Methods

By investigating and analyzing the theories of operating speed prediction and evaluation methods currently used wordlwide, such as the IHSDM in the United States, safety evaluation guidelines for highway projects in China, etc., through the studies and analysis of these basic theories, modeling methods and calculation steps, modeling methods and experimental steps that can be used to generate continuous operating speed prediction calculations are proposed.

(2) Road Section Research and Data Collection

For low-pressure hypoxic environment and geological features of the Qinghai-Tibet Plateau, researches are carried out on the project area, and basic environmental features of the plateau are collected. By surveying and collecting geometry parameters, traffic management status, and operating speed on the following roads and areas, which are: similar roads in different areas (1~2 similar linear sections in Plain area), different roads in similar areas (1~2 other linear design or grade of the sections in Plateau area), and similar roads in similar areas (1~2 built freeways or engineering application highways in Qinghai-Tibet Plateau region), the data simulation analysis of operating speed of Qinghai-Tibet Plateau freeway can be conducted.

Survey collections are made mainly on the alignment design, main model of vehicles and corresponding proportion, and to compare the differences between highways in different regions and at different levels, while data collections are mainly continuous speed collection along the line and operating speed collection at the feature cross-section. When performing feature cross-section selection, factors such as design speed, terrain features, plane alignment, longitudinal alignment, cross-sectional alignment are comprehensive considered, types of section should include straight lines + transition curves + circular curves, S-shaped curves, convex vertical curves + flat curves, concave vertical curves + flat curves, convex vertical curves + straight lines and concave vertical curves + straight lines. The number of comprehensive collection cross-sections should be over 100, and number of each vehicle model per section should be over 200.

(3) Building of Operating Speed Prediction Model

After parsing and data collection of traditional operating speed model, to analyze the factors, which influence the driving condition and speed, generated by special environmental features and vehicle power performance changes which may exist in the Qinghai-Tibet highway area, mainly include excessive vehicle speed caused by environmental uniformity at roadside, and engine output road attenuation and vehicle acceleration performance attenuation that are caused by low pressure and oxygen deficient environment. Performances of the vehicle are based on the study results of typical vehicle performances and safety feature changes under low pressure and oxygen deficient environment, correction are made to acceleration, terminal speed, climbing performance, etc. in the current vehicle operating speed model, so as to build an operating speed prediction model applicable to freeways in Qinghai-Tibet Plateau region.

(4) Model Validation and Engineering Applications

Field operating speed observation are carried out on freeways that can characterize the Qinghai-Tibet highway, meanwhile, operating speed prediction model are used to calculate according to the design information and environmental features of the road section, and compare the differences between observed and predicted values of operating speed. When large differences are found, analysis shall be carried out in terms of road environment, vehicle factors and model foundation theory, and making the corresponding corrections. When the model cannot significantly characterize the operating speed features of this validation section, try to modify the model from the basic parameters of modeling, to ensure the objective validity of the model.

After establishing an effective operating speed prediction model, to carry out corresponding operating speed prediction in engineering application sections, to calculate and evaluate the speed distribution along the road, in order to provide the basis and recommended data for the following design of highway routes and traffic safety facilities, etc.

1.3.2 *Research on Traffic Features and Speed Control of Freeways in Qinghai-Tibet Plateau Region*

The researches are mainly three parts.

(1) **Research on the Features of Traffic Accidents and Traffic Safety Service Level in High-altitude Hypoxic Areas**

(1) Analysis of the features of Road Traffic Accidents on the Qinghai-Tibet Plateau

Combining road information, demographic and economic information, and weather features of the Qinghai-Tibet Plateau, to carry out statistical analysis of the spatial and temporal distribution, types and patterns of road accident features on the Qinghai-Tibet Plateau, and summarize factors of traffic accidents, and on the basis of the analysis of the main causes of traffic accidents under the conditions of hypoxic in the plateau and the main patterns of accidents, according to the features of the distribution of traffic accident locations along the route and the frequency distribution of traffic accidents, to select high traffic accident locations for observation for data collection, to further explore the interaction between road geometric alignment, geographical environment features, traffic operation status and traffic safety.

(2) Correlations between Different Oxygen Levels and Traffic Accident Features in Plateau Area

According to the relationship between oxygen content and altitude and air pressure, to carry out oxygen level survey of the studied road. Taking the road elevation division method as a reference, grading roads according to different oxygen levels, indicators such as 10,000-vehicle accident rate, 100 million vehicle-kilometer accident rate, 10,000-vehicle fatality rate and 100 million vehicle-kilometer fatality rate in plateau area are selected to reflect the traffic accident rate, to establish association of different oxygen content roads between traffic accident rates and accident types respectively, types of accident include severity and morphological features of the accident.

(3) Research on the Features of Traffic Accidents in Uninhabited Plateau Area

Summarize the special features of the uninhabited plateau area, such as alpine, windy, climate variability and extreme rapidity, and combined with road linear features, accident information data in uninhabited plateau area, comparative analyze the features of traffic accidents and traffic accident rates between uninhabited roads and roads near residential areas at the same elevation, to carry out statistical analysis of the impact of geographical features of uninhabited plateau area on the morphological features of traffic accidents, traffic accident rates, etc.

Select accident rate statistics, features (severity, form, time) and causes of traffic accidents of non-uninhabited roads at the same elevation, compare the features of traffic accidents on uninhabited roads with those on non-uninhabited roads, and to analyze the features of traffic accidents in uninhabited areas.

Based on linear features and accident data in uninhabited plateau area, combined with the geographical weather features and causes of accidents, to analyze the influence of road environment features on pattern features (severity, form, time) and rate of traffic accident.

(4) Research on Traffic Safety Service Level in High Altitude Hypoxic Area

Based on the above traffic survey and study of the relationship between road elevation, road environment and traffic accidents, determining the set of key factors influencing road traffic safety on the Tibetan plateau, and through correlation analysis, its relationship with traffic accident rate and road traffic accident prediction model in plateau area were established. Meanwhile, according to reviews of domestic and international researches, traffic accident rate statistics, determine the grading of traffic safety service level on roads in Plateau area, and then combined with the road traffic accident prediction model in Plateau area, to carry out forecasting on level of service for road safety on the Tibetan plateau, and establish quantitative standards for the level of traffic safety services on Qinghai-Tibet highway.

(5) Traffic Accident Prediction Model for High Altitude Hypoxic Area

Based on the probability distribution pattern of the number of traffic accidents or accident rates, to carry out studies on the relationship between road elevation, road environment and traffic accidents by analyzing existing traffic accident data, and considering the relationship between various independent variables (alignment, traffic flow, environment) and the dependent variable (frequency of accidents), and to determine a set of key influencing factors for road traffic safety on the Tibetan plateau, then establish prediction model of traffic accidents in plateau area by statistical regression method.

(6) Quantitative Standards for Safety Service Level in High Altitude Hypoxic Areas

By analyzing the accident rate and traffic volume in plateau area, based on the above established freeway traffic accident prediction model, to discuss its grading criteria by using probability and statistical theory, and selecting safety service level grading indicators to predict road traffic safety service level on the Qinghai-Tibet Plateau, so as to improve the quantitative standards research of traffic safety service level in plateau areas.

(2) **Research on the Design Technology of Dynamic Speed Control (Speed Limit) of Freeway in Special Environment**

(1) Relationship between Speed Limit and Traffic Safety in High Altitude Hypoxic Area

In Qinghai-Tibet Plateau, the traffic volume is low and with good visibility, the topographic conditions along the highway is fine, generally with higher alignment indicator, roads are free of lateral disturbance in permafrost areas, therefore, drivers usually expect to go through the high altitude and hypoxic section at higher speeds. And the overall performance of the vehicle decreases in the low pressure and oxygen

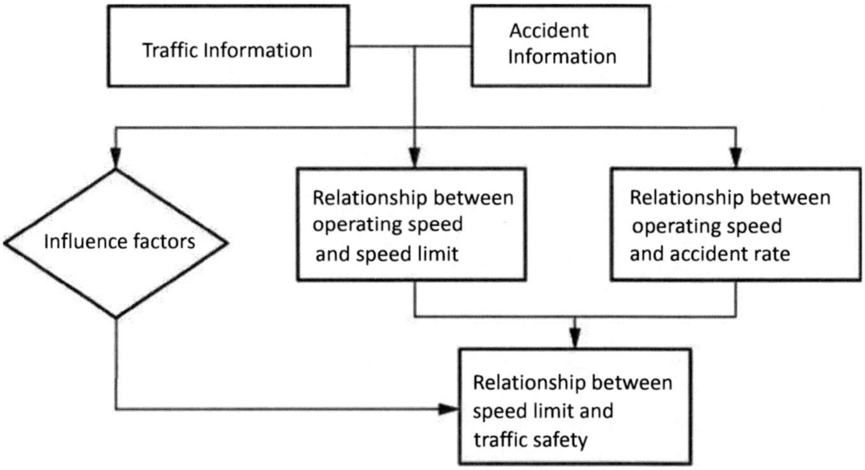

Fig. 1.4 Technical road map for research on the relationship between speed restriction and traffic safety in high altitude hypoxic areas

deficient environment, and driving stability declines sharply due to long-term snow and ice accumulation, drivers are hard to concentrate at high speeds and with reduced vehicle dynamics, in order to ensure the driving safety, speed limits are imposed on special sections of road while satisfying comfortable driving.

Based on the relationship between restricted speed and operating speed on different special sections such as the Qinghai-Tibet Plateau with good visibility, faster driving expectation, and reduced vehicle performance, by studying the relationship between operating speed and accident rate on the Qinghai-Tibet Plateau, and to establish a research relationship between speed limit and traffic safety on the Qinghai-Tibet Plateau (Fig. 1.4).

(2) Multi-objective Decision Model for Speed Limit Value in High Altitude Hypoxic Area

Comprehensively considering the influence of low-pressure hypoxic, alpine freezing, scenery monotony and other effects on the driver's physiology and psychology, as well as on the power performance of the vehicle, and based on the objective constraint of integrated traffic safety and operational efficiency, a decision model with safety objective and integrated efficiency impact is established for the speed limit value of Qinghai-Tibet Plateau highway. It is proposed the vehicle operating speed control standards and arrangement basis at different altitude, alignment features and terrain conditions in high altitude hypoxic area, and to evaluate the effect of engineering applications, and using speed control technology to improve the safety factor of highways in Qinghai-Tibet Plateau.

(3) Research on Dividing the Length of Speed Limit Zone of High Altitude Hypoxic Area Road

Due to the hypoxic content in the Qinghai-Tibet Plateau, the driver's driving behavior contrasts greatly with that of at lower altitudes, generally the higher the altitude the greater the driving workload for the driver to work normally. At the same time, it is often long distance running in plateau area, drivers are prone to be fatigue and nervous, and driver's reaction time increases as altitude increases, and driving agility decreases as a result. Due to the long-term impact of snow and ice, the vehicle power performance is weakened, the road is slippery, and vehicle braking effect is reduced.

Based on the operating speed variation features, design specifications, and technical standards of Qinghai-Tibet highway, considering the impact of plateau environments such as hypoxic content, high altitude, snow and ice on roads, drivers, vehicle performance, analyzing and obtaining the factors influencing the length of the speed limit zone on the Tibetan Plateau, and establishing the minimum length of the speed limit zone under different circumstances, and and to determine the length of the speed limit zone division combined with safety, economic and other principles.

(4) Study on the Transition of Speed Limit Area in High Altitude Hypoxic Area

When a highway transits from a speed limit zone with a higher speed limit value to a zone with lower speed limit value, if the speed limit difference exceeds a certain range, in order to make drivers slow down safely, smoothly and comfortably, it is suggested to arrange a certain length of transition section between the two speed limit areas, and arrange the corresponding prompt signs.

The transition of the speed limit zone mainly considers the distance required for the visual recognition of traffic cues, the distance required for running speed adjustment and driving length needed to maintain stable driving for drivers on the Tibetan plateau. Based on design specifications, technical standards, etc., considering the factors in the Qinghai-Tibet Plateau under alpine and hypoxic, such as wind and snow or ice on the windshield, the vehicle's power performance is weakened, the road friction coefficient is reduced, the driver's responsiveness is weakened, and the driver's vision is obstructed, and establish a model for the transition length of the speed limit zone on the plateau and determine the length of the transition section.

1.3.3 Research on Main Technical Indicators of Freeway Alignment in Low Pressure and Oxygen Deficient Environment

Based on the specialized research results on operating speed, vehicle performance and driving features in high altitude hypoxic area, combined with the special construction environment along the Qinghai-Tibet freeway and the functional features of the Qinghai-Tibet freeway, etc., and to determine the main technical indicators and parameters using thresholds of the Qinghai-Tibet freeway from a safety perspective, implementing as the following parts.

(1) **Research on the Main Geometric Indicators and Parameters of Qinghai-Tibet Freeway**

(1) Systematically Summarize the Previous Research Results

Through the research topic of characterization of typical vehicle performance and safety in low pressure and oxygen deficient environment, vehicle dynamic features model and vehicle braking performance model are established, calculation and analysis are made for the corresponding performance values of the vehicle, and obtained the variation of vehicle features at different altitudes, and analyzing vehicle safety in long straight lines, small radius circular curves, and longitudinal slopes.

By studying the features of physiological and psychological changes of drivers in low pressure and oxygen deficient environment, to establish the correlation of the parameters such as plateau natural environment, routes and traffic environment between the physiological and psychological indicators of drivers.

And to systematically summarize the research results and provide theoretical support for the selection of geometric indicators and parameters of Qinghai-Tibet freeway.

(2) Research on Building A Mobile Driving Simulation System, Propose Thresholds for Relevant Indicators Based on Simulation Tests and Comparative Validation

Based on the previous research results such as vehicle dynamics research results, driver's psycho-physiological change rules, to carry out studies on the relationship between vehicle performance and driver's psycho-physiological changes under the conditions of single indicator and different alignment combinations of highway, and to establish the relationship model between operating speed, vehicle performance, driving features indexes and road single route indicators and different alignment combinations at different altitudes.

Taking mobile driving simulation test system as a research platform, to achieve multiple superposition in the simulated driving test system, such as weather conditions, surrounding environment and any complex road geometry, to reproduce the impact of different meteorological conditions and altitude on traffic safety, and to realize the functions of simulated driving, operating speed analysis, visibility and safety sight distance detection on roads with complex linear conditions.

Finally, study and propose the main geometric indicators and parameters suitable for the freeway under low pressure and oxygen deficient environment in the Qinghai-Tibet Plateau area, such as the length of the section divided by the same design speed, stopping sight distance value, maximum and minimum longitudinal slope value, average longitudinal slope value, longitudinal grade reduction, etc.

(2) **Study on Cross Slope of Crown and Pavement Superelevation**

(1) Systematically Summarize the Previous Research Results

Through the topic of changes in performance and safety features of typical vehicles under low pressure and oxygen deficient environment, the study concludes the change pattern of power performance and braking performance of vehicles under different natural conditions in the plateau, and to provide theoretical support for the presentation of relevant technical indicators.

(2) Proposed Values of Cross Slope of Crown and Pavement Superelevation Combining with Experimental Tests

Aimed at the security problem regarding the technical parameters of roadbed cross-section of Qinghai-Tibet freeway, taking the Tibetan highway as a sample, conducting research on the correlation between pavement performance parameters and traffic accidents on Tibetan roads, and carry out on-site testing and inspection on cross slope of crown and pavement superelevation by relying project.

Considering the serious ice and snow disasters affecting local sections of the Qinghai-Tibet highway, to carry out the distribution features of dangerous road sections with icy, frosty and snowy roads in plateau areas based on the analysis of the detection data of slippery road conditions, combining with the features of the distribution of traffic accidents under ice and snow conditions, so as to determine the maximum ultra-high critical value of road traffic safety under the ice slippery degree of freeway pavement in high altitude areas. G214 to Jiegu section is selected for testing, combined with the vehicle speed and driving stability analysis, the road arch cross-slope thresholds for icy and slippery road surfaces under the condition of ensuring driving safety are proposed.

(3) **Study on Reasonable Spacing of Service Facilities**

(1) Systematically Summarize the Previous Research Results

Discuss the possible effects of the hypoxic environment on the human body in terms of recognition, judgment and operation ability through the analysis of the features of human changes in the hypoxic environment. Through the driver's visual recognition, judgment and operation test under hypoxic environment, to establish a model of the relationship between plateau hypoxic environment, driver perception and operational ability. Finally, the change pattern of driving fatigue features under low pressure and oxygen deficient environment is obtained.

(2) Determining Reasonable Spacing of Service Facilities

Base on the previous theoretical analysis and related dependent projects, by detecting driving fatigue at different altitudes and under a certain length of continuous driving, to carry out study on the relationship between factors such as altitude, continuous driving time and fatigue, analyze the change pattern of fatigue time with altitude, continuous driving time, etc., and establish the corresponding model.

By collecting data on road traffic accidents in Tibetan areas, the correlation between driving fatigue and traffic accidents in the plateau environment and the mechanism of driving fatigue induced traffic accidents are analyzed, so as to provide references for taking values of reasonable spacing of service areas.

Combining the above factors, to propose the best time and mileage for drivers to rest and relieve fatigue on the way in different plateau hypoxic areas, and to finalize the reasonable spacing of service facilities on Qinghai-Tibet freeway and some specific requirements for the design of service facilities, etc.

1.3.4 Study on Technical Standards and Design Guidelines for Qinghai-Tibet Freeway

(1) **Systematically Summarize the Previous Research Results**

To establish an effective operational speed prediction model through the research topic of freeway operational speed model in Qinghai-Tibet Plateau region. Calculate the vehicle operating speed distribution along the highway, and to provide basis for highway geometric design, selection of technical standards and other aspects. At the same time to summarize research results related to vehicle performance change pattern and driver's psycho-physiology in the plateau area, to propose theoretical support for technical standards.

(2) **Determine the Selection Methods and Principles for Technical Standards**

To determine the selection method and principles of technical standards on the basis of the preliminary study. Due to the diverse altitude, topography, and climate change on the Tibetan Plateau, it is impossible to choose the same technical standard for the design of the whole freeway on the Plateau, different design criteria are required depending on the actual situation.

Principles and methods for determining the technical standards are based on the following four main factors, which are, **i.** Consistency and coordination of operating speed, that is, the operating speed difference or the gradient of operating speed of adjacent sections should be less than a certain critical value, so that the transition between different operating speeds can be as smooth as possible. **ii.** Natural conditions such as topography, landform, altitude and its climatic conditions along the freeway, that is, the topographic and terrain features along the Qinghai-Tibet freeway are complex and changeable, vehicle power, tire and braking performance models also vary at different altitude conditions, therefore, in the selection of technical standards, different natural conditions of the surrounding should be considered. **iii.** Depending on the effectiveness of the implementation of the project, that is, carry out comparison of road sections of Hua-Da freeway and Gonghe-Jiegu of G214 for technology demonstration application, evaluate the application effect before and after the implementation of the demonstration project, and summarize the experience in the application of technical standards. **iv.** Selection of technical criteria also requires to considerate the adoption of route geometry indicators.

Based on the research results related to operating speed, vehicle performance and driving features in high altitude hypoxic areas, combined with the special construction environment along the Qinghai-Tibet freeway and the functional features, and to finally determine the methods and principles of dynamic deployment of technical standards, and the minimum length of use of the same technical standard.

(3) **Compilation of *Safety Design Guide for Qinghai-Tibet Plateau Freeway***

Summarize the research results of on highway operating speed, vehicle performance, driver's features, highway geometric alignment indicators and design standards, traffic engineering and design essentials of facilities along the Qinghai-Tibet Plateau under low pressure and oxygen deficient environment, and carry out compilation of *Safety Design Guide for Qinghai-Tibet Plateau Freeway*, for instructing the overall design of the Qinghai-Tibet Freeway route.

1.4 Technical Difficulties and Research Methods

The key technical issues of this study including the study of vehicle performance and driver's psychophysiological load and fatigue behavior under low pressure and oxygen deficient environment, the study of main technical indicators and application technologies of freeway routes, and the study of key technologies for traffic safety assurance of Qinghai-Tibet Plateau freeway.

i. Aiming at many problems that low pressure and hypoxia environment affect the safe operation of driver and vehicle in Plateau area, a mobile driving simulation system is built on-site through vehicle test calibration, and the test research on driver-vehicle coupling environment operation is carried out. A correlation model between altitude and driver response, altitude and vehicle performance is built to study the technical indicators of highway route in Plateau area. The parameters are supported by methods and safety design techniques.
ii. According to the impact of traffic characteristics, vehicle performance, and driver's psychophysiological characteristics on highway geometric alignment indicators in high-altitude hypoxia areas, an inherently safe route indicator system for Qinghai-Tibet Freeway is proposed to adapt to environmental requirements.
iii. By analyzing driving behavior, fatigue time, and accident characteristics at high-altitude hypoxic regions, we regressed to derive a model of the relationship between driving distance and drivers' psychophysical characteristics, and then proposed a reasonable spacing of service facilities along the Qinghai-Tibet Freeway.

iv. By studying the key influencing factors of high altitude road traffic safety service levels, we establish a prediction model for high altitude traffic accidents based on the characteristics of altitude and oxygen content, and propose a grading standard for high altitude road traffic safety service levels.
v. Based on the traffic safety goal, comprehensive study of the interaction relationship between rate limiting and traffic accidents, operation efficiency, and establish the control standard and rate limiting decision model of high altitude road speed in cold and low oxygen condition.
vi. It was developed to be suitable for the special climate condition of snow and winds, with strong crash proof ability, flexible and energy absorbing, and easy to conserve on freeway road side safety facilities.

The main research methods are as follows: first, we conducted an on-site and data survey analysis to analyze the road conditions, the characteristics of the traffic situation and the vehicle type, and the mechanism of the traffic accident due to the low oxygen environment of the Qinghai-Tibet Freeway. On site measurement of on-site detection and integrated on-board detection and sensing equipment has been autonomously developed and integrated to obtain parameters of power characteristics and brake performance and tire characteristics of vehicles under low oxygen environment, measurement indicators of drivers' psychophysiological status, drivers' perception and operation ability, and establish a relationship model using mathematical and statistical analysis methods. Through a temporary construction of a virtual reality simulation platform in the survey area for a number of comparative verification: according to the changing law of the vehicle power and brake performance, the parameters distribution of the driving condition of the road surface are the flat, longitudinal and cross design technical parameters of the road in this area using the value domain and the application method to provide a technical basis; using the results of the analysis of drivers' psychophysiological status and perception and operation capability, the traffic engineering facilities such as service area spacing and landmark marks were further optimized in combination with the impact relationship with traffic safety facilities. Ultimately, indicators for geometric design on the Qinghai-Tibet Freeway targeting the hypoxic areas of plateau will be determined based on research on road condition safety, vehicle dynamic characteristics, and drivers' psychophysiological characteristics, to determine the reasonable spacing of service facilities along the Qinghai-Tibet Freeway and the principles and points of traffic engineering and arrange facilities along the road.

At the same time, under the guidance of both the research results and theories of speed control technology, from the protection of road traffic safety in special environments, the study proposes freeway dynamic speed control (rate limiting) technology and related specific methods, indicators and requirements in special environments to guide the design and construction of the Qinghai-Tibet Freeway route and traffic engineering; To investigate the prevention and treatment technologies for ice and ice disasters on freeways wordlwide, aiming at the problems of relatively severe impacts of ice and ice disasters on the local section of the Qinghai-Tibet Freeway, studies propose or develop applicable, economical, and feasible targeted safety and security technologies and equipment.

By conducting a field demonstration that considers the equipment of partial protection technologies under basically the same driving conditions as the Qinghai-Tibet Freeway and summarizing the research results, we establish a system of main technical indicators of freeway routes under low pressure and oxygen deficient environments on the Qinghai-Tibetan Plateau and propose a dynamic design speed segmentation method, speed control techniques, and Transportation Engineering with facility arrangement along the Qinghai-Tibetan plateau that are suitable for the geographical environment, traffic driving characteristics, Thus forming a key technology system for freeway safety design under special circumstances, and finally proposing a technical standard for Freeway Engineering in the Qinghai-Tibetan region and compiling a *Safety Design Guide for Qinghai-Tibet Plateau Freeway.* The technical flow are shown in Fig. 1.5.

1.5 Scientific and Technological Achievements and Innovation

The author of this book carries out indoor and outdoor test and driving simulation to obtain basic information under low pressure and oxygen deficient environment at high altitude, builds relevant models through theoretical analysis, and then determines relevant technical indicators and parameters for geometric design of Qinghai-Tibet Freeway. Combining the current application status and trend of freeway operation safety management and accident prevention technology, the expresswa driving safety assurance technology under special weather conditions is studied to prevent accidents from the angle of improving intrinsic safety and strengthening active intervention of the system. The main scientific and technological achievements include the following 6 aspects and 24 achievements.

(1) **Psychophysiological Characteristics of Drivers and Vehicle Performance at Different Altitudes under Low Pressure and Oxygen Deficient Environment**

   i. Analyzing the traffic accident distribution law and characteristics of G214 and G109 in Qinghai-Tibet Plateau from the aspects of space, time, cause, form, traffic mode, weather and road condition of traffic accident. According to the analysis results of traffic accident risk degree, the risk of speeding, fatigued driving and improper braking is the highest. The major influencing factors of traffic accident risk include trucks, snow and ice pavement conditions, fog and wind weather conditions, steep bends and steep slopes.
   ii. According to the natural environment of low pressure and hypoxia environment in Plateau and the traffic composition characteristics in Qinghai-Tibet region, the truck with 30t load, 8.3 kW/t power-to-weight ratio, 49 T load-to-weight ratio and 5.6 kW/t power-to-weight ratio is used as the test

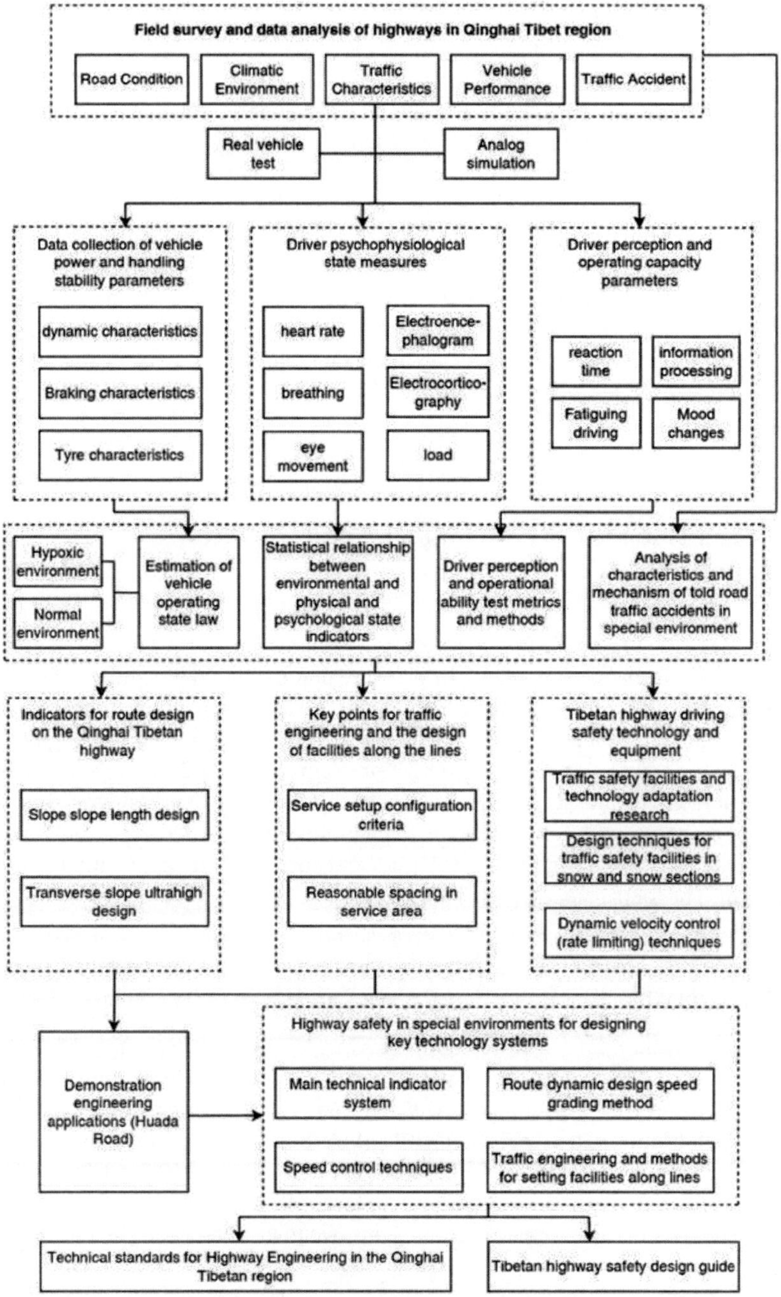

Fig. 1.5 Technical flow chart

vehicle to conduct field tests at altitudes of 3000, 3500, 4000, 4500, 5000 and 5500 m, respectively. External engine characteristics/braking characteristics models of trucks (power-to-weight ratio 8.3 kW/t) and articulated trains (power-to-weight ratio 5.6 kW/t) under full load at high altitude and engine driving/braking torque changing with altitude were established. The research results show that overall vehicle power performance decreases with the increase of altitude, engine torque decreases by 7–10% for each 1000 m elevation.

iii. According to the results of driver's psychophysiological indoor and outdoor tests, the driver's heart rate and heart rate growth rate obviously increased with the elevation of altitude, the oxygen saturation obviously decreased, and the respiratory tidal volume had a slight upward change characteristic. The factors that influence the change of Heart Rate indicators at altitude are mainly elevation, alignment indicators, and speed, of which the greatest extent of the effect is observed at altitudes, with mild hypoxic elevations starting below 4000 m of altitude and severe hypoxic elevations beginning above 4000 m of altitude. When a driver travels at a certain speed, the higher the elevation, the worse the alignment indicator, that is, the smaller the radius of the flat curve, the greater the value of alignment combination, the greater the heart rate growth rate, and the greater the tension during the driver's driving.

iv. According to the results of simulated driving simulation trials at different altitudes, the earlier the time point the driver developed fatigue as the elevation rose. Fatigue time points were approximately 27.8% earlier for large car drivers and 25.2% earlier for small car drivers at 4600 m altitude compared with 3540 m. The higher the altitude, the less oxygen and the more likely fatigue will occur.

(2) **Value Criteria for the Indicator Limit Values of the Flat Aspect Geometric Technique for Freeways on the Qinghai-Tibetan Plateau Region**

i. By studying the four-variable model of heart rate growth rate, altitude, linear indicator (radius of flat curve, combined value of line shape) and vehicle speed, the dividing point of psychological and physiological reaction, i.e. R'and N', is put forward for drivers at 3500–4700 m elevation when driving on combined sections of flat curve or bend. In the range of 3567–3957 m above sea level, the minimum radius of flat curve is 300 m. The minimum radius of flat curve is 350 m in the range of 4100–4702 m above sea level.

ii. Based on the external characteristics of a typical representative vehicle model under different altitude conditions to characterize the change of climbing motility and brake resistance in sections with different slopes, this study proposes that the equilibrium speed (satisfying different design speeds and lowest allowable speeds) and the maximum slope impairment in sections with different slope and ultimate slope length design parameters are between 0 and 6% (graded every 0.5%) slope.

iii. According to the results of the driver reaction time test, the driver's reaction time was prolonged by about 70 ms on average for every 0.5 h increase in driving duration, and the driver reaction time interval at high altitude was larger than that at low altitude, indicating that the higher the altitude, the greater the driver's psychophysical load, and prolonged driving was more likely to cause fatigue. However, according to the test results, the perceived time (maximum 0.75 s) of drivers at high altitude in the event of unpredictable events is less than the use criterion (1.5 s) for the perceived time in our current standard specification, so the recommendation to take the value of parking distance at high altitude is still to adopt the value standard value in the current standard specification.
iv. From the aspects of highway function, operation safety, scale effect of frozen subgrade, etc., the reasonable width of each component element of freeway cross-section in Qinghai-Tibet Plateau area is systematically analyzed, and the recommended form of cross-section suitable for special environment of Qinghai-Tibet Plateau and the minimum width of each component element are put forward.
v. By carrying out a force analysis on the vehicle's moving condition in the ultrahigh segment, the lateral friction coefficient of the vehicle when the road surface is most unfavorable under ice and ice conditions is used to calculate the safety net of the vehicle while moving in the ultrahigh segment at 40, 60, and 80 km/h conditions, respectively, taking the intermediate value between the maximum and minimum ultrahigh values as the most desirable ultrahigh value, Finally, reasonable values of ultrahigh and radius at different velocity conditions were obtained.
vi. Based on the mobile driving simulation equipment, driving behavior tests under low pressure and oxygen deficient environment were conducted at different altitudes, and the electrocardiographic changes during driving were analyzed, comprehensively considering the main indicators of elevation, driving time length, beat to beat interval change rate, driver class, and so on, a multivariate logit regression model was used to establish a multi index discrimination model of driving fatigue, driver fatigue probability distribution curves and fatigue cut-off points at different altitudes were obtained.
vii. The service facilities are divided into three categories based on the actual needs of vehicles and drivers in the plateau area, and a comprehensive consideration of the five main aspects, including the vehicle fueling needs, the physiological and safety needs of drivers in the high-altitude environment, the truck needs, the emergency needs of natural disasters, and the observational needs, will be conducted to coordinately arrange the intervals between the three categories of service facilities, and to study the distribution spacing, as well as the configuration criteria, of service facilities in different areas.

(3) **Key Design Technology and Application Demonstration of Traffic Safety Guarantee of Qinghai-Tibet Freeway**

i. Using a large number of measured data from Metro Count Traffic Tester, this paper explores the influence of oxygen content in air on the speed difference between trucks and passenger cars in straight and longitudinal sections. Finally, it is concluded that the speed difference increases gradually with the decrease of oxygen content in the straight section. When the altitude is higher than about 4350 m, the speed difference between trucks and passenger cars is more than 20 km/h. Through the research on speed difference and operating speed characteristics of vehicles with different oxygen content on Qinghai-Tibet Freeway, the influence of oxygen content on its variation law is obtained. Considering the effects of low pressure and hypoxia environment, cold freezing and monotonous scenery on driver's physiology, psychology and vehicle dynamic performance, a speed limit decision-making model of freeway under special environment of low pressure oxygen deficiency is established.

ii. According to the number and characteristic distribution of road traffic accidents in Qinghai-Tibet Plateau region under oxygen deficient environment, a highway traffic accident prediction model in plateau area is proposed, which is related to the terrain along the line, vehicle composition, regional oxygen content and other indicators. At the same time, according to safety service level classification standard of σ-theory, with the predicted value of the accident mean equation as the center line, the classification standard of highway traffic safety service level in plateau area is put forward.

iii. Establish simulation analysis model and method of traffic sign wind load CFD, establish index of efficiency of wind load reduction, analyze and compare the effect and efficiency of different measures to reduce wind load in separate use and comprehensive application, put forward and complete verification of low wind load two-pillar sign structure of two layout designs, which are location indication sign and large vehicle right-hand warning sign respectively.

iv. Research and development of two new freeway side guardrails (cable guardrail, steel pipe guardrail) suitable for high altitude wind, ice and snow conditions and easy maintenance. The snow-proof cable guardrail and two low wind load marking facilities have passed simulation test analysis and crash test, and have been demonstratively applied on Hua-Da Freeway relying on the project.

v. Carry out freeway trial design and safety evaluation in the demonstration project of Huashixia-Dawu section of Chengdu-Xiangride freeway, formulate speed control plan of demonstration section, design and application of new low wind load sign and snow barrier, and complete the implementation of demonstration engineering application proposal.

(4) **Optimized Technology of Highway Route Safety Design in Qinghai-Tibet Plateau**

i. By analyzing the stress state of trucks on the uphill section, the equilibrium speed of corresponding gradient at each elevation is obtained. Based on the theory of equivalent gradient at different elevations, the conversion method of equivalent gradient corresponding to actual gradient at different elevations is calculated, and the modification and application method of operating speed prediction model in high elevation area is proposed. According to the model, the error between the predicted operating speed and the observed results is 3–7%.
ii. Based on the influencing factors such as highway function, topographic and geological conditions, and distribution of operating speed, the section technology of freeway design speed in Qinghai-Tibet Plateau area is put forward.
iii. Put forward the technology and process of route safety optimization design applicable to freeway in Qinghai-Tibet Plateau area.
iv. Combining the conversion method of actual and equivalent gradient of highway longitudinal slope in high altitude areas and brake temperature prediction model considering altitude factors, the safety evaluation technology for continuous long and steep longitudinal slope section above 3000 m is put forward.

(5) **Selection Principles and Methods of Technical Standards for Qinghai-Tibet Freeway**

According to the special geographical and climatic environment of Qinghai-Tibet Plateau, taking full account of the influencing factors such as vehicle performance, driver's psychophysiology, operating speed distribution, traffic volume and its composition, engineering economy, etc. in the high-cold and low pressure and oxygen deficient environment, the selection principles and methods of technical standards such as design speed, design vehicle, cross-section form and width, interchange spacing, traffic engineering and facilities along the freeway are put forward.

(6) **Development of Test Equipment and Analysis Software**

i. Based on various sensors and test electronics, the vehicle performance test system in plateau area has been independently integrated, and has been tested and used in the performance test of six-axle truck and four-axle large truck under low pressure and oxygen deficient environment.
ii. Independently developed and integrated mobile driving simulation test platform, with SMI eye tracker and BIOPAC multi-channel physiological monitor, can meet the test requirements of route geometry parameters, safety facilities and traffic operation characteristics in the coordinated operation environment of the human-vehicle environment simulation system.

iii. Integrating highway aerial photography, remote sensing, traffic accident and other data in Qinghai-Tibet Plateau area, the secondary development was carried out based on ArcGIS software, and the "Tibet Highway Network Basic Information Platform V1.0" was independently developed to realize the functions of traffic accident statistics, map browsing and design document correlation.

The author of this book has made breakthroughs in the field of technical standards for freeway geometric design and safety design in high-altitude areas, and creatively proposed service facilities and reasonable spacing for different service functions of freeway in high-altitude areas. The research results generally reach the leading international level. The main innovations in the research include the following three aspects:

i. The geometric design indicators of highway, such as minimum radius of circular curve, maximum grade and slope length, parking sight distance, cross-section form, width of each element and superhighway, which meet the traffic operation safety requirements in high-altitude areas, are systematically studied by synthesizing the influencing factors such as driver's psychological and physiological characteristics, vehicle performance variation regularity, skid resistance of ice and snow asphalt pavement, physical characteristics of Frozen Subgrade and so on under different altitudes in Qinghai-Tibet Plateau area. Freeway technical standards and general design methods meeting the requirements of safe operation in cold, low pressure and oxygen deficient environment are established. Key technical indicators and standards for freeway geometric design in Qinghai-Tibet region are put forward for the first time, which fills the gap of main technical indicators in high-altitude areas in current *Highway Engineering Technology Standards* in China and provides a solid foundation for highway safety design in high-altitude areas.
ii. According to different service functions, three types of freeway service facilities and their allocation standards under high altitude and oxygen deficient environment, including driver's physiological needs, fatigue time, vehicle refueling and emergency rescue requirements for ice and snow disasters, are put forward for the first time according to different service functions. It provides basis for supplementing and perfecting service facilities arrangement standards in highway industry standard specifications.
iii. According to the characteristics of long mileage and complex construction conditions of highway in Qinghai-Tibet Plateau area, the speed model of freeway under different altitudes is developed by combining highway functions, topographic and geological conditions, and distribution characteristics of operating speed. A speed-limiting decision model considering safety and operating efficiency in special environment of plateau is established, and sectional speed control technology of freeway in Qinghai-Tibet Plateau area is put forward for the first time. The problem of speed control such as wide-open highway section, large speed difference between trucks and passenger cars has been solved, and

the technology of route safety optimization and highway speed control has been further supplemented and improved.

1.6 Application of Results

Some of the research results in this book have been applied in the design, safety evaluation and speed control of freeway in Qinghai-Tibet region. During the application process, the geometric design indicator and design method of freeway in high altitude and low pressure and oxygen deficient area proposed by stages were inspected and corrected continuously, which finally laid a good foundation for the establishment of technical indicator system of the project and the promotion of research results. At the same time, the research and engineering demonstration work cleared the technical barriers for the successful construction of freeway in high-altitude areas. The main applications of the research results are as follows:

(1) **Application in Highway Design**

Based on the trial design study of the expansion project of Huashixia-Dawu Freeway (hereinafter referred to as Hua-Da Highway), and using the proposed speed model of Qinghai-Tibet Freeway, the indicator such as operating speed and space visual distance of the supported project are calculated and evaluated, and the route plan, modification and adjustment suggestions of relevant indicators are put forward. The optimized route plan is evaluated by experienced technical experts, which significantly improves the coordination and consistency of the alignment indicators.

(2) **Application in Highway Safety Assessment**

Through safety evaluation of high-grade highway in G318 Linzhi-Lhasa Section (hereinafter referred to as Lin-La Highway), operating speed calculation model considering the impact of equivalent gradient correction in high-altitude areas is adopted to get the operating speed distribution of trucks and passenger cars along the highway. On the one hand, the accuracy of relevant research results is further verified by combining the observation results of actual operating speed on site. On the other hand, the possible influencing factors of traffic safety operation in the project are further discovered and the corresponding traffic safety improvement countermeasures are put forward, which play an important role in ensuring efficient and safe operation of Lin-La Highway.

(3) **Application in Speed Control Management**

Through the research on speed limit plan of Hua-Da Highway Expansion Project supported by the project, and adopting comprehensive decision-making model of highway speed limit in high altitude area, the main factors influencing speed limit value of Hua-Da Highway are analyzed, including continuous long longitudinal slope section, small radius flat curve section and interchange section. The speed limit value by vehicle type and corresponding traffic safety improvement countermeasures are put forward, which provide scientific basis for speed control and management of Hua-Da Highway.

1.7 Economic and Social Benefits

The author of this book carries out research on geometric indicator parameters, selection principles and methods of technical standards and traffic safety guarantee technology of freeway in high altitude areas on the basis of research and revealing comprehensive performance changes of large-scale typical freight vehicles at high altitudes, summarizing psychological and physiological changes of driving behavior in freeway driving in Plateau areas, and developing new safety protection facilities suitable for snow and ice environment. Integrating key technology of freeway safety design in special plateau environment, and carrying out project demonstration and application, provide basic test, research and project demonstration support for ensuring freeway driving safety in plateau area and realizing breakthrough of "zero" freeway in Tibet area. The completion of this research will bring considerable economic and social benefits in improving the quality of freeway design and the intrinsic safety of highway in Qinghai-Tibet region. Detailed below:

(1) **Promote Highway Technical Standards in Qinghai-Tibet Region to Promote the Safety and Smoothness of Transportation**

At present, the transportation on the Qinghai-Tibet Plateau is mainly carried by the lower grade two-lane highway. Taking Qinghai-Tibet Highway as an example, Qinghai-Tibet Highway is located in a smooth terrain with good visibility. There is no lateral interference in the permafrost areas for many years. In the section of Kunlun Mountain Pass-Tanggulan Nananduo County, drivers generally expect to cross the cold and oxygen deficient areas at a high speed. However, as the overall technical of Qinghai-Tibet Freeway is relatively low, the running environment is poor, the horizontal and longitudinal alignment indicator of the section crossing mountains is low, and the combination section of continuous longitudinal slope and bend slope has a great influence on driving safety, especially on braking safety of trucks and driving stability of snow-covered sections in winter. In the permafrost area, the horizontal and longitudinal alignment indicator is high, the running speed of vehicles is relatively high and the speed difference between trucks and passenger cars is large. However, as trucks and passenger cars can't be driven by separate lanes, vicious accidents such as rear-end collision and frontal collision are likely to occur. As trucks and passenger cars drive on the same lane, the overall traffic capacity of the highway is also reduced. According to the statistical results of traffic volume of Qinghai-Tibet Freeway in recent years, the service level of the vehicles has approached the lower limit of service level required by the second-class highway (*Highway Engineering Technology Standard* 2014). At the same time, because the Qinghai-Tibet Freeway is not completely closed at present, some livestock (such as yaks), wildlife (such as Tibetan antelope) often travel on the highway, as well as some high fill or cliff section of the road side has poor safety, which has a greater impact on the safety of high-speed vehicles.

From the perspective of transportation demand and national strategy, an freeway is urgently needed in Qinghai-Tibet Plateau to make up for the shortcomings of regional economic construction. Therefore, through the study of the author of this book, it can provide technical basis for the construction of high-speed, first-class and other high-grade highways in the Qinghai-Tibet Plateau region, and provide support for the efficient, safe and stable development of regional transportation. At the same time, the freeway safety design technology, speed control design technology and traffic security facilities proposed by the author of this book improve the intrinsic safety of freeway design in Plateau area, and provide a solid guarantee for traffic safety and smooth operation after highway operation.

(2) **It is helpful to Promote the Humanized Design of Highway and Improve the Life Support Ability of Highway**

The construction of freeways should reflect the design concept of "person centered, safety first," and the authors of this book fully considered the comfort as well as the survival needs of drivers for driving on the plateau areas during the process of developing a study on the safety design technology of freeways on the Tibetan Plateau. In the study of design indicators of flat longitudinal geometry, driver's mental states such as discomfort and fear caused by the change of heart rate growth rate should be fully considered; In the study of service facility spacing, the effects of driver physiological needs, fatigue characteristics are fully considered, and the requirements of arrange life support facilities such as emergency rescue, medical treatment and so on are also considered in the choice of service facility functional type. The above research results can contribute to the improvement of the service quality of freeways and the improvement of the satisfaction level of flight attendants, and make important contributions to the promotion of road construction and social harmony development.

(3) **Further Supplemented and Improved the Standard Specification System of Highway Engineering in China**

The Qinghai—Tibet Plateau, with its complex geo-climatic environment and special transportation characteristics, has different technical requirements for highway construction in this area from that in plain areas. The current standard system of highway engineering technology has some differences in implementation and use across the country. The current national highway industry standard and norm system plays an important role in highway construction in Qinghai-Tibet Plateau area, but it is difficult to fully meet a series of specific requirements for highway design and construction in the special geographical and climatic environment of the area.

Through carrying out research on freeway safety design technology in complex environment of high cold, low pressure and oxygen deficiency in Qinghai-Tibet Plateau area, the author of this book proposes highway horizontal, longitudinal and transverse geometric design indicators and their parameters, and service facilities arrangement spacing requirements, which are suitable for highway service function, topographic and geological conditions, vehicle performance and driver's psychological and physiological characteristics in this area, in order to supplement and improve the relevant provisions of current highway industry standard specifications. Establishing local standard for highway construction in Qinghai-Tibet Plateau provides reliable technical support.

The research results have broad application prospects, can effectively improve intrinsic safety of freeway design in high-altitude areas, improve traffic safety and service level of highway construction projects, and then drive economic development and improve people's livelihood in Qinghai and Tibet provinces and regions, with remarkable economic and social benefits. It will be directly applied to freeway construction projects in High Altitude Area, which can reduce project risks, speed up project progress, reduce management costs and save project investment. It has broad market demand and will generate huge economic benefits.

Chapter 2
The Characteristics of Traffic Operation in High Altitude Areas

2.1 Characteristics of Traffic Accidents in Special High Altitude Environments

High altitude

Traffic accidents are caused by the interaction of four factors: people, vehicles, roads, and the environment. The four factors together constitute a complex people-vehicles-roads-environment system. In this system, people, the driver as the main body of the system, the vehicles refer to various types of vehicles driving on the highway, the roads mainly refer to the highways, and the environment refers to the natural environment and climatic conditions of the area where the highway is located. By analyzing the distribution characteristics of traffic accidents, such as space, causes, and forms, the relationship between traffic accidents and the four factors of people, vehicles, roads, and the environment and the main impacts on traffic safety are clarified. It provided the basis for the research on the operating characteristics of people, vehicles, roads and environments in the low pressure and oxygen deficient environment and high altitude areas.

2.1.1 Basic Data on Traffic Accidents

The collected traffic accident data includes G109 and G214 traffic accident data in Qinghai areas from 2010 to 2015, and G109 traffic accident data from 2012 to 2014 in Tibet areas. Traffic accident data includes any number of indicator statistics such as accident category, accident type, and accident cause.

In order to study the distribution characteristics of traffic accidents, the author's research team carried out secondary development based on ArcGIS software, and independently developed the "Basic Information Platform for Highway Network in Tibet V1.0" (Fig. 2.1). The information platform integrates the data of highway aerial

J. Liu, *Technical Indicators and Safety Design of Freeway in High Altitude Area*,
https://doi.org/10.1007/978-981-99-0620-8_2

Fig. 2.1 Basic information platform of highway network in Tibet V1.0

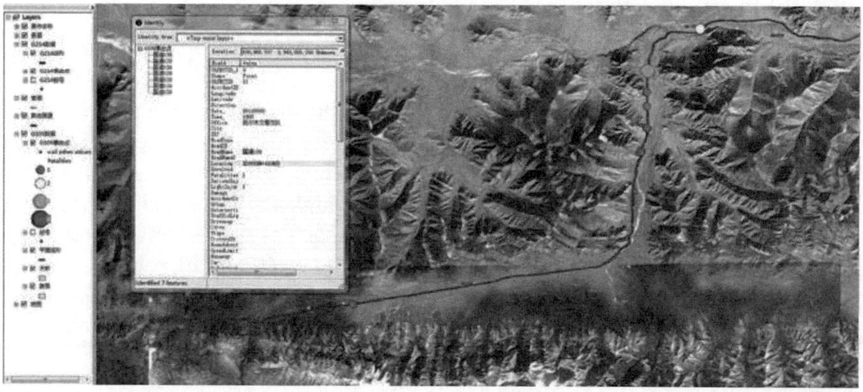

Fig. 2.2 Display of query function of traffic accident basic information (G109 Xidatan section)

photography, remote sensing, traffic accidents in the Qinghai-Tibet Plateau Region to realize traffic accident statistics, map browsing, and design file associations. It aims to help users quickly inquire and count comprehensive data such as the stake number, geometric alignment, location, time, cause, and shape of traffic accidents in the Qinghai-Tibet region and provide technical support for rapid and accurate analysis of the traffic safety status of the highway network in the region (Fig. 2.2).

2.1.2 *The Indicators and Methods of Traffic Accident Analysis*

Due to the complexity of traffic accidents, in order to reflect the overall quantitative characteristics of traffic accidents better, statistical methods are used to analyze a series of indicators, thereby reflecting the quantitative characteristics of various aspects of the accident and revealing the overall internal regularity of the accident.

2.1.2.1 Indicators of Road Traffic Accidents

The indicators of road traffic accidents mainly including absolute indicators, relative indicators, average indicators and dynamic indicators.

(1) **Absolute Indicator**

The absolute indicator including the number of accidents, the number of injuries, death toll and economic losses, which are the basic indicators reflecting the situation of traffic accidents. The absolute indicator of traffic accidents can reflect the scale, total amount, and level of traffic accidents in a certain region in a certain period as well as clearly reflect its development trend. However, absolute indicator are almost static and isolated, which cannot reflect the impact caused by the difference of actual road and traffic conditions on accidents.

(2) **Relative Indicator**

The relative indicator introducing some related factors as the basis for comparison. These factors are related to accidents directly or internally, therefore the accidents have better comparability relative to these related factors. Relative indicator includes accident rates per kilometer, accident rate and death rate per 100 million vehicles/km, accident rate, death rate, and accident fatality rate per 10 thousand vehicle/km, etc.

i. Accident rate per kilometer, namely the average number of accidents per kilometer, is also called accident frequency. It considered the length of the road segment to make the number of accidents more comparable:

$$R_L = \frac{A}{L} \tag{2.1}$$

where,

R_L Per kilometer accident rate;
A Number of statistical accidents of the road segment (times);
L Length of road segment (km).

ii. Accident rate per 100 million vehicle/km, in which takes both the road length and traffic volume into account. It refers to the average number of traffic accidents (or casualties) obtained by the sum of kilometers of all motor vehicles in a year in a certain area, usually calculated by the accident rate of 100 million vehicle km:

$$R_K = \frac{A}{K} \times 10^8 \tag{2.2}$$

where,

- R_K Number of accidents or casualties per 100 million vehicle km in a year;
- A Number of traffic accidents or casualties throughout the year;
- K Total driving vehicle kilometers throughout the year (the vehicle km is usually calculated by multiplying the length of the road by the annual volume of traffic on the road, or the annual traffic volume is calculated from the annual average daily traffic).

iii. The accident fatality rate is an indicator to evaluate the degree of injury and rescue level, and also reflects the severity of various traffic accidents:

$$d = \frac{D}{W + D} \times 100\% \tag{2.3}$$

where,

- d Fatality rate (%);
- D Death roll (person);
- W Number of injured (person).

(3) **Average Indicator**

The average indicator including the average of death roll per accident, the average accident rate 100 million vehicle in each section of the road, the average number of accidents, etc.

Here, the average indicator introduces the concept of traffic accident risk coefficient, which refers to the ratio of death roll caused by a certain type of traffic accident to the number of such accidents each year (unit: person/accident)

$$\theta_d = \frac{N_{\mathrm{P}}}{N_{\mathrm{T}}} \tag{2.4}$$

where

- N_P —Accident fatality (person/year)
- N_T —Number of accident (No./year)

This indicator reflects the proportion of fatalities caused by certain type of traffic accident, that is the probability of death of the person involved after a traffic accident under a specific condition. In terms of national and regional statistics, it is not possible to ignore the impact caused by mass fatalities and injuries; while for the traffic accident risk factor of a certain highway in a certain period, considering the chance and uncommonness of mass fatalities and injuries, it is necessary to exclude such accidents to avoid excessive interference with the analysis results.

(4) **Dynamic Indicator**

Dynamic indicators include growth volume, accident development rate, accident growth rate, average development speed, average growth rate, etc., which are used to reflect the process and trend of traffic accident development and change.

Among the above-mentioned indicator, absolute indicator are the basis, and relative, average and dynamic indicator are determined through absolute indicator; in turn, relative, average and dynamic indicators reflect the accident patterns more precisely that are difficult to be reflected through absolute indicator. By using accident indicator, the features and patterns of accident distribution are studied to achieve the purpose of lowering the number and the severity of accidents.

2.1.2.2 Analysis Methods for Traffic Accident

Analysis methods of traffic accidents mainly including statistical analysis, classification analysis, graphical method, etc.

(1) **Statistical Analysis Method**

The statistical analysis method is to use the statistical data of traffic accidents to explore the patterns of occurrence and changes of traffic accidents from a macroscopic perspective, and to propose suggestions and methods to improve traffic and reduce traffic accidents on this basis. The statistical analysis method is based on data that can objectively and comprehensively reflect the original face of traffic accidents, to reflect the original state of the accident accurately and comprehensively, and carry out scientific reasoning and judgment, so as to reveal the patterns contained in the data.

(2) **Classification Analysis Method**

The classification analysis method is both an important method for handling data and a basic method for analyzing the causes of traffic accidents. Its purpose is to give a clear, intuitive, and regular concept by classifying the data of different nature and the intricate causes of traffic accidents clearly.

(3) **Graphical Method**

The graphical method is a more visual representation of the analysis results of the statistical and classification analysis methods in the form of graphs. The three main methods include scale diagrams, coordinate diagrams and analysis diagrams.

In order to analyze the distribution pattern of traffic accidents in high altitude areas and according to the form of accident data information, this book mainly adopts absolute, relative, and average indicators as accident indicators, and applies statistical analysis method, classification analysis method and graphical method to analyze the traffic accident pattern.

2.1.3 General Distribution Characteristics of Traffic Accidents

(1) **Spatial Distribution of Accidents**

According to the accident statistics, the number of accidents near stake number G214 is counted, including Xining and Jinghong. The initial stake number ranges from K107 to K992, 885 km in total. 188 accidents totally occurred, with an accident rate of 0.21 per kilometer. The specific stake number distribution is shown in the Fig. 2.3.

It can be seen from Fig. 2.4 that the accident rate of stake number G214 ranges from K100 to K200 is high, 0.38Nos./100 km, and the overall trend decreases with the increase of stake number. From the perspective of traffic accident risk coefficient, the accident rate is the highest from stake number number K200 to K300, 1.77Nos./ person, and then gradually decreases. In general, the accident rate is high and the accident risk is serious in the range of K100 to K300.

According to the accident data of G109 Qinghai section, the statistics of the accidents occurred near the stake number number show that the initial stake number number range is K1844~K2736, totally 892 km, 388 accidents occurred, and the accident rate per kilometer is 0.43. The specific stake number distribution is as Fig. 2.5.

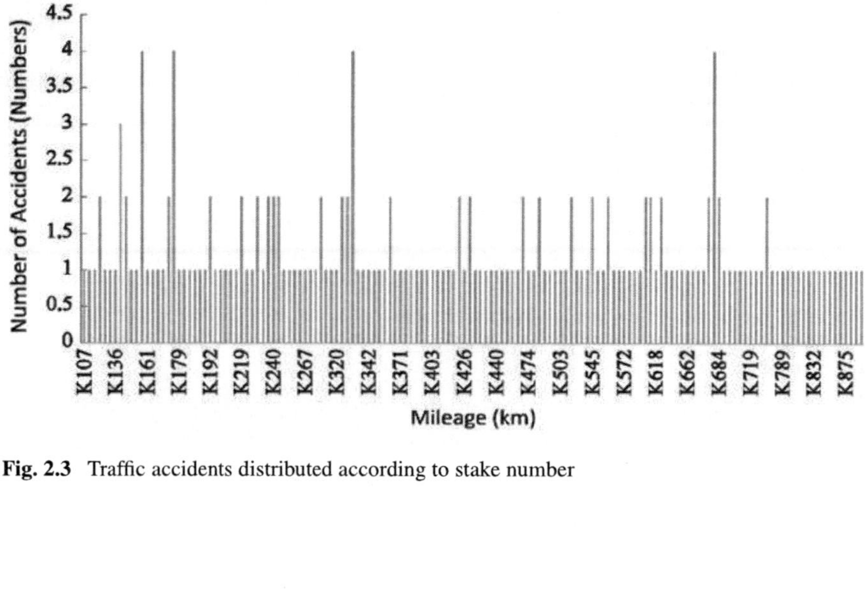

Fig. 2.3 Traffic accidents distributed according to stake number

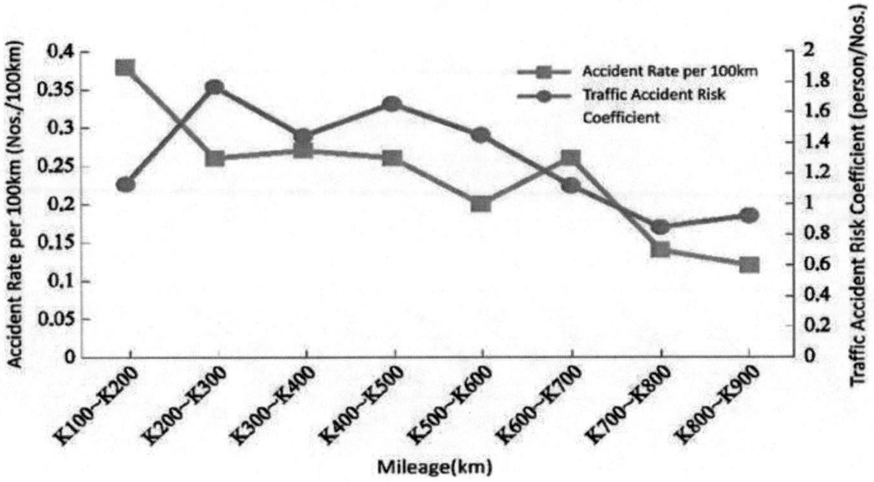

Fig. 2.4 Accident rate per 100 km & distribution of accident risk coefficient

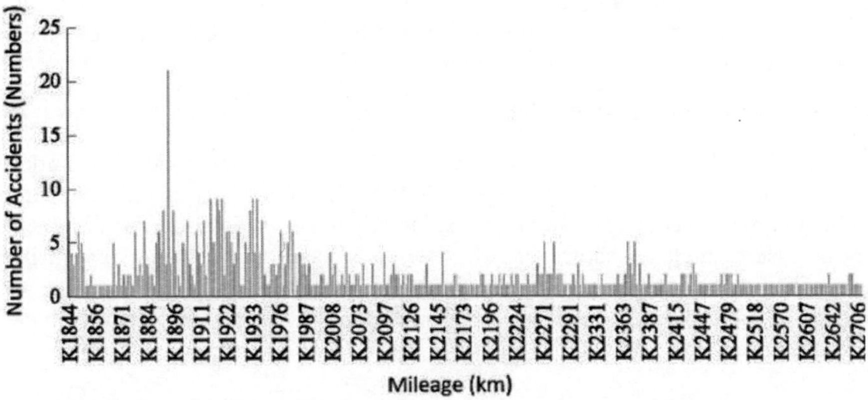

Fig. 2.5 Traffic accidents distributed according to stake number of G109 Qinghai section

In general, the accident rate of 100 km from K1900 to K2000 is the highest, reaching 0.6. From the perspective of accident risk, the severity of traffic accidents between K2000 and K2100 is relatively high. Combining these two aspects (Fig. 2.6), it can be concluded that the accident rate and severity are relatively high within the range of K1900 to K2100. From the statistical data, 94 accidents have occurred in this range in the past five years, causing 144 deaths, these two parts accounted for 24.85% of the total accidents and 26.52% of the total deaths.

According to the accident statistics, the number of accidents occurred near the stake number of G109 Tibet section was counted, including Golmud to Lhasa. The stake number starting from K2773 to K3876, totally 1103 km, 113Nos. of accidents

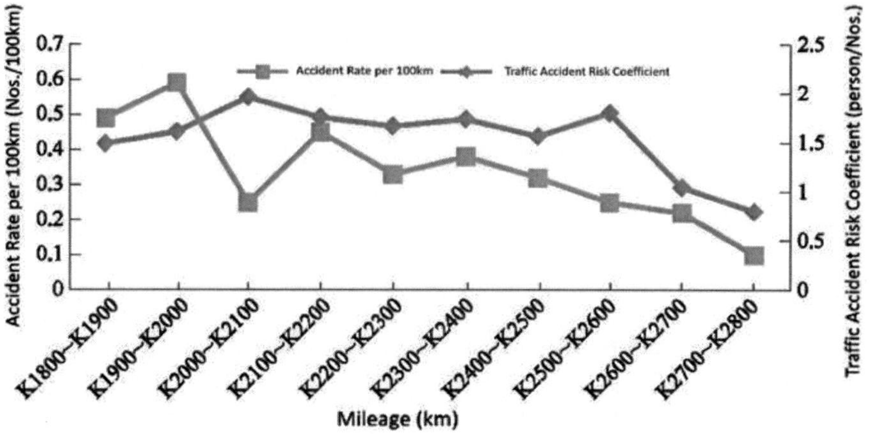

Fig. 2.6 Accident rate per 100 km & traffic accident risk coefficient of G109 Qinghai section

occurred, with an accident rate of 0.1 per kilometer. The specific stake number distribution is shown in Fig. 2.7.

From Fig. 2.8, the accident rate per 100 km from K3600–K3800 is the highest, where is also the most dangerous section.

(2) **Accident Cause Distribution**

China's Road Traffic Safety Department defines the cause of accident according to five major categories, the motor vehicle violation, motor vehicle non-violation fault, non-motor vehicle violation, pedestrian or passenger violation, road problem. In the case of multiple causes, it is done usually based on the major responsible part, which means the cause of the accident decided by the major responsible part, if equal responsibility, then refer to sequence of motor vehicle, non-motor vehicle, pedestrians, others.

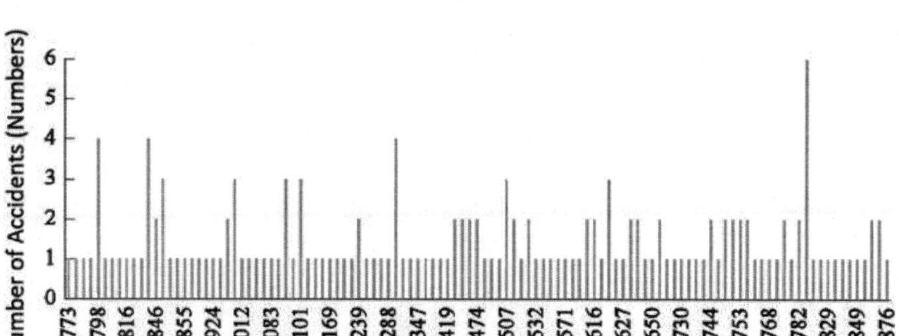

Fig. 2.7 Traffic accidents distributed according to stake number of G109 Tibet section

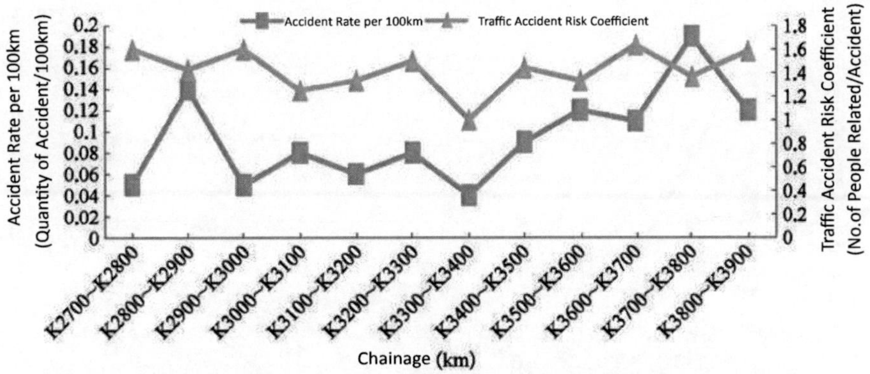

Fig. 2.8 Accident rate per 100 km and risk coefficient distribution of G109 in Tibet area

According to the G214 Qinghai area traffic accidents, as shown in Fig. 2.9, 94.68% of the accidents were caused by motor vehicle illegal behavior, and the number of accidents caused by non-motor vehicle violations or pedestrian/passenger violations is 0.

By further subdividing the various types of motor vehicle violations, ignoring the "others", we can get a descending order distribution of detailed motor vehicle violations in G214 Qinghai area, as shown in Fig. 2.10. If seeing these figures, the distribution characteristics of each specific cause of violations have basically the same pattern in terms of accidents number, the fatality and the injured. Among all the causes, Speeding is the major one of traffic accidents, accounted for 43.09%, 37.05% and 47.88% of the three aspects, respectively. And the proportion of Illegal Cross Lane Driving, Fatigue Driving and other serious violations are also listed in top reasons. The accidents caused by these three violations takes 66.49% of the total number of accidents.

The number of accidents and fatality caused by different reasons, as well as the death toll are not the same. Above the causes are high in accident rates. While the most

Fig. 2.9 Distribution characters of the major cause of different accidents

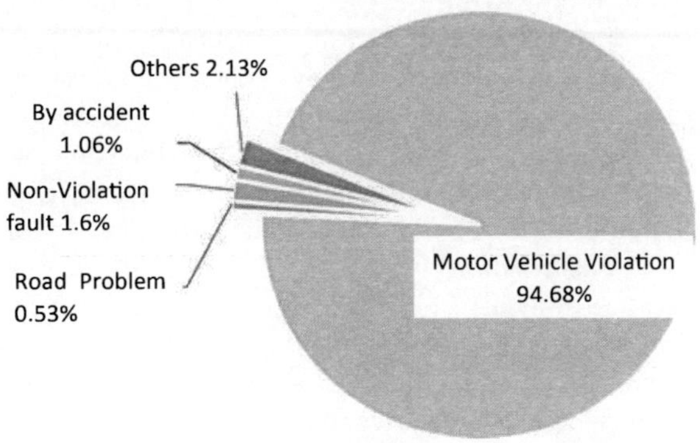

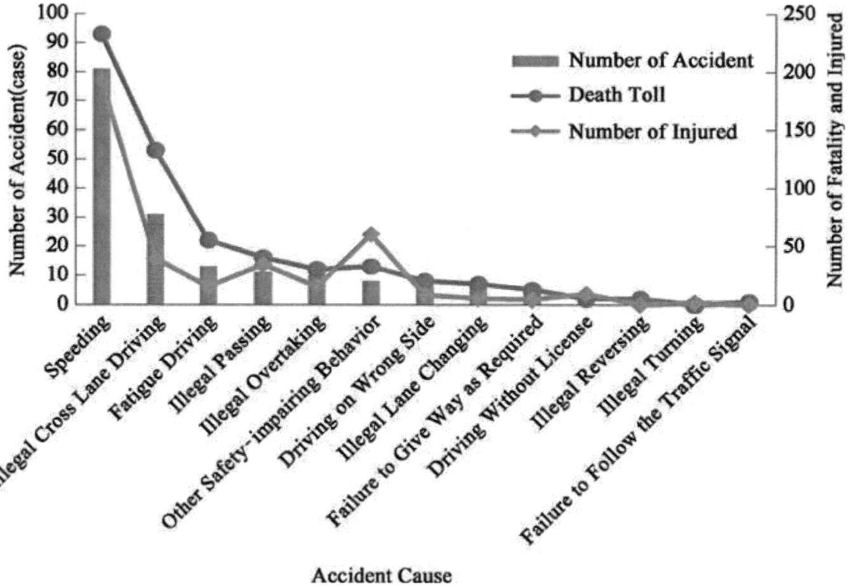

Fig. 2.10 Distribution of accident causes in G214

severe accidents are caused by " Illegal Cross Lane Driving", then by Fatigue Driving, Illegal Passing, and Wrong Side Driving etc., without considering the "others". The fatality rates of these accidents are over 50% (Fig. 2.11).

According to the traffic accidents in G109 Qinghai area, 96.72% of the accidents were caused by motor vehicle violations, followed by non-violation fault, 2.03%. The total of both accounts for 98.75% of all accident causes, as shown in Fig. 2.12.

As Fig. 2.13, in descending order, the major causes of accidents are Speeding, Other Safety-impairing Behavior, and Failure to Give Way as Required, together accounting for 55.03% of the total number of accidents, of which Speeding accounts for 37.06%.

According to different accident causes of traffic accident risk coefficient statistics (Fig. 2.14), "Improper Braking" risk coefficient is the highest. In G214 high altitude area, traffic accidents caused by car brake failure count for 0.23% of the total number of traffic accidents. There are many long and steep slopes in high altitude area. During driving, trucks need to brake constantly in long time, which will lead to overheating of brake system, brake failure, then cause accidents. Coupled with the high probability of double accident with other vehicles, the severity of the accident is obvious.

The distribution characteristics of the causes of traffic accidents in the G109 Tibet area are shown in Fig. 2.15. It can be seen that Improper Operating and Illegal Lane Occupancy are the main causes of accidents, of which 23.73% of the total accidents caused by Illegal Lane Occupancy. Improper Operating usually means that in an emergency, the driver does not take reasonable and effective measures such as emergency braking or direction control, or mis-operating, which lead to accident.

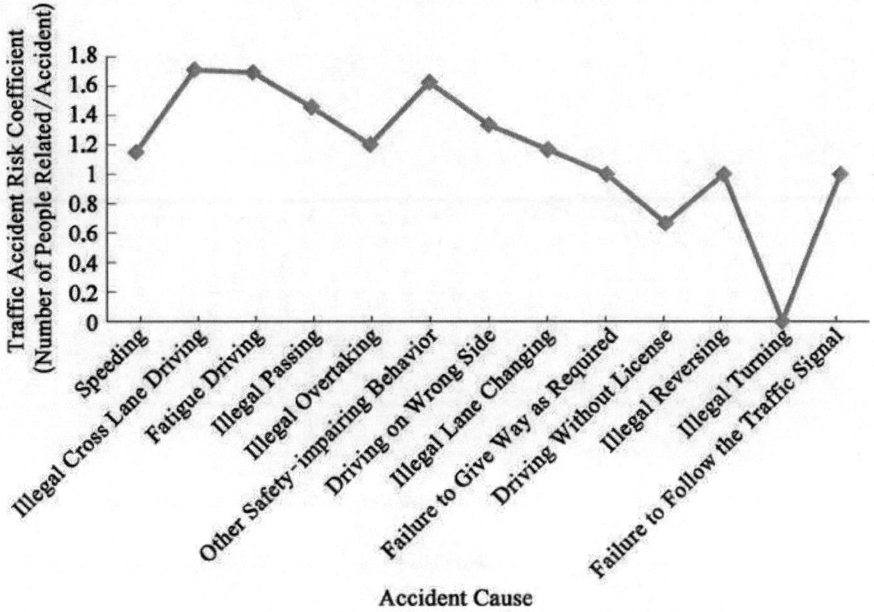

Fig. 2.11 G214 Distribution of risk of different accident causes in G214

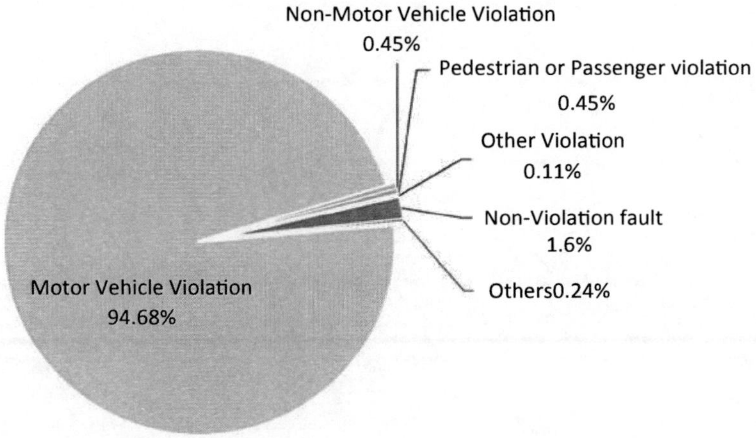

Fig. 2.12 Distribution different causes of accident in G109 Qinghai area

According to the statistics of traffic accident risk coefficients for different accident causes (Fig. 2.16), Speeding has the highest risk. And from the analysis, the number of accidents due to Speeding was 8 in total, but caused 15 deaths and 11 injured.

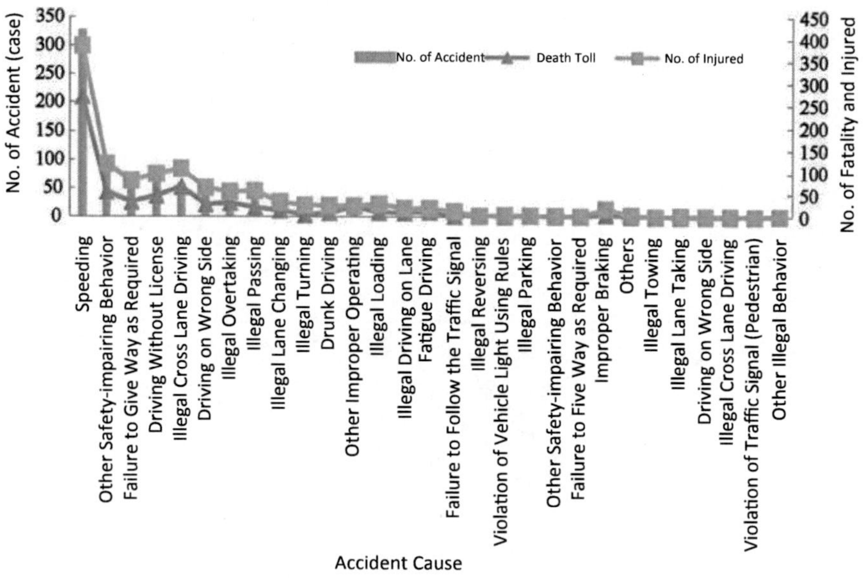

Fig. 2.13 Distribution of traffic accident causes in G109 Qinghai area

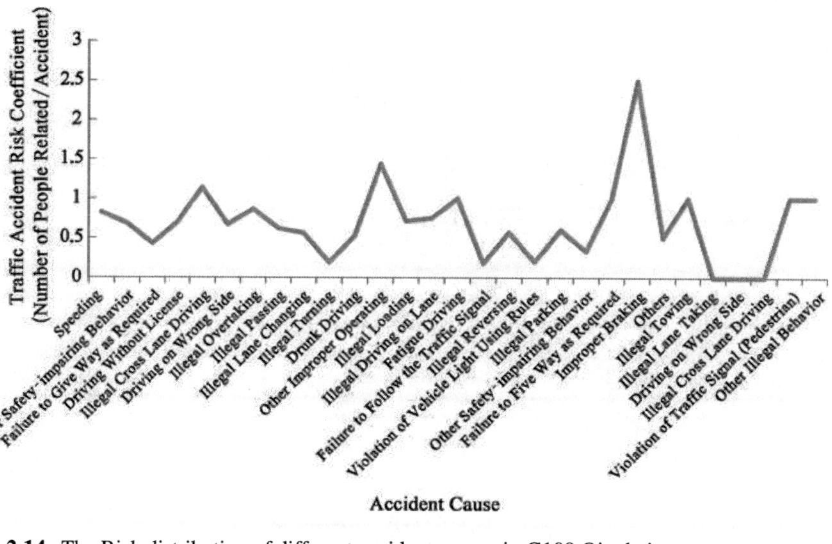

Fig. 2.14 The Risk distribution of different accident causes in G109 Qinghai area

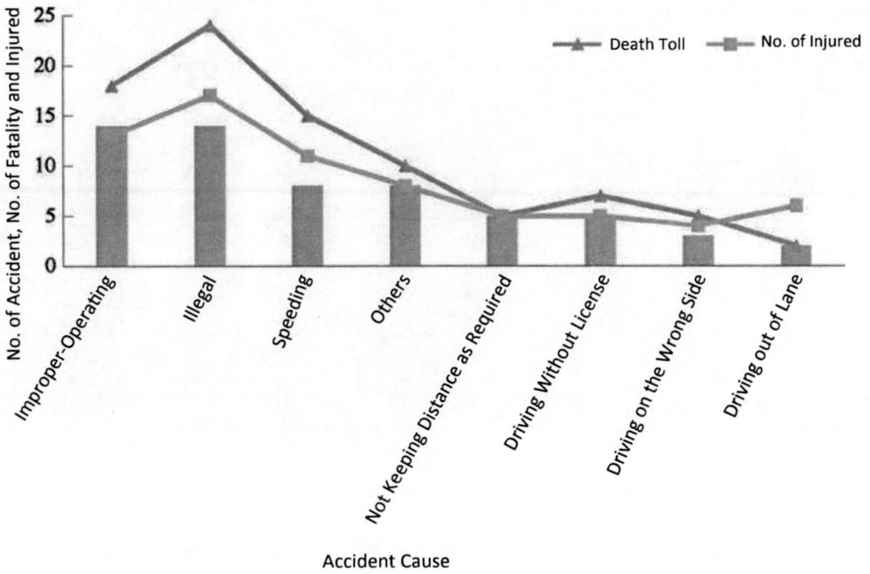

Fig. 2.15 Distribution characters of accident causes in G109 Tibet Area

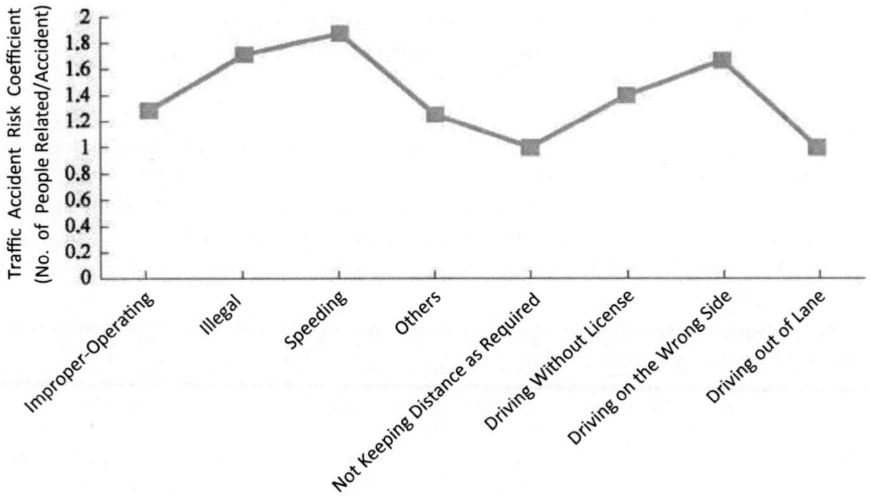

Fig. 2.16 Risk distribution of accident causes in G109 Tibet area

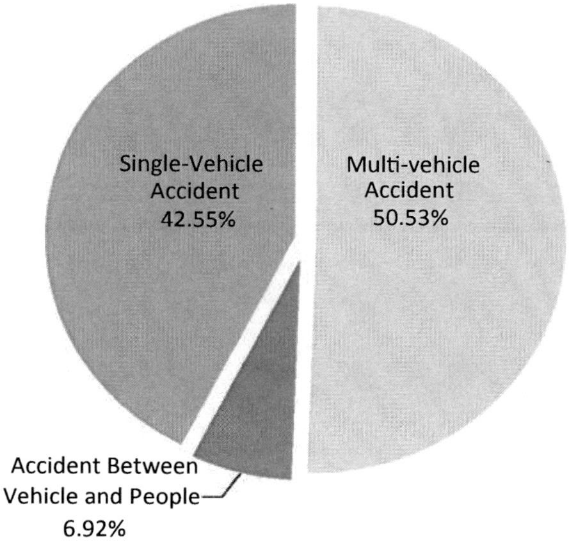

Fig. 2.17 Accident types distribution of G214 Qinghai area

(3) **Accident Forms Distribution**

Traffic accident form refers to the specific situation manifested by the conflict between traffic participants or their own uncontrolled accidents. According to the relevant provisions of China's Road Traffic Management, road traffic accidents are mainly divided into collision, crushing, scraping, overturning, crashing, fire, and other accident forms. According to the statistics, in the past five years, the high altitude area G214 Qinghai area road accident type distribution are shown in Fig. 2.17.

As is shown in Fig. 2.17, multi-vehicle accidents as the main type of accident in G214, accounting for 50.53% of the total accident type, followed by single-vehicle accidents, accounting for 42.55%. And the proportion of accidents happen between vehicles and people is low.

As is shown in Fig. 2.18, in the last five years, among the G214 high altitude area traffic accidents, collision-type accidents are 51.06% of total number of accidents; the fall accident 41.49% of the total number of accidents.

There are many low-grade Freeways in high altitude areas. And some Freeways have not arrange central separation facilities, which are prone to collision accidents when overtaking. In addition, there are lots of mountain roads, when the vehicle turns, it is easy to collide with the guardrail then overturn and fall.

So, collision accident (including rear-end collision, Head-on collision, hitting fixed objects and standing vehicles, etc.) and overturning (rollover, rollover sideways) are the main forms. According to the accident risk coefficients of different accident forms (Fig. 2.19), the "Other Accidents Between Vehicles " has the highest risk, 9 persons/accident, while the rests share the same risk.

According to the accident statistics of G109 Qinghai area in the past 5 years, the distribution of accident types is shown in Fig. 2.20.

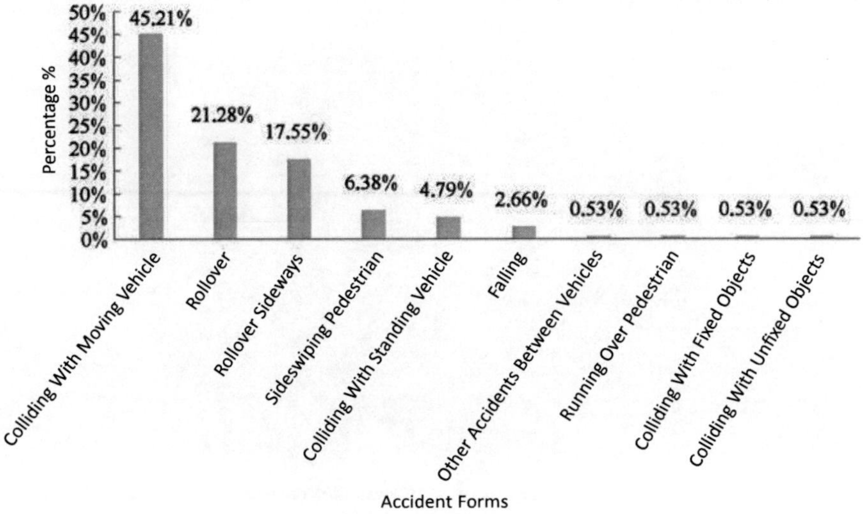

Fig. 2.18 Accident forms distribution of G214 Qinghai area

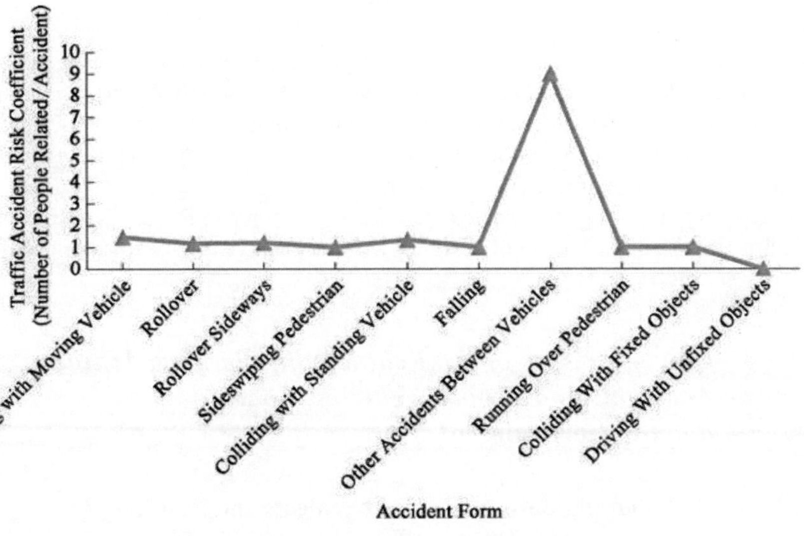

Fig. 2.19 Risk Distribution of different accident forms of G214 Qinghai area

It can be seen in Fig. 2.20, the main accident type is multi-vehicle accidents, accounting for 62% of the total number of accidents, followed by accidents between vehicles and people, accounting for 22% of the total, and single-vehicle accidents are not reported, so the statistics are low, accounting for only 16%. Figure 2.21 shows the

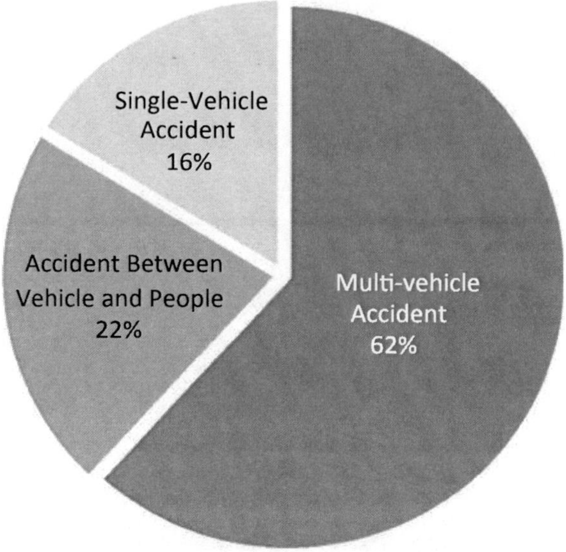

Fig. 2.20 Accident types distribution of G109 Qinghai area

different accident forms for each accident type. As can be seen from Fig. 2.21, multi-vehicle accidents are dominated by collisions with moving vehicles, accounting for 56. 05% of the total accident forms, while single-vehicle accidents are dominated by collisions with fixed objects, accounting for only 2.15% of the total accident pattern, but 59. 4% of the single-vehicle accidents.

The severity of the different accident forms is shown in Fig. 2.22. As can be seen, falling has the highest level of severity, with an average of 2 deaths per accident.

2.1.4 Characteristics of the Relationship Between Traffic Accidents and People, Vehicles, Roads and the Environment

2.1.4.1 Relationship Between Traffic Accidents and People

People isthe main body of the road traffic system, and the car driver must keep abreast of the vehicle, road and traffic changes in the process of driving, and constantly make correct judgments and reactions. The control of the vehicle is achieved through the accelerator pedal, brake and steering wheel, manipulating the direction, and controlling the travel speed to adapt to the dynamic operating process of this system. Compared to plain areas, in a high altitude, low pressure and oxygen deficient environment, drivers are more likely to experience plateau reactions such as headache,

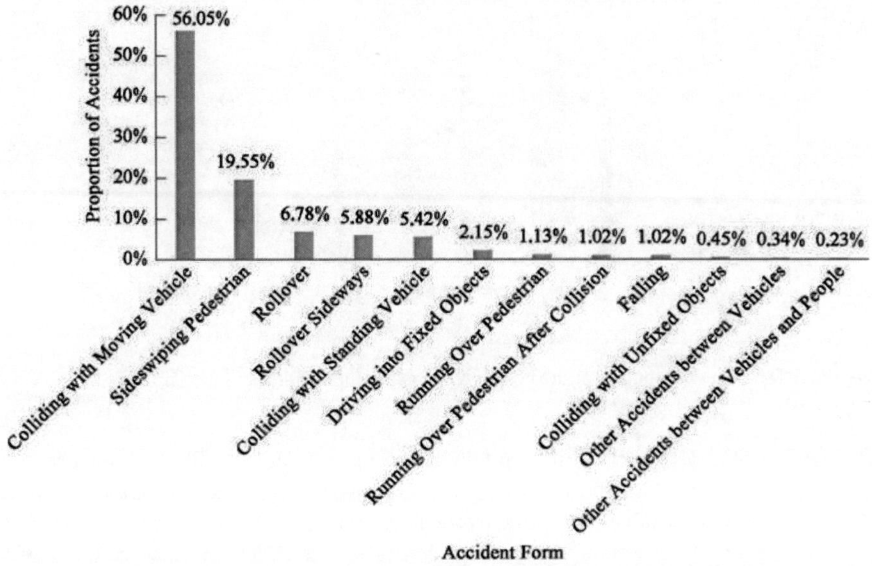

Fig. 2.21 Accident forms distribution of G109 Qinghai area

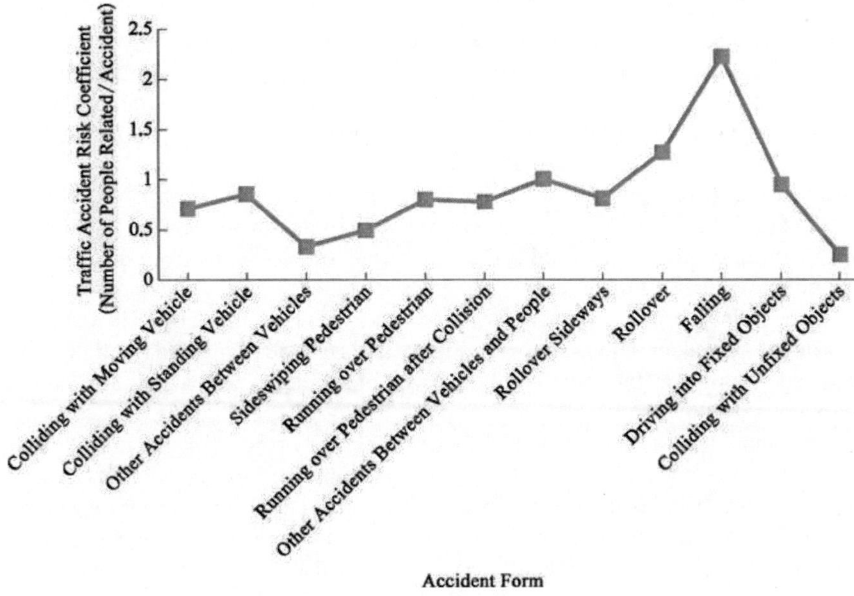

Fig. 2.22 Severity level distribution of accident forms

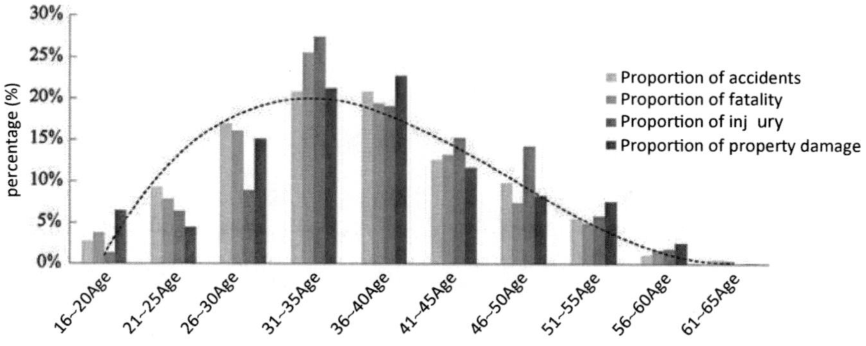

Fig. 2.23 The age distribution of the people responsible for accidents in the G214

fatigue and breathing difficulties, and their driving state is more likely to be affected by the environment and show abnormal driving behaviors. Therefore, the distribution characteristics of drivers in road traffic accidents in high altitude areas should be analyzed to clarify the main influencing factors which affects human driving status in traffic accidents.

(1) **Age and Driving Age Distribution of Drivers**

Figure 2.23 shows the age distribution of the people responsible for accidents in the G214 Qinghai **areas** from 2010 to 2015. The age of the person responsible for the accident is basically consistent in the distribution of the number of accidents, the number of injuries and fatalities, and property damage, and the distribution characteristics are basically normal.

According to accident statistics, the age distribution of people involved in accidents in the G109 Qinghai areas is shown in Fig. 2.24. It can be seen from the figure, the age group of 21–40 years old is the main group of people involved in accidents, accounting for 64% of the total.

The age distribution of persons involved in accidents in the G109 Tibetan areas according to accident data is shown in Fig. 2.25. It can be seen from the figure, the 21–40 age group was the main group involved in accidents, accounting for 58% of the total.

In summary, as the composition of the driver age group corresponds to that of the socially active age group, the 26–40 age group is both the most active and the most accident-prone. The 26–40 age group is the most active and accident-prone group.

In the current accident information collection data system, driving age is an **Indicator** that can objectively reflect the experience and level of drivers. Figure 2.26 shows the distribution of the age of drivers responsible for accidents according to G214. The age of the person responsible for the accident is consistent in the number of accidents, the number of injuries and fatalities, and the four indices of property damage. Drivers with 6–10 years of driving experience are the most frequent victims of traffic accidents. The percentages of accidents, injuries and fatalities, and property

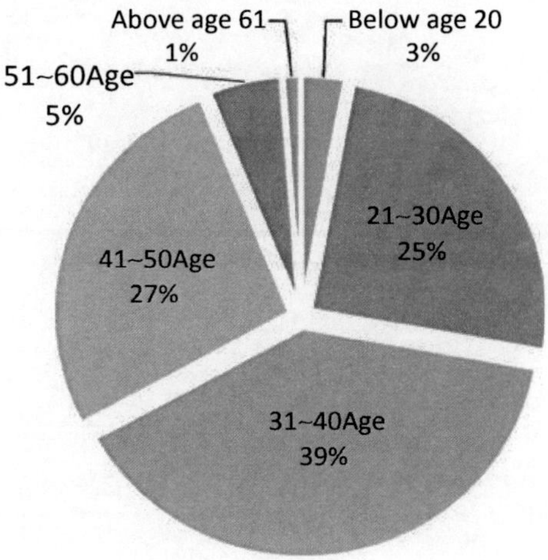

Fig. 2.24 Age distribution of people involved in accidents in the G109 Qinghai areas

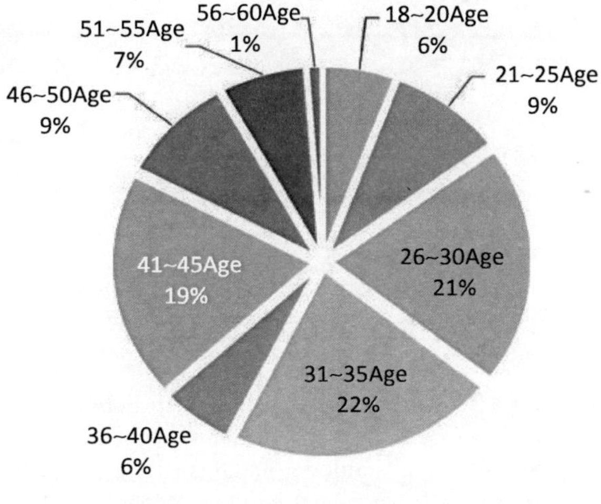

Fig. 2.25 Age distribution of persons involved in accidents in the G109 Tibetan areas

damage among drivers with less than 6–10 years of driving experience were 25%, 27%, 15% and 20.23% respectively.

From the above analysis, drivers with 6–10 years of driving experience account for a larger proportion of overall accidents, which may be related to the higher proportion of drivers with 6–10 years of driving experience among road participants, which is not necessarily the highest in terms of risk, and is therefore characterized by the traffic accident risk factor, as shown in Fig. 2.27.

Drivers with less than 4 years of driving experience have a higher risk factor in traffic accidents. Those responsible for accidents with 4 years of driving experience have a risk factor of 2.0, much higher than of other driving age groups.

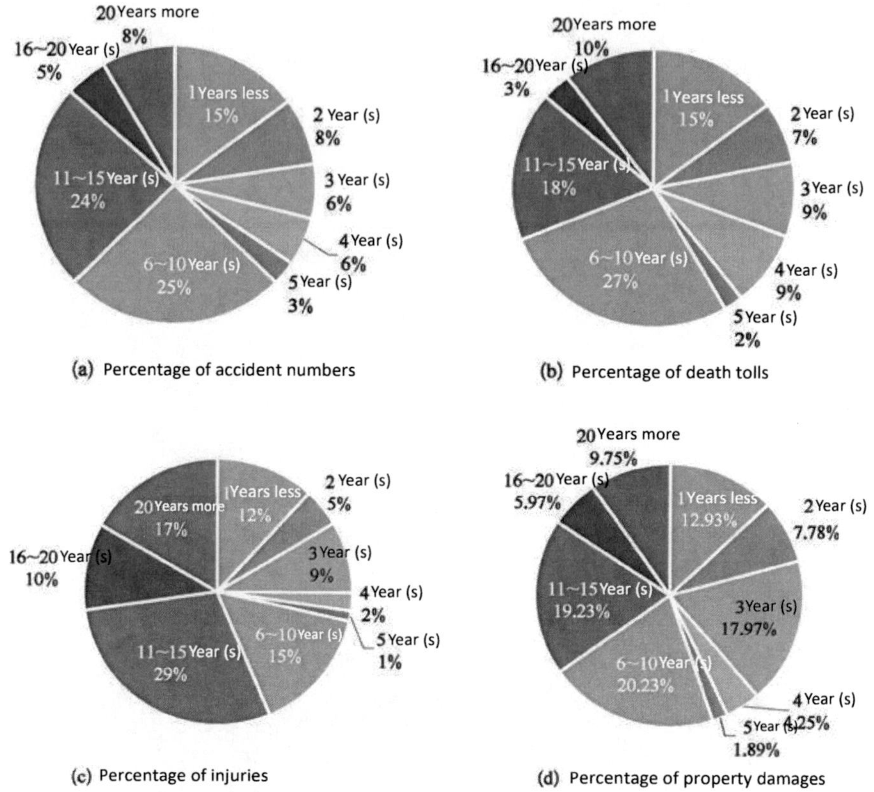

Fig. 2.26 G214 characteristics of age distribution for persons of accidents in Qinghai areas

Figure 2.28 shows the statistical distribution of the age of those responsible for the G109 accident. The age profile of accident victims is consistent in terms of the number of accidents, injuries and fatalities, and the distribution of data confirms that drivers with 6–10 years of driving experience are the most frequent victims of traffic accidents. The percentages of accidents, injuries and fatalities, and property damage among drivers with less than 6–10 years of driving experience were 28%, 29%, and 27% respectively.

Figure 2.29 shows that drivers with less than 4 years of driving experience have a higher risk factor in traffic accidents. The risk factor for drivers with 4 years of driving experience is 0.85, which is much higher than the distribution of other driving age groups.

The findings of the driving age analysis on the one hand indicate that the overall potential risk of traffic safety is indeed higher for low driving age drivers; on the other hand, whether there are more scientific, simple, and operational indicators or methods to evaluate the actual driving ability of drivers is the question before every road. On the other hand, whether there are more scientific, simple, and operational indicators

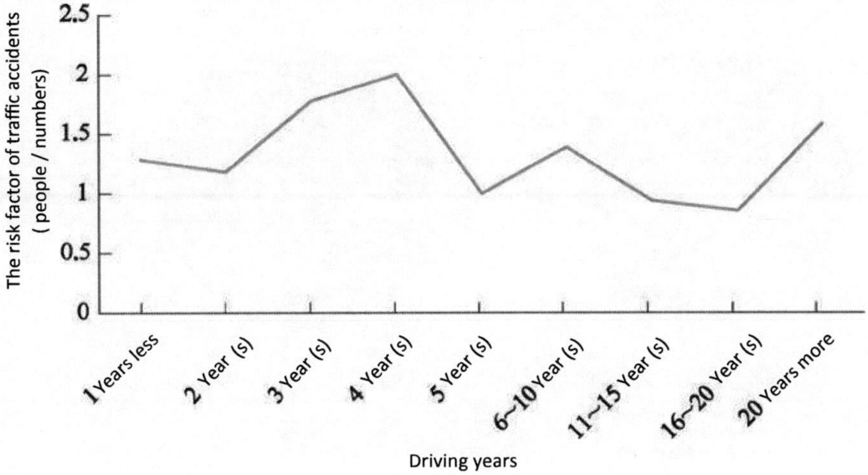

Fig. 2.27 The Risk factor of traffic accidents under different driving age conditions

or methods to evaluate the actual driving ability of drivers is another question for every road safety worker.

(2) **Analysis of Driver Influence Factors in Traffic Accidents**

According to the analysis results in Sect. 2.1.3 of this book, the causes of traffic accidents related to drivers on high altitude areas roads are the highest risk of traffic accidents caused by speeding and fatigue driving. In addition to some mountain ranges (such as Bayankara Mountain, Tanggula Mountain, Kunlun Mountain, Fenghuo Mountain, etc.) in the Qinghai-Tibet Plateau region, the road lines are more complicated and there are more steep slopes and sharp curves, but the rest of the road sections are generally flat and straight, so the road conditions for high-speed driving are mild. The situation of speeding is more prominent because drivers prefer to go through the high altitude area faster to reduce the side effects of physiological status caused by the low pressure and deficiency of oxygen. In addition, because of the long road stake number in Qinghai-Tibet Plateau area, the driving time is generally long, and the road environment landscape is single, so fatigue driving is also more likely to occur.

According to the study results from Doctor Mr. Cui Jianhua from the Institute of Alpine Distress of the No. 18th Hospital of the Chinese People's Liberation Army, factors such as excessive fatigue and upper respiratory tract infection before entering the plateau will increase the incidence of plateau reactions, among which the incidence of plateau cerebral edema is 0.05% to 2%, and the incidence of plateau cerebral edema increases with the increase of altitude and labor intensity. Therefore, from the perspective of ensuring traffic safety, priority should be given to reducing the impact of fatigue driving on drivers, such as by setting up necessary parking and rest facilities along the route and optimizing the landscape along the highway. At the same time,

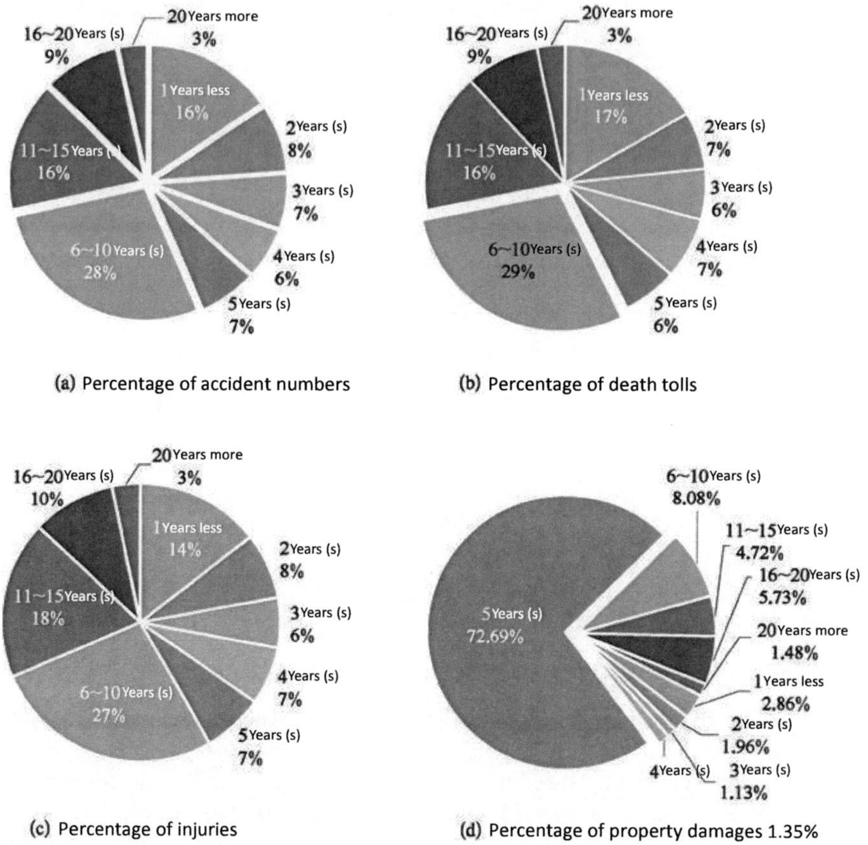

Fig. 2.28 Age distribution of persons involved in accidents in the G109 Qinghai areas

speed limit facilities should be reasonably set up to effectively control the running speed of vehicles and reduce the unpleasant effect of speeding to reduce some unsafe driving behaviors.

2.1.4.2 Relationship Between Traffic Accidents and Vehicles

The type of vehicle in the traffic flow also has an impact on traffic accidents, which is related to the power performance, speed, external dimensions, climbing ability and load degree of various vehicles, especially in the high altitude area. In terms of traffic flow characteristics, the more types of vehicles, the greater the range of speed differences. Under this circumstance, more overtakes occur, which will increase the possibility of traffic accidents.

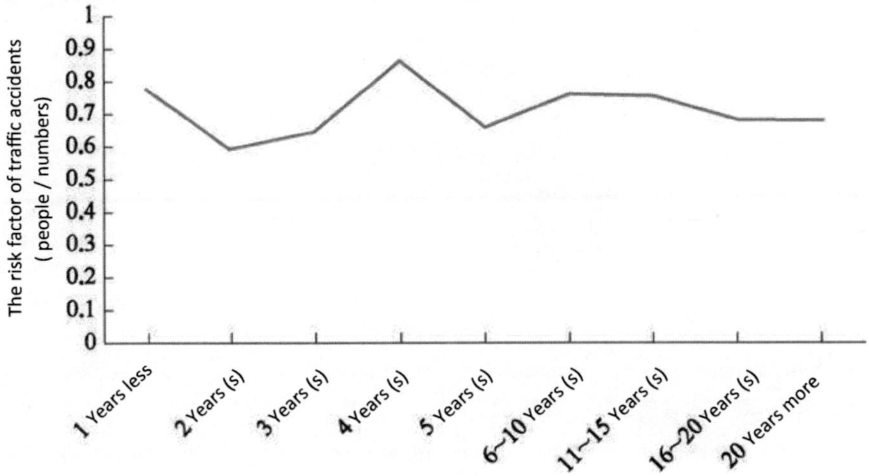

Fig. 2.29 The risk factor of traffic accidents under different driving age conditions

Figure 2.30 shows a diagram of the proportion of fatalities caused by driving different motor vehicles in G214. In the traffic accidents among the different modes of transportation, driving vehicles occupies the absolute majority. For example, out of the 243 traffic deaths, 235 were caused by driving a car, which is 96.71%.

According to statistics of the accident data, vehicles involved in traffic accidents were classified into four categories: buses, trucks, motorcycles, and other models, as shown in Fig. 2.31. According to the statistics of the traffic accident in G214 Qinghai areas in the past five years, buses accidents are the main type, which accounted for

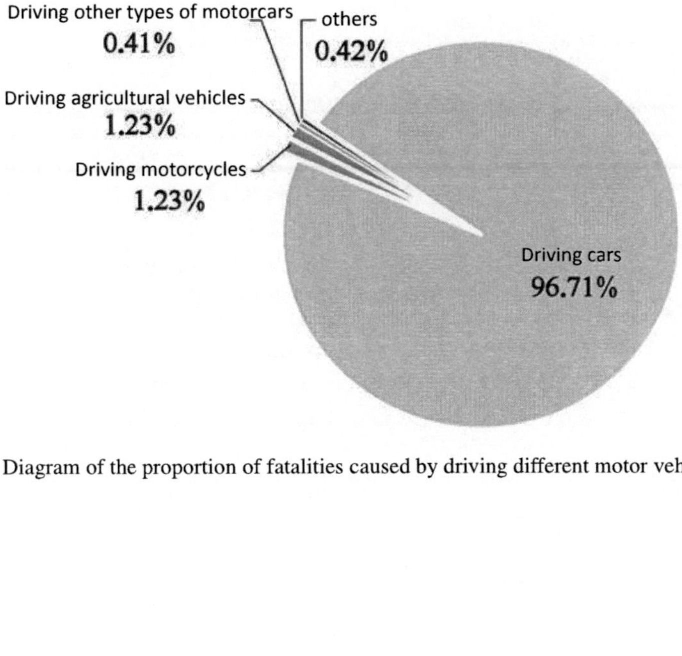

Fig. 2.30 Diagram of the proportion of fatalities caused by driving different motor vehicles

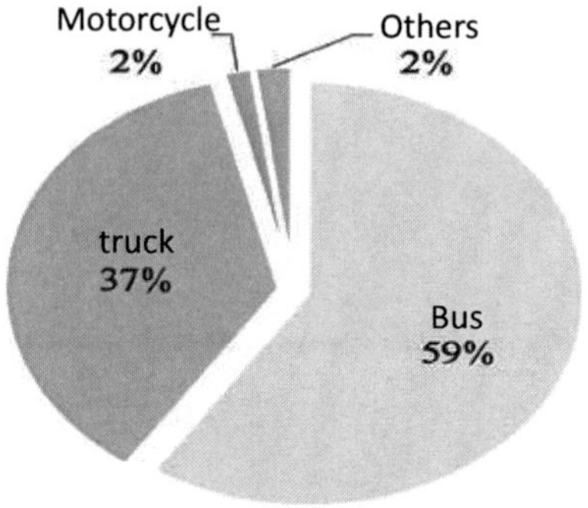

Fig. 2.31 Proportion of vehicles involved in traffic accidents

59% of the total accident models; The accident involving trucks accounted for 37%, motorcycles and other vehicles involved in fewer accidents.

The following four statistical indicators of traffic accidents of different car models are used to represent their distribution characteristics, as shown in Fig. 2.32.

If further subdivision of the accident vehicle types, from Fig. 2.32, the accident vehicles are mainly passenger cars and heavy trucks. The accidents caused by small passenger cars accounted for 46.29% of the total number of G214 accidents and heavy trucks accounted for 26.29%.

From Fig. 2.33, mini-passenger buses and mini-trucks, heavy trucks tend to suffer a higher degree of severity, an average of 2 deaths per accident. Due to the higher speed of those vehicles, more passengers, poorer truck mechanical performance and braking performance, as well as easily overloaded, once the accidents happen, the consequences are severed.

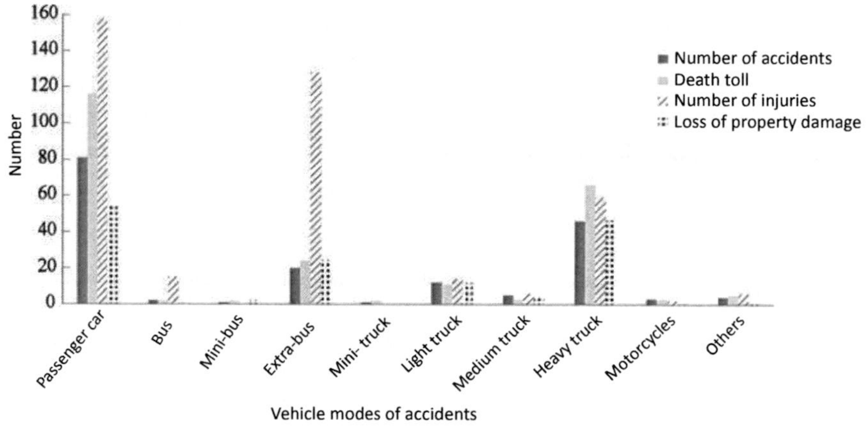

Fig. 2.32 In the G214 the distribution characteristics of transportation modes of people responsible for the accidents

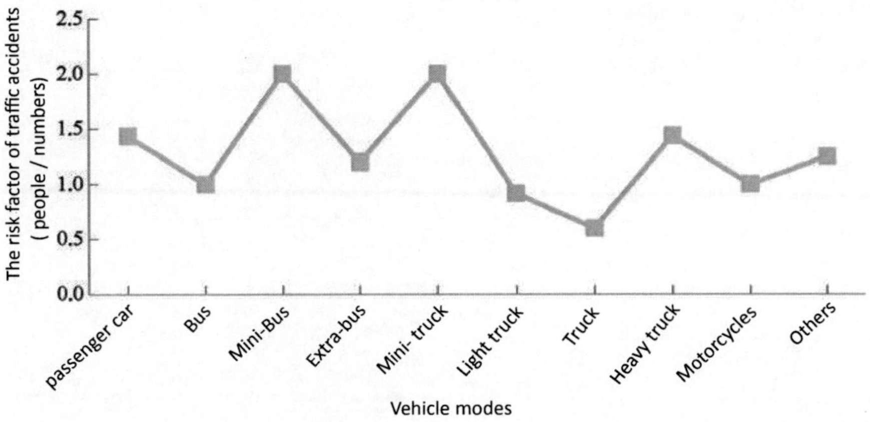

Fig. 2.33 The Risk factor of different vehicles

Figure 2.34 shows a schematic diagram of the proportion of fatalities caused by driving different vehicles in traffic accidents in G109. Among the different modes of transportation, driving a car occupies the absolute majority. For example, out of 655 traffic deaths, 570 were caused by driving a car, which is 87.02%.

According to the accident statistics of G109, the vehicles involved in the traffic accidents were classified into five categories: buses, trucks, motorcycles, semi-trailers, and other models, as shown in Fig. 2.35. Among them, the bus is the main type of vehicle involved in the accident, accounting for 49% of the total number of models; followed by the truck, accounting for 31% of the total, and both account for 80%.

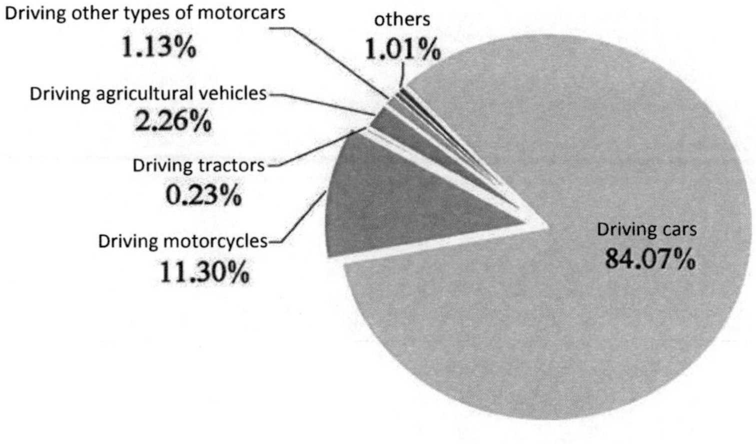

Fig. 2.34 G109 diagram of the proportion of fatalities caused by driving different motor vehicles

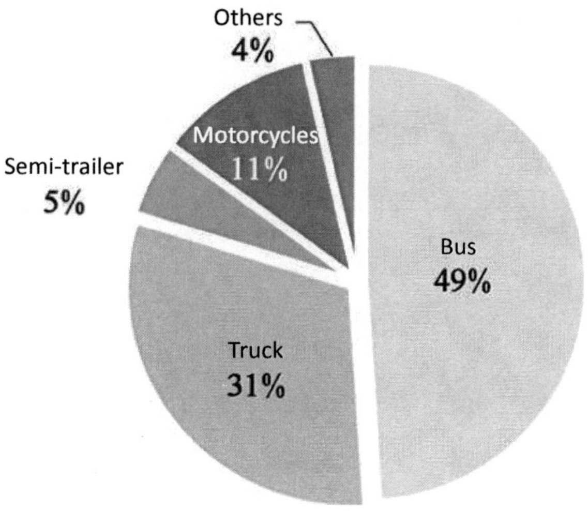

Fig. 2.35 G109 proportion of vehicles involved in traffic accidents

From Fig. 2.36, the vehicles involved in the accident are mainly passenger cars and trucks. The accidents involving passenger cars accounted for 41.39% of the total number of G109 accidents. The accidents involving passenger cars accounted for 41.39% of the total number of accidents in G109; The accidents involving trucks accounted for 17.23% of the total number of accidents in G109.

From the perspective of accident risk, it can be seen from Fig. 2.37, the danger coefficient of the semi-trailer is higher, the average number of fatalities caused by the vehicle is 1.5 people per accident. Reasons can be concluded that the semi-trailer

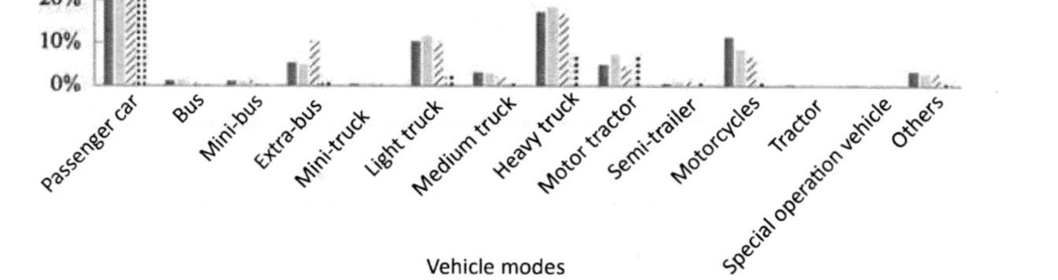

Fig. 2.36 G109 the distribution characteristics of transportation modes of people responsible for the accidents in Qinghai areas

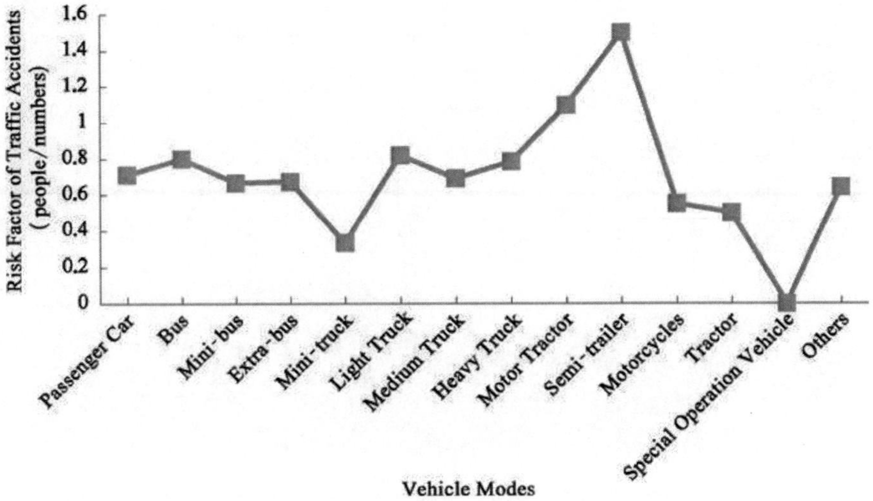

Fig. 2.37 G109 the risk factor of different vehicles in Qinghai areas

volume tonnage is larger, easier to be overloaded, higher power, especially when in high altitude areas, the overall vehicle performance is greatly reduced. When an accident occurs, it will often cause serious casualties and property damage.

In Fig. 2.38, we can see that the fatality rate of buses and the semi-trailers exceeds 40%. From the above analysis, semi-trailers are more dangerous, while the buses, because they carry more people, will cause serious casualties in one accident, with the highest fatality rate, over 50%.

According to the accident information statistics of G109 Tibetan areas, there are 108 valid accident samples with complete vehicle types. The distribution of the accident types is shown in Fig. 2.39. The main vehicle modes of accidents are passenger car and Extra-bus, with 78% of the total number of accidents, among which 53 accidents occurred in minibuses, accounting for 49% of the total. The total number of accidents involving passenger car is 53, accounting for 49% of the total; 31 accidents involving extra-trucks, accounting for 29% of the total.

From the perspective of the accident risk factor (Fig. 2.40), the risk level of heavy trucks is the highest. The average death per accident is 1.6 people.

From the G214, G109 traffic accident distribution of the type of vehicles involved, the proportion of traffic accidents caused by heavy trucks such as heavy trucks, trailers are higher. According to data from the G109 in Qinghai areas, the risk of traffic accidents caused by improper braking caused is the highest. Trucks must brake continuously for a long time because there are more long downhill sections in high altitude areas, which easily cause the brake system overheating and brake failure. Besides, the probability of secondary accidents with other vehicles is high, and the severity of accidents is high.

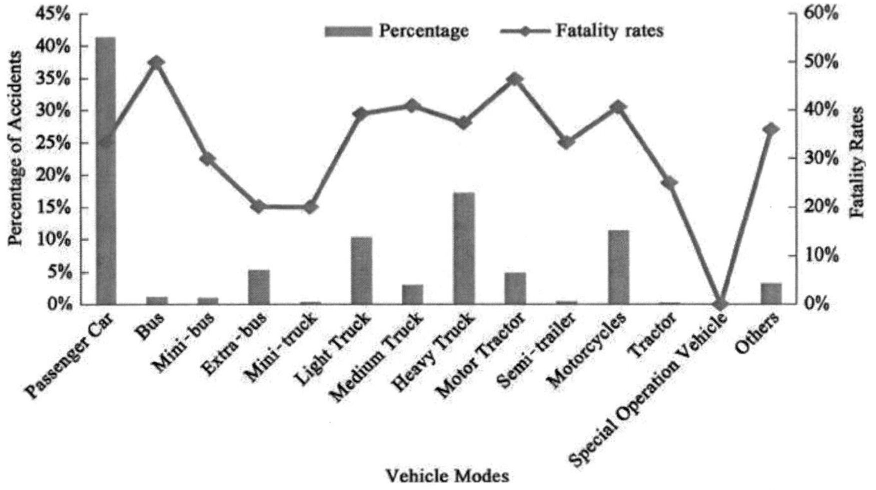

Fig. 2.38 G109 characteristics of fatalities and accidents caused by different vehicles in Qinghai areas

Fig. 2.39 G109 proportion of vehicles involved in traffic accidents in Tibet areas

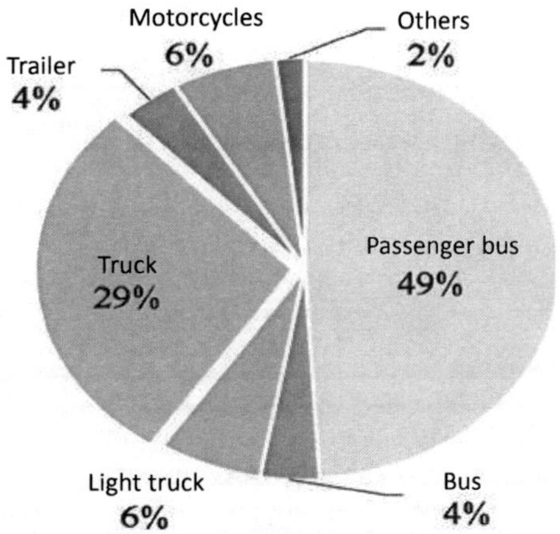

It can be seen that the dynamic performance and braking performance of trucks are very vulnerable to the comprehensive impact of high altitude, low-pressure, hypoxia environment and complex terrain conditions, thus affecting the driving safety. Therefore, when studying the vehicle performance in high altitude areas, we should focus on the performance analysis of heavy trucks, and grasp their change rules and the impact mechanism on traffic safety.

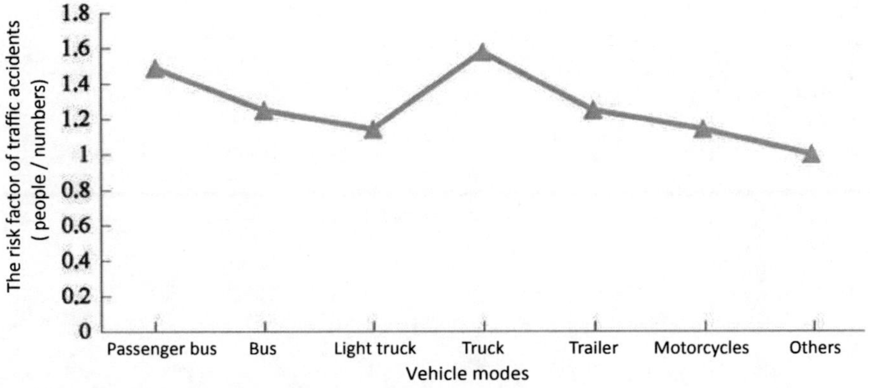

Fig. 2.40 G109 the risk factor of different vehicles in Tibet areas

2.1.4.3 The Relationship Between Traffic Accidents and Roads Design

(1) **Alignment Distribution of Accident Sections**

According to accident statistics, the alignment of highways in high altitude areas is divided into nine types: straight, general curves, general slopes, and sharp curves. Among them, there is no accident data for the steep slopes and sharp curves. Figure 2.41 shows the number of traffic accidents and the distribution of casualties in each road alignment. It is seen from Fig. 2.41 that the number of accidents, the number of casualties and the number of injuries in the straight sections and general curved sections accounted for a relatively high proportion. Of these, the three indicators in the straight sections accounted for 42.25%, 28.80% and 35.41%, and the three indicators of general curved road sections accounted for 26.20%, 37.20% and 32.67%, respectively.

In Fig. 2.42, from the traffic accident risk coefficient statistics, the traffic accident risk coefficient of the curved road section is higher than that of the straight road section, and especially the continuous downhill road section is at a higher level.

Figure 2.43 shows the accident distribution of different alignment road sections in G109 Qinghai areas. The accident rate of the straight section and the continuous downhill section is higher, and the three indicators of the straight section are 85. 08%, 84. 89% and 83. 25%.

From the perspective of accident severity, the accidents under different alignments are analyzed. It can be concluded from Figs. 2.44 that the accident risk of the sharp bend and steep slope is relatively high, and this is due to the poor sight distance of the sharp curve and steep slope. If the vehicle exceeds the limit speed or overloads, it is very easy to cause accidents such as overturning, brake failure and crashes.

(2) **Distribution of Accident Road Surface State**

The degree of wetness of the road surface is an essential factor affecting road driving safety, and the wet road surface will directly lead to the decrease of friction coefficient.

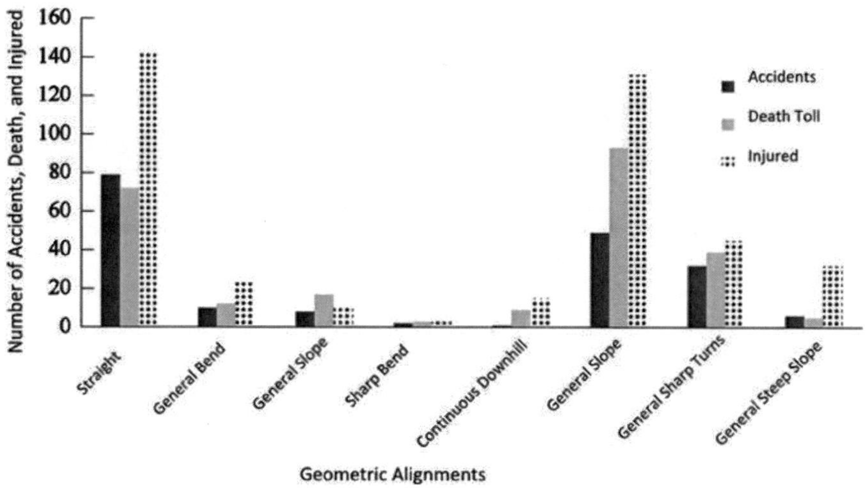

Fig. 2.41 Distribution of accidents and death toll on different road alignment in Qinghai G214

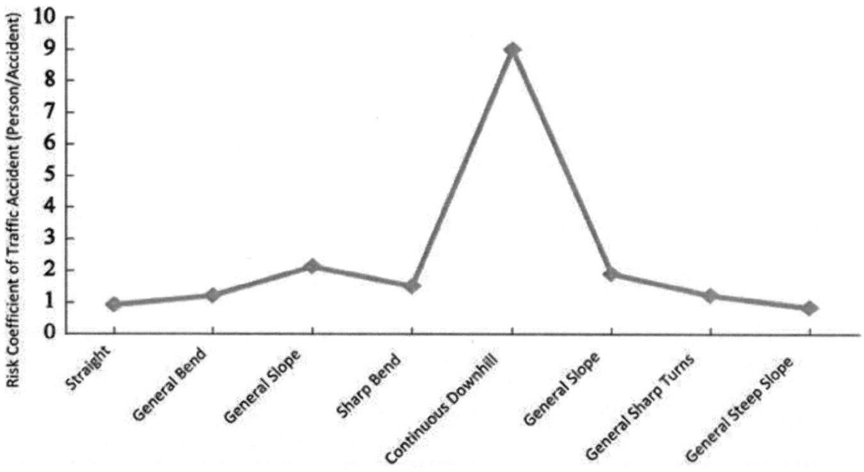

Fig. 2.42 Changes in the risk coefficient of traffic accidents in different alignment sections

The degree of wetness of the road surface can be divided into dry, wet, stagnant, flooded, ice and snow, muddy, and other conditions. According to the accident data of G214 in Qinghai in the past five years, the statistical results are shown in Fig. 2.45.

In general, the accidents that occurred under dry road conditions accounted for the most significant proportion, with 142 accidents accounting for 75.53% of the total number of accidents, followed by snow and ice conditions, with 22 accidents accounting for about 11.70%. The two accounted for 87.23% of the total number of accidents.

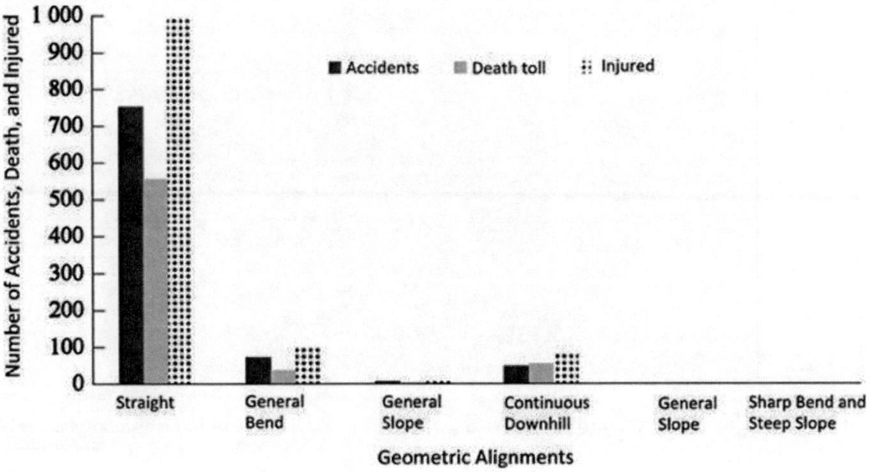

Fig. 2.43 Distribution of accidents and death toll on different road alignment in Qinghai G109

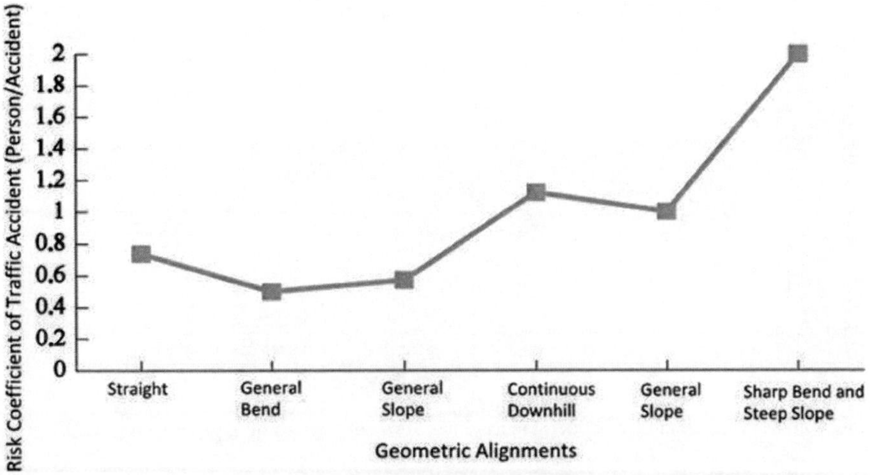

Fig. 2.44 Changes in the risk coefficient of traffic accidents in different alignment sections

According to the statistics of traffic accident risk coefficients under different degrees of road surface wetness (Fig. 2.46), the degree of risk is higher in a flooded state, reaching 2.5 persons per accident, and the risk of other road surface states is generally similar.

According to the statistics of traffic accidents in G109 Qinghai areas under various road surface conditions in the past five years, the results are shown in Fig. 2.47.

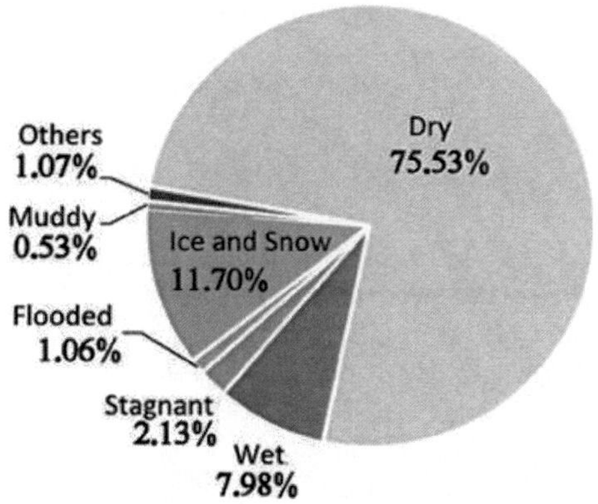

Fig. 2.45 Accident proportion distribution under different degrees of road surface wetness in G214

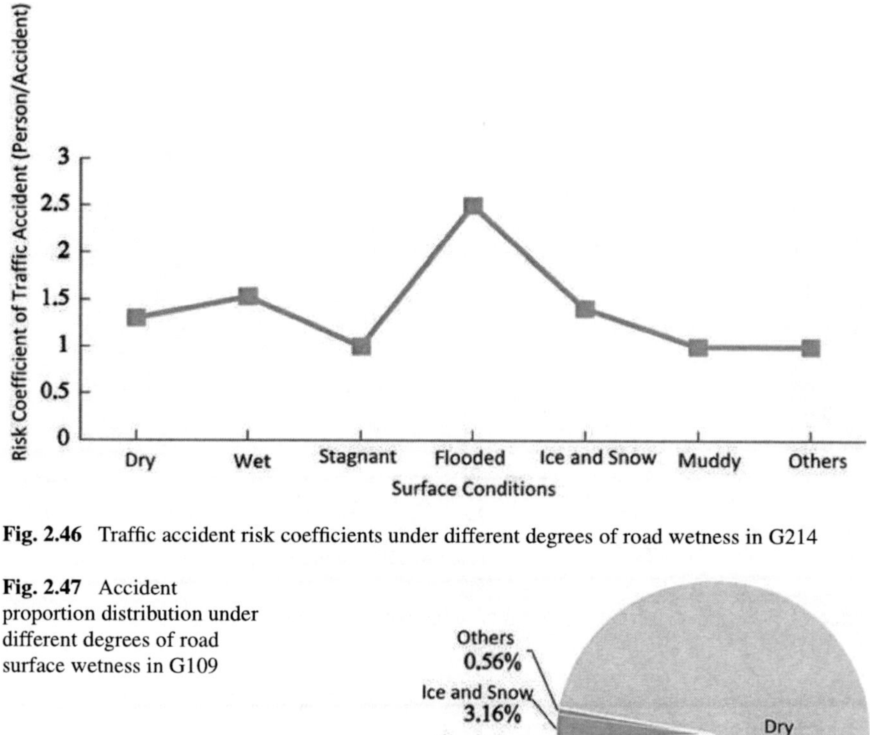

Fig. 2.46 Traffic accident risk coefficients under different degrees of road wetness in G214

Fig. 2.47 Accident proportion distribution under different degrees of road surface wetness in G109

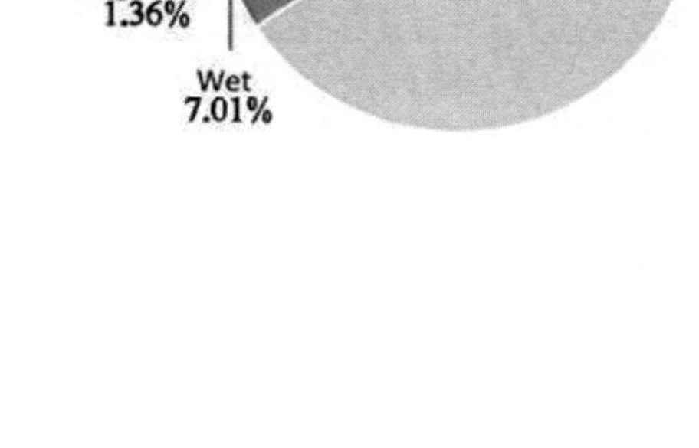

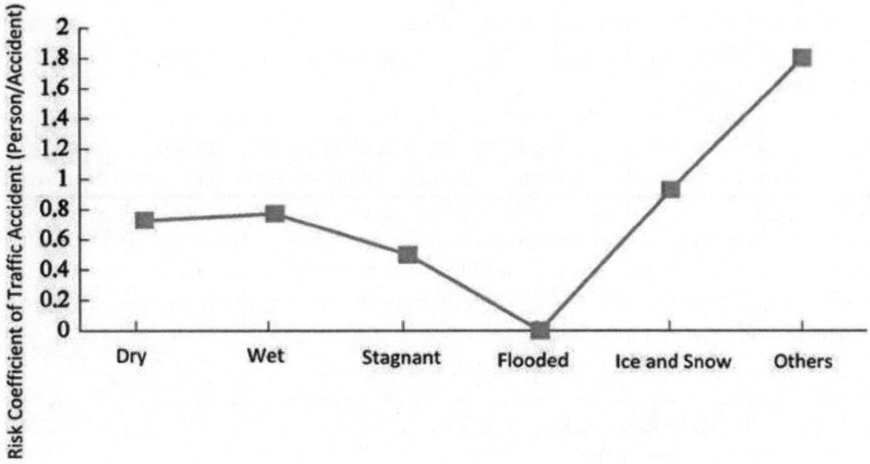

Fig. 2.48 Traffic accident risk coefficients under different degrees of road wetness in G109

In general, the accident rate is higher under dry conditions, accounting for 87.80% of the total number of accidents, followed by wet conditions, which account for 94.81% of the total accidents.

Judging from the accident risk coefficient (Fig. 2.48), excluding "other" road conditions, the accident risk is the highest under ice and snow conditions. This is because under ice and snow conditions, the adhesion coefficient between the vehicle and the road surface is reduced, and extremely dangerous traffic accidents such as side-slip, rollover, and excessive braking distance are prone.

(3) **Analysis of road influence factors in traffic accidents**

According to the analysis results of road conditions in traffic accidents, the number of traffic accidents on straight road sections is the largest. However, the danger of traffic accidents caused by sharp bends and steep slopes is more serious. The alignment of highways in high altitude areas is generally smooth and straight, and some sections even have a continuous 10 km long straight line. In addition, the roads along the highway are relatively barren, and the landscape is monotonous, which can easily cause driver fatigue when driving for a long time. Moreover, when driving on a long straight section, it is very easy to cause the driver to Overspeed, causing an accident. Most traffic accidents are distributed on the straight road section, which indicates that the smooth and monotonous road alignment condition is an essential incentive for speeding and fatigued driving. When the alignments are more complex, the driver generally needs to perform large steering operations and frequent braking and acceleration, which significantly impacts the driver's physical load and requires the body to provide more physical energy by increasing the heart rate. The high rate of traffic accidents under complex alignments indicates that the driver's psychological and physiological state is easily affected when road conditions change, and the change in the psychological and physiological state significantly impacts the severity of traffic accidents. It is necessary to research drivers' psychological and physiological states under complex alignments and put forward reasonable horizontal and vertical alignment indicators.

Most traffic accidents occurred under dry and ordinary road conditions, but when the road is icy or snow-covered, it can easily lead to more serious traffic accidents. Under the conditions of ice and snow, the lateral and longitudinal friction of the road surface is reduced, and the vehicle is prone to skidding, causing the driver to lose control of the vehicle and resulting in serious accidents. Therefore, the influence of snow and freezing conditions should be considered in the design of geometric indicators such as the radius of the circular curve, the longitudinal slope, the super-elevation, and the sight distance.

2.1.4.4 The Relationship Between Traffic Accidents and the Environment

(1) **The Distribution of Accident Weather**

According to the statistics of 188 traffic accidents in G214 Qinghai area in the past five years under various weather conditions, the results are shown in Fig. 2.49. It can be seen from the figure that the proportion of accidents and the proportion of casualties have the same distribution characteristics. Traffic accident varies in different climatic conditions. The ratio of sunny and non-sunny accidents was 130:58. Nearly 69.15% of accidents and 65.74% of deaths occurred in good weather. Weather conditions such as sunny days account for the highest proportion of the year, and the number of accidents is also higher.

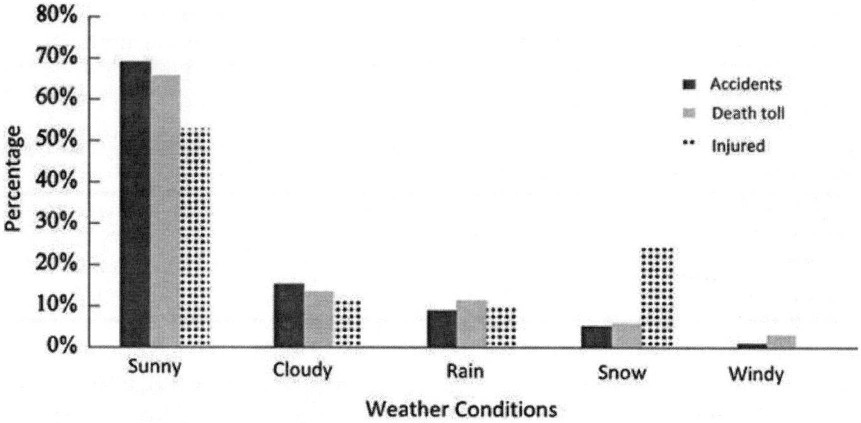

Fig. 2.49 Proportion distribution of traffic accidents under different weather conditions in Qinghai G214

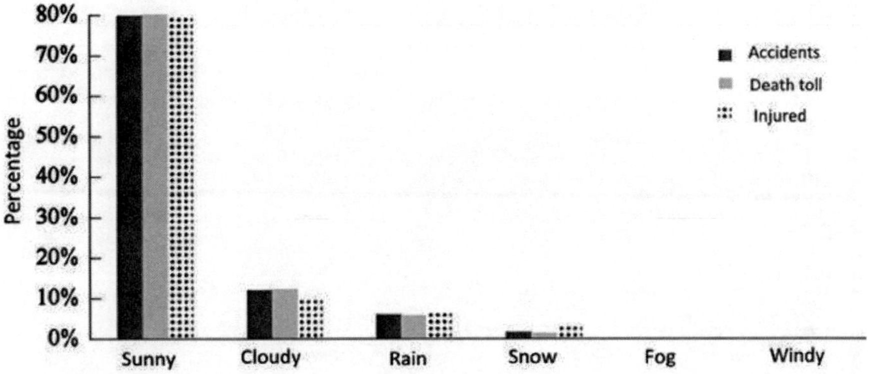

Fig. 2.50 Proportion distribution of traffic accidents under different weather conditions in Qinghai G109

On sunny days, due to good driving conditions, fast speed, and driver's arbitrary, it is difficult to take specific measures to avoid a collision in an emergency. Due to the enormous collision energy, it will inevitably cause severe accident casualties.

According to the statistics of 885 traffic accidents in the G109 Qinghai area in the past five years under various weather conditions, the results are shown in Fig. 2.50. It can be seen from the figure that the proportion of accidents and the proportion of casualties have the same distribution characteristics, and the traffic accident under different climatic conditions is different. The ratio of sunny and non-sunny accidents was 707:178. Nearly 79.89% of accidents and 80.15% of casualties occurred in good weather.

In G109, sunny days accounts for the highest proportion of the year, and the number of accidents is also higher. The accident severity under different weather conditions is analyzed from the accident risk coefficient, as shown in Fig. 2.51. It can be seen from the figure that in fog and intense wind weather, the accident risk is higher, with an average of 1 casualty per accident.

The visibility on foggy days is low, which affects the driver's sight distance, and is prone to dangerous accidents. The impact of strong winds on driving safety is mainly reflected in the impact on driving stability, especially crosswinds, which can easily lead to vehicle slippage and endanger driving safety.

(2) **Accumulative Correlation between Different Oxygen Content and Traffic Accidents**

The oxygen content level is mainly related to the local altitude, climate, and vegetation coverage, so its regional characteristics are apparent. Oxygen content is not the main factor affecting accidents as an environmental factor. However, the level of oxygen content has a more significant impact on the driver's state and motor vehicle performance, making traffic accidents more likely to occur. At the same time, oxygen content also impacts the traffic environment. Low oxygen content and in-habitation usually lead to less population and low traffic. Therefore, areas with

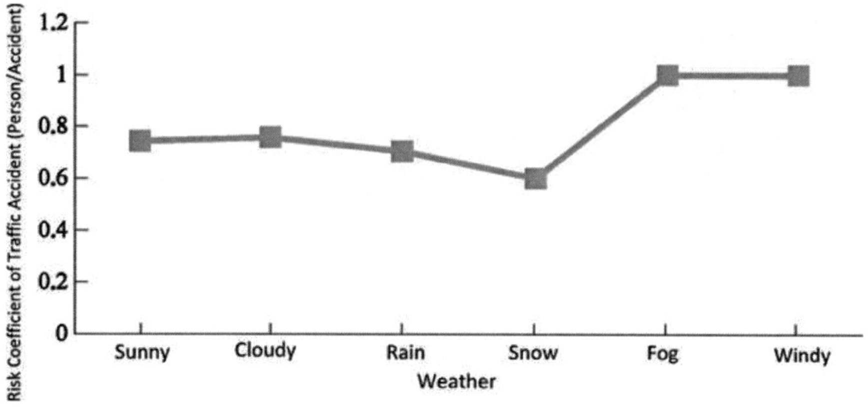

Fig. 2.51 Accident risk distribution under different weather conditions

different oxygen content will have a specific correlation with the distribution of accident characteristics.

Figure 2.52 shows the changes in the forms of traffic accidents in areas with different oxygen content. It can be seen from the figure that the proportion of single-car accidents decreases gradually with the increase in oxygen content. Car-to-car accidents and car-to-person accidents are on the rise overall.

Figure 2.53 is a line figure of the average number of casualties per accident with the change in oxygen content. Overall, the casualties in traffic accidents gradually decrease with the increase in oxygen content. At this critical point, the number of deaths was above 1.0, and above this critical point, the number of deaths was only about 0.5.

From the above two perspectives, it can be concluded that in areas with low oxygen content, the proportion of single-car accidents is relatively high, and the

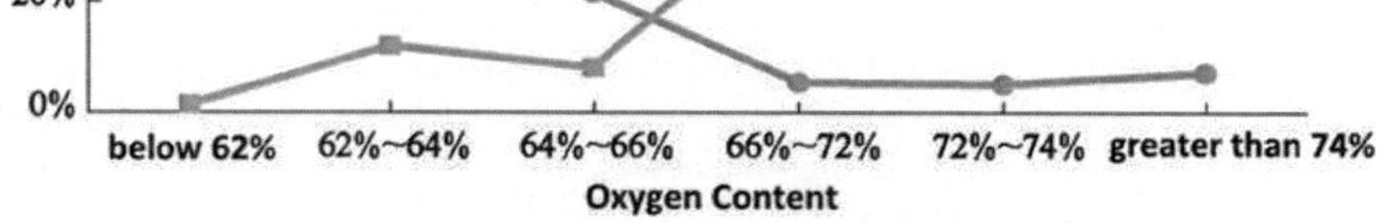

Fig. 2.52 Forms of traffic accidents in areas with different oxygen content

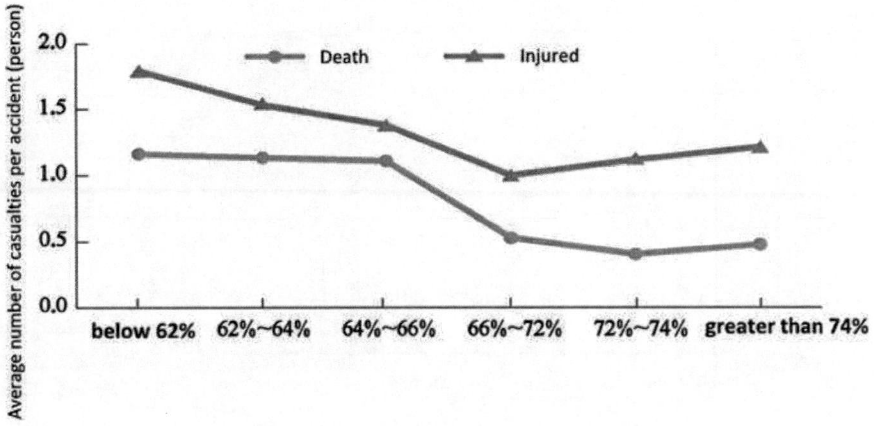

Fig. 2.53 Traffic accident casualties in different oxygen-containing areas

casualties of accidents are severe. With the increase in oxygen content, the proportion of single-car accidents gradually decreased, the car-to-car accidents and the car-to-person accidents increased accordingly, and the accident casualty level also gradually decreased.

(3) The Relationship between the Geographical Features of Uninhabited Areas and the Characteristics of Traffic Accidents

The survey shows that traffic accident characteristics under different geographical features in high altitude areas are analyzed and summarized. The following analysis is mainly from two aspects: the accident rate per million vehicle kilometers and the casualties.

As shown in Fig. 2.54, the accident rate per 100,000 vehicle kilometers of highways in villages and towns is much higher than that in the other three areas, which is consistent with the previous analysis, in which the accident rate between vehicles is three times or more than the other three areas. Meanwhile, the accident rate of vehicles and pedestrians is more than six times, and the accident rate for single-vehicle is lower compared to other areas. The other three non-village roads have similar accident rates per 100,000 vehicle kilometers. However, the roadside environment in uninhabited mountainous areas is usually poor, and cliffs and many longitudinal slopes can cause rollover and collision accidents. The single-car accidents and passenger cars accidents shown in the figure are more prominent. The multiple-car accidents in the tourist areas are slightly more severe than those in the uninhabited mountainous areas, and the car-to-person accidents are more prominent compared with the uninhabited mountainous areas. Due to the characteristics of a monotonous roadside environment, less lateral interference, and less traffic volume in the Gobi inhabited region, the accident rate is low.

It can be seen from Fig. 2.55 that the death level of roads in villages and towns is only 1/3 or less of that in other areas. Villages and towns have mixed traffic, large flow,

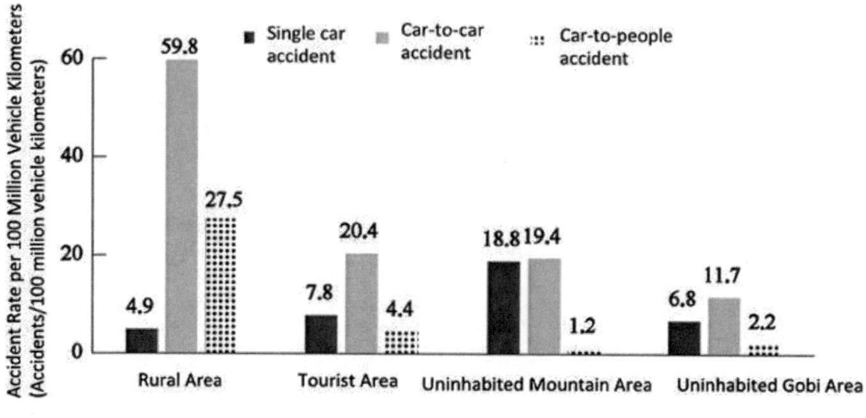

Fig. 2.54 Distribution of accident rate per 100,000 vehicle-kilometers by accident form under different geographical features

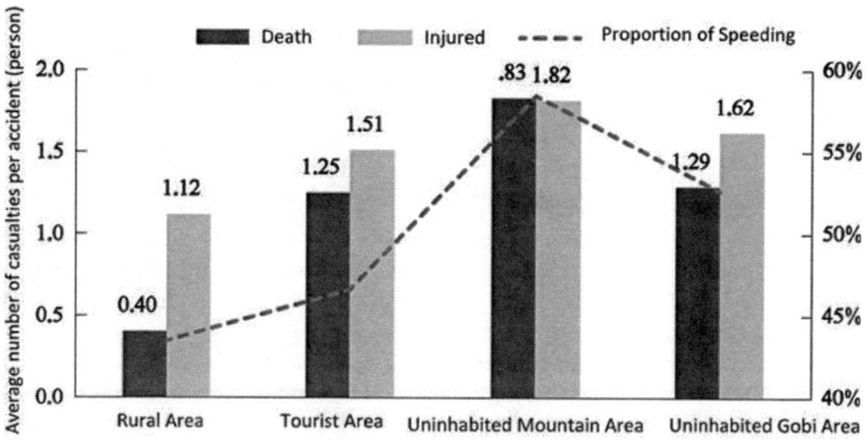

Fig. 2.55 Accident casualties in different geographical features

large accident base, and many simple collision accidents. The most severe casualties are in uninhabited mountainous areas, with an average death toll of 1.8 per accident. The curve in the figure represents the proportion of accidents caused by over-speed to the total accidents. The data can show a positive correlation between the average number of casualties and the proportion of over-speed accidents.

From the above analysis, the characteristics of the roads in the villages and towns are consistent with the previous study, and the high accident rate, car-to-car accidents and car-to-person accidents are all prominent. The road sections in uninhabited areas have obvious terrain characteristics, and the roadside environment and lateral interference are the main factors affecting the accident.

(4) **Analysis of Environmental Influence Factors in Traffic Accidents**

From the distribution of traffic accidents under different weathers, most traffic accidents occurred in good weather, and there was little difference in the degree of danger of traffic accidents under various weather conditions. The occurrence type and severity of traffic accidents are related to oxygen content. In areas with low oxygen content, the proportion of single-car accidents is relatively high, and the degree of casualties is severe, indicating that the low pressure and oxygen deficient environment has a more significant impact on drivers and vehicles themselves. It is easy to cause a traffic accident due to the driver's driving state or vehicle performance.

The number and severity of traffic accidents vary in different traffic environments. The number of traffic accidents in villages and towns is high. However, the casualties in mountainous and uninhabited areas are higher, which is related to the high speed of vehicles. The mountainous and uninhabited areas are sparsely populated and have harsh environmental conditions. Drivers generally have the requirement to pass quickly. Therefore, the running speed is fast and accessible to over-speed, and the traffic accidents caused by over speeding are more serious.

To sum up, from the analysis of the location, causes, and forms of traffic accidents, the general rule of changes of traffic accidents on G109 and G214 highways in the Qinghai-Tibet Plateau region was obtained, and the relationship between the four elements of people, vehicles, roads, and the environment and traffic accidents was determined respectively. The main factors affecting the occurrence of traffic accidents are obtained.

2.2 The Characteristics of the Traffic Flow in Special High Altitude Environment

2.2.1 Basic Data Collection

2.2.1.1 Equipment Overview

According to the research, collecting relevant primary data such as vehicle speed, oxygen volume fraction, altitude, and temperature is necessary.

(1) **Vehicle Speed Detection Equipment**

The traffic density on the Qinghai-Tibet Highway is low, and the equipment must be exposed to the field for a long time to collect enough samples. In order to ensure the power supply of the equipment and the safety of the field test, this test adopts the Metro Count traffic flow detection system, which is powered by dry batteries that have the advantages of long working time, no need for real-time connection to the

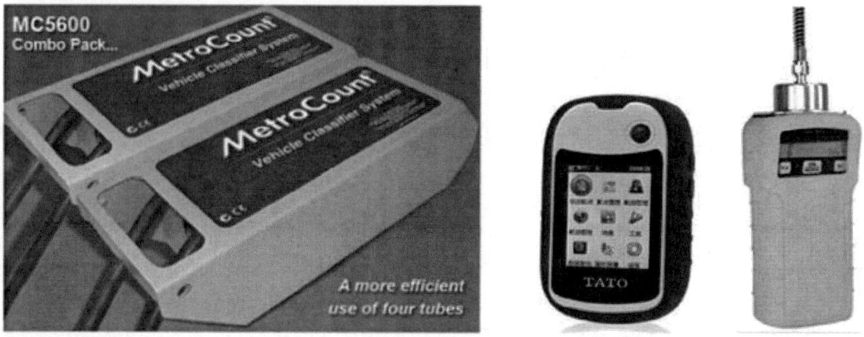

Fig. 2.56 MetroCount traffic flow detection system, GPS and gas detector

computer, and strong field operation reliability. When the vehicle passes the detection section, the system obtains the air pressure pulse signal through the rubber air pressure pipe on the road surface. It forms the axle electrical signal in the roadside unit to realize real-time traffic flow detection. The detected data includes the information of each vehicle passing through the section, such as date of arrival, time, maximum wheelbase, vehicle speed, headway, time interval, number of axle groups, number of axles, and vehicle type.

(2) **High altitude Environmental Testing Equipment**

In order to obtain the oxygen content of the test section, a hand-held GPS recorder, a hand-held gas detector (Fig. 2.56), and a thermometer is used to detect the altitude, oxygen volume concentration and temperature of the test site.

The data measured by the gas detector is the volume fraction of oxygen, that is, the volume of oxygen in a unit volume of air. Due to the thin air, the oxygen volume fraction measured by the gas detector cannot reflect oxygen content level in high altitude areas. Through the gas volume fraction and mass concentration conversion formula, Laplace formula, and using the measured oxygen volume fraction, altitude and other data, the oxygen mass concentration that can reflect the oxygen content is calculated:

$$C_m = \frac{N}{22.4} \times 10^{-6} \times \frac{273.15}{273.15 + T} \times \frac{P_1}{P_2} \tag{2.5}$$

$$Z_2 - Z_1 = 18410 \times (1 + \frac{T}{273.15 + T}) \times \lg \frac{P_1}{P_2} \tag{2.6}$$

In the above two formulas,

C_m is the oxygen mass concentration;
N is the molecular weight of the gas, where the molecular weight of oxygen is 32;
T temperature;

p1, p2 —air pressure, *p2* is the standard atmospheric pressure;
Z1, Z2 —altitude, *Z1* is the measured altitude of the inspection section, and *Z1* is the sea level.

In the following, the percentage of the detected section's mass concentration and oxygen's mass concentration in the plain's standard air is used to study oxygen content.

2.2.1.2 Data Overview

Two data surveys (23 days in total) were conducted on parts G214 and G109. Meanwhile, the G214 Gonghe-Maduro section, G109 Minhe-Dulan section (including the G6 Huangyuan-Daotanghe section), and G109 Golmud-Lhasa section road attribute data, traffic flow characteristics, traffic environment data and traffic accidents data were collected in 32 survey points and 91 survey sections. The road attribute data includes road alignment, transverse and longitudinal slope, and section spacing. The traffic flow characteristic data includes the data detected by Metro Count, such as speed, flow, vehicle type, head distance, wheelbase, etc. Traffic environment data includes speed limit, altitude, and oxygen content. The data collection overview is shown in Table 2.1.

2.2.2 *Traffic Volume and Composition*

2.2.2.1 Traffic Volume Data Collection

In order to study the distribution and composition characteristics of highway traffic in the Qinghai-Tibet Plateau region, the research team collected the traffic volume data of G109 (Qinghai-Tibet Highway) and G317 (Sichuan-Tibet Highway) in Qinghai-Tibet Plateau region. Since the traffic volume of G109 is relatively large and the data is complete in high altitude areas, G109 is the main study section. The primary data

Table 2.1 Data collection overview

National Highway	Section	Route length	Inspection point	Observation section	Altitude range (m)	Number of vehicle samples
G214	Gonghe-Maduo	342	8	20	2911–4378	33,882
G109	Minhe-Dulan section (including G6 Huangyuan-Daotanghe section)	534	10	28	1800–3563	132,917
G109	Golmud-Lhasa	1162	14	43	3300–5231	119,094

of the G109 Naijigou checkpoint, Wudaoliang section, Amdo checkpoint, and Xijiao checkpoint from 2012 to 2014 were collected. The essential information is shown in Tables 2.2 and 2.3.

2.2.2.2 Traffic Volume Analysis

Traffic volume is an important basic data to study the road traffic situation and its composition change pattern. Traffic volume changes all the time, it usually takes the average value of a certain time period to represent traffic volume of that time period. By the length of the time period from which the average is taken, average traffic volumes commonly used include annual average daily traffic (AADT), monthly average daily traffic (MADT), and weekly average daily traffic (WADT), among which, the annual average daily traffic volume is a vital controlling indicator in highway engineering, its calculation formula is:

$$\mathrm{AADT} = \frac{1}{365}\sum_{i=1}^{365} Q_i$$

Where Q_i——Daily traffic volume in each specified time period (vehicle/d).

By summarizing the average daily traffic data and 24 h observation data of Qinghai-Tibet highway for each quarter and each month from 2012 to 2014, the annual average daily traffic volume data by vehicle type for each year is obtained (Table 2.4). The data can reflect the traffic volume of Qinghai-Tibet highway and its changing pattern of time distribution for traffic composition in the last three years.

The distribution pattern of daily average traffic volume for three consecutive years can be obtained from Table 2.4, in addition, the quarterly average daily traffic volume and the 24hoursaverage traffic volume can be further analyzed. The average daily traffic volume per quarter is calculated by averaging the traffic volume (natural numbers) for three consecutive years, and 24 h traffic volumes are averaged over 2014 traffic volumes (natural numbers), finally, the traffic volume distribution is shown in Fig. 2.57.

The data in Fig. 2.58 are mainly from Naijigou checkpoint, Wudaoliang section, Amduo checkpoint and Xijiao checkpoint along the G109. Based on the annual average daily traffic volume in the observation area of each checkpoint, the overall spatial distribution characteristics of the traffic volume and its composition of the Qinghai-Tibet highway can be derived. Figure 2.59 shows the distribution of traffic volume at each checkpoint on Qinghai-Tibet highway in 2014.

Further analysis of the distribution pattern of the traffic composition of each road section can be obtained as shown in Fig. 2.60.

According to the graphical analysis, from the time distribution of traffic volume, the traffic volume of G109 in the past three years has shown an upward trend, and types of passenger car and truck has shown a yearly growth trend with an average annual growth rate of about 6.2%, while the agricultural vehicles such as tractors

Table 2.2 The traffic volume sample data of the continuous observation station at the Amdo checkpoint for 24 h in October 2014

Time	Vechicle											
	Car									Minivan	Motocycle	Total mixed vechicle
							Tractor		Total			
	Small truck	Medium truck	Large truck	Small passenger car	Large passenger car	Trailer	Small tractor	Large tractor				
21:00-22:00	9	6	0	50	0	20	0	0	85	0	12	97
22:00-23:00	4	0	1	43	0	27	0	0	75	0	11	86
23:00-0:00	5	1	3	45	0	16	0	0	70	0	10	80
0:00-1:00	1	1	0	15	0	30	0	0	47	0	8	55
1:00-2:00	5	3	0	20	0	33	0	0	61	0	5	66
2:00-3:00	0	4	0	9	0	42	0	0	55	0	3	58
3:00-4:00	4	4	3	4	0	43	0	0	58	0	2	60

Table 2.3 Annual average daily traffic volume in the first quarter of 2012

Route Number	Observation Mileage(km)	Annual average daily traffic volume (vehicle/day)							
		Motor Vechicle		Car					
		Total Equivalents	Total of Natural Numbers	Total Equivalents	Total of Natural Numbers	Small Truck	Medium Truck	Large Truck	Extra Large Truck
A	1	2	3	4	5	6	7	8	9
G109	76.87	3215	1678	2822	1550	275	150	207	194
Route Number	Annual average daily traffic volume (vehicle/day)						Driving volume (10,000 vehicle kilometers/d)	Adapt to Traffic Volume (vehicle/d)	Traffic Congestion
	Car			Motocycle	Tractor				
	Container Car	Small and Medium Passenger car	Large Passenger Car		Total Equivalents	Total of Natural Numbers			
A	10	11	12	13	14	15	16	17	18
G109	53	587	84	40	353	88	173	15,000	0.21

Table 2.4 Annual average daily traffic volume of qinghai-tibet highway from 2012 to 2014 (Vehicle/d)

Year	Automobiles (Vehicle/d)									Natural number of motorcycles	Subtotal of Tractors		Total of motor vehicles	
	Minivan	Medium truck	Large truck	Extra-large truck	Container truck	Small and medium sized bus	Bus	Subtotal of automobiles			Equivalent number	Natural Number	Natural Number	Equivalent number
								Natural number	Equivalent number					
2012	315	174	281	454	59	522	66	1869	4087	71	345	86	2027	4504
2013	244	169	196	439	40	792	41	1922	3854	120	68	17	2059	4042
2014	207	169	193	493	29	955	58	2105	4171	152	29	7	2264	4352

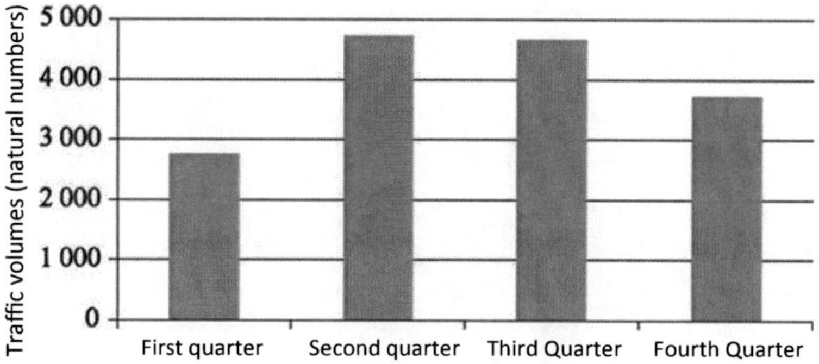

Fig. 2.57 Distribution of traffic volume of qinghai-tibet highway by seasons

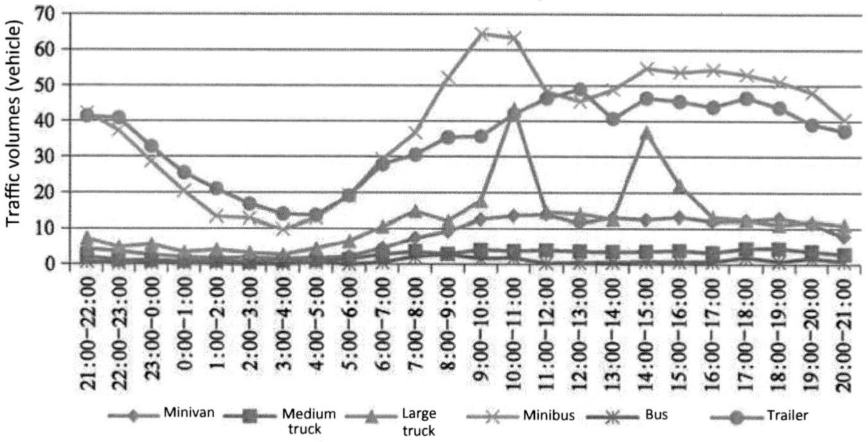

Fig. 2.58 24-hour traffic volume distribution by vehicle type of Qinghai-Tibet highway in 2014

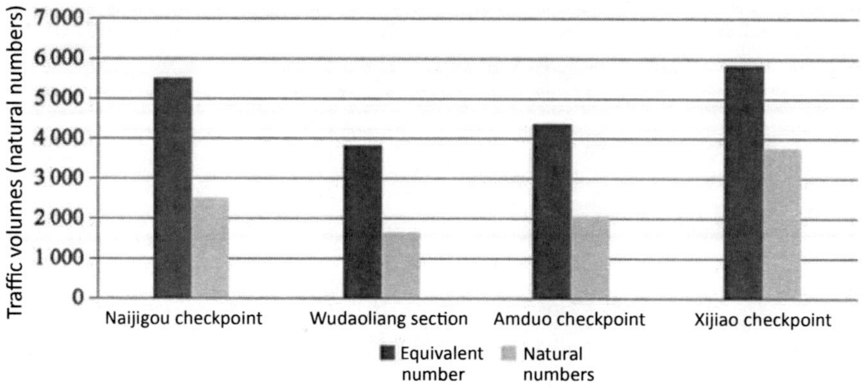

Fig. 2.59 Traffic volume distribution by section of Qinghai-Tibet highway in 2014

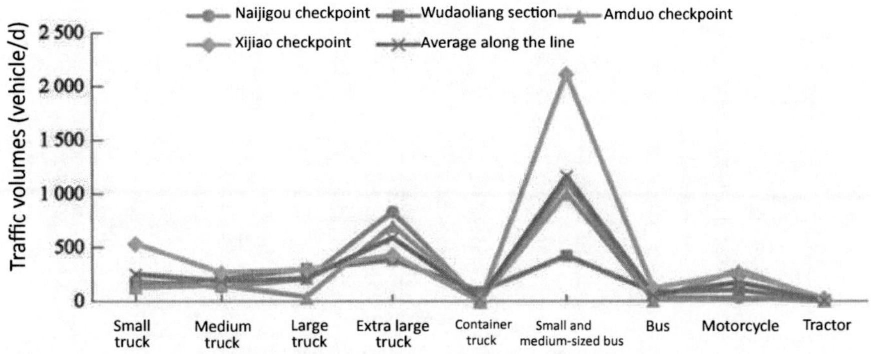

Fig. 2.60 Traffic volume distribution by section and vehicle type of Qinghai-Tibet highway in 2014

have decreased year by year, it reflects that the leading vehicle type of Qinghai-Tibet highway is developing in a faster and motorization way. In terms of traffic volume of each vehicle type, the leading vehicle types on Qinghai-Tibet highway are passenger car and trucks, and only these two types of vehicles show continuous growth, which reflects a polarized development.

From the spatial distribution of traffic volume, the distribution of traffic volume along the Qinghai-Tibet highway shows the characteristics of high at both ends and low in the middle, among which Naijigou checkpoint and Xijiao checkpoint are in Golmud City of Qinghai Province and Lhasa of Tibet Autonomous Region respectively, which is consistent with the regional population distribution characteristics along the highway. As for the distribution of vehicle types along the road, the vehicle type composition of the Golmud section is mostly oversized trucks, while vehicle type composition of the Lhasa section is mostly small and medium-sized buses, reflecting that the Golmud section along the Qinghai-Tibet highway mainly undertakes the function of freight input and output, while the Lhasa section mainly undertakes the function of passenger input and output.

2.2.2.3 Spatial Distribution Features of Traffic Volume

According to the vehicle traffic flow detection instruments laid in G109 and G214, the traffic flow information of each stake point is derived, and the spatial distribution maps of traffic flow in G109 and G214 are made respectively.

(1) **Spatial Distribution of Traffic Volume of G109**

There are 24 stake points along G109, and the instruments of two of them are malfunctioning, so the data of 22 stake points are valid. The spatial distribution of traffic volume of G109 was made based on the traffic volume information of these 22 stake points, as shown in Fig. 2.61.

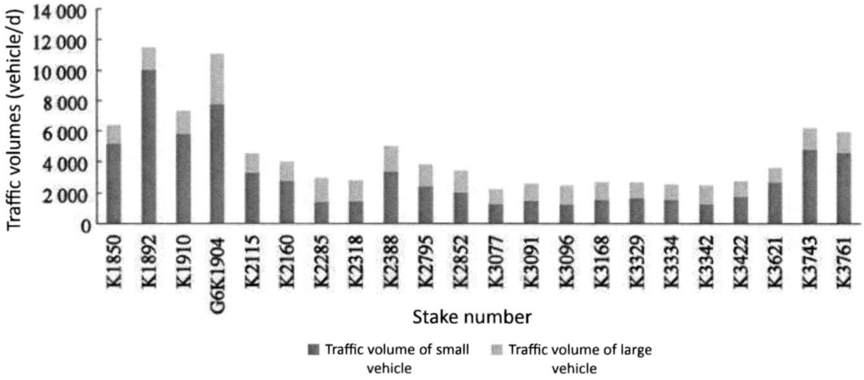

Fig. 2.61 Spatial distribution of traffic volume of G109

From Fig. 2.61, higher traffic volume is gathered between stakes K1850 and K1910, near the town, among which K1892 has a large traffic flow through the town; K1850 and K1910 are between the town, so the traffic volume is relatively small; K1910 is located at the diversion of G214 and G109, which is a class-I road (other points are all class 2 roads). K1910 is located at the diversion of G214 and G109, which is a class-I road (all other points are second-class roads) with two lanes in both directions, the traffic volume is large. Between K2115 and K3621 is a non-urban area with low traffic volume, among which the traffic volume of stake number K2388 is larger because it is close to the county town of Turan. The traffic volume between K3743 and K3761 becomes larger than before, because this area is close to Lhasa and has more traffic flow.

(2) **Spatial Distribution of Traffic Volume of G214**

There are 8 stake points laid along G214, which are class-II roads, two-way two-lanes, its data are valid, according to the traffic information of these 8 stake points to make traffic volume spatial distribution map of G214, as shown in Fig. 2.62.

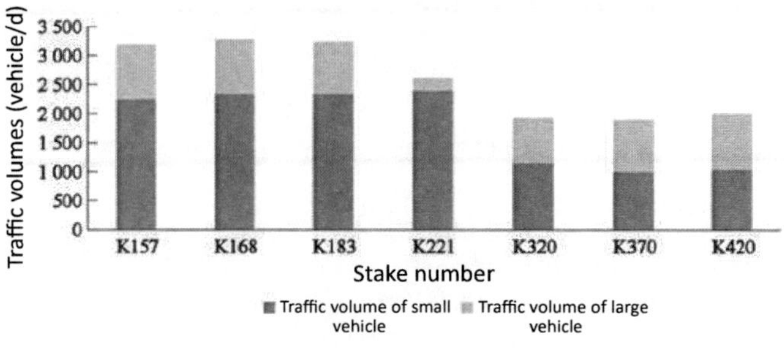

Fig. 2.62 Spatial distribution of traffic volume of G214

From Fig. 2.62, the observed traffic volumes at these stakes are concentrated between 1,800 and 3,300 vehicles/d. It can be seen from the figure that the traffic volume is higher between K157 and K183 because they are in towns and have high traffic volume, while piles K320 and K420 are non-towns, so the traffic volume is smaller.

2.2.2.4 Spatial Distribution Coefficient

According to the vehicle traffic flow detection instruments laid at G109 and G214, the traffic flow are divided into two directions: Xi'an-Lhasa and Lhasa-Xi'an, to make direction distribution coefficient maps for G109 and G214 in two directions receptively (Figs. 2.63 and 2.64).

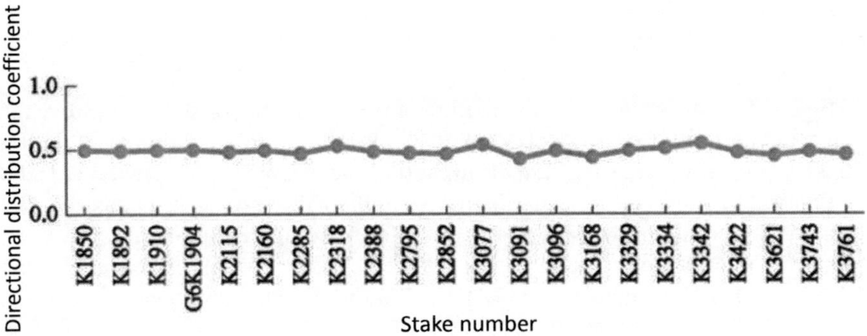

Fig. 2.63 Directional distribution coefficient of G109 in Xi'an-Lhasa direction

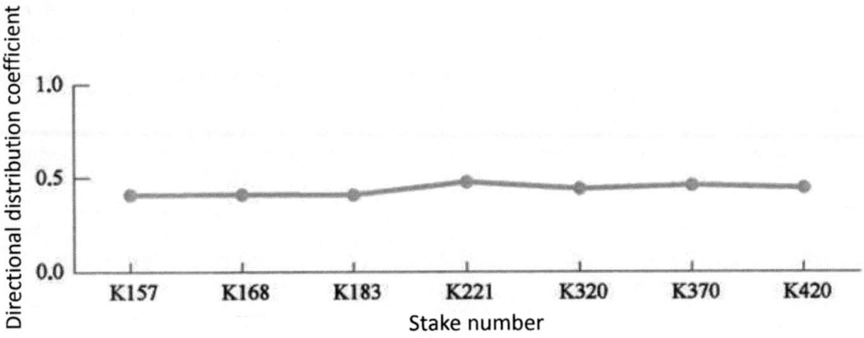

Fig. 2.64 Directional distribution coefficient of G214 in Xi'an-Lhasa direction

(1) **Directional Distribution Coefficient of G109 in Xi'an-Lhasa Direction**

From Fig. 2.63, the distribution coefficients in Xi'an-Lhasa directions of G109 fluctuate up and down at 0.5, indicating that there is not much difference between the traffic flows of G109 in both directions.

(2) **Directional Distribution Coefficient of G214 in Xi'an-Lhasa Direction**

From Fig. 2.64, the distribution coefficients in Xi'an-Lhasa directions of G214 is less than but very close to 0.5, indicating that there is not much difference between the traffic flows of G214 in both directions.

2.2.2.5 Time Distribution Features of Traffic Volume of G109

To study the time distribution features, several representative stakes of G109 were selected, which roadside environment includes towns, mountains, and Gobi, etc. The following are the traffic time distribution maps of these stake points (Figs. 2.65, 2.66, 2.67, 2.68, 2.69, 2.70 and Fig. 2.71).

Figure 2.65 shows the time distribution of traffic volume for stake K1850, which is located between the towns of Minho. The trend of traffic volume from 0:00 to 16:00 is gradually increasing, and the highest traffic volume is 260 vehicles at 16:00.

Figure 2.66 shows the time distribution of traffic volume for stake number K1892, which is between the towns, it can be seen from the figure that the traffic volume from 0:00 to 5:00 is very little, within 100 vehicles, then it sharply increases from 5:00 to 7:00, and changes smoothly from 7:00 to 20:00, with about 620 and 830 vehicles, and then decreases.

Figure 2.67 shows the time distribution of traffic volume for stake number K2115, which is located near Qinghai Lake with a plain on the road side. From the figure, the traffic volume from 0:00 to 6:00 is less than 100 vehicles, and the traffic volume increases from 7:00 to 10:00, and the traffic volume changes smoothly between 10:00 and 19:00, between 250 and 300 vehicles, and then the traffic volume decreases.

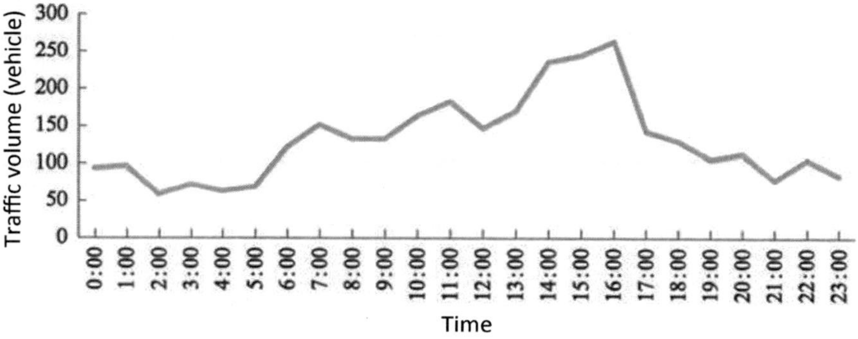

Fig. 2.65 Time distribution of traffic volume at stake K1850

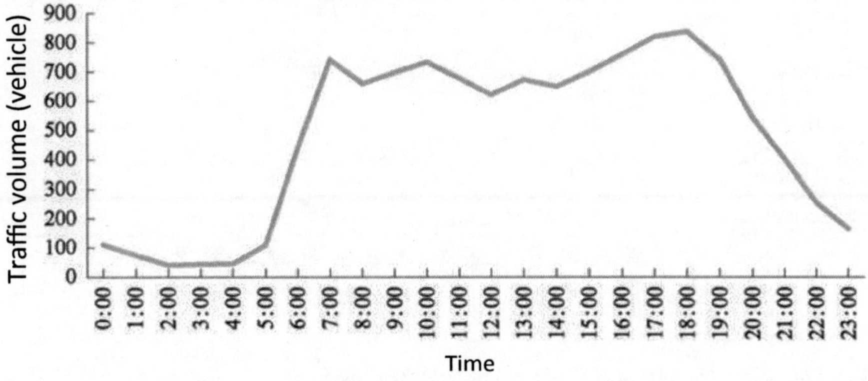

Fig. 2.66 Time distribution of traffic volume at stake K1892

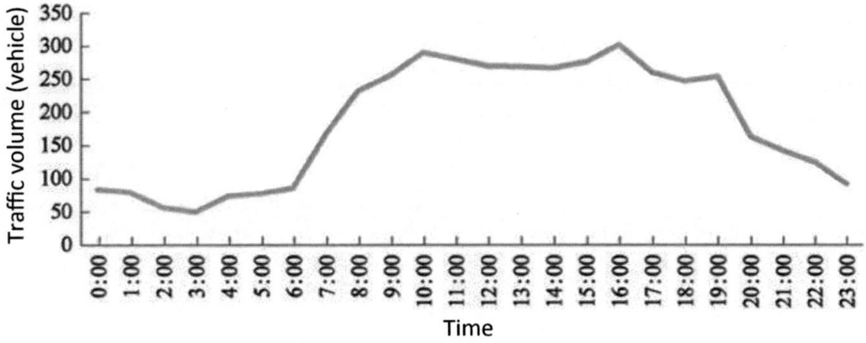

Fig. 2.67 Time distribution of traffic volume at stake K2115

Figure 2.68 shows the time distribution of traffic volumes for stake number K2285, which is in a mountainous area instead of town area. It can be seen from the figure that the traffic volume here is low, maximum 180 vehicles.

Figure 2.69 shows the time distribution of traffic volume for stake number K3168, which is located in an uninhabited area instead of town area. The traffic volume from 0:00 to 10:00 is less than 100 vehicles, and the traffic volume increases from 10:00 to 18:00 with a maximum of 250 vehicles, and after 18:00 the traffic volume decreases.

Figure 2.70 shows the time distribution of traffic volume for stake number K3342, which is located near Tanggula Mountain. From the figure, the traffic volume fluctuates, from 0:00 to 8:00 is below 100 vehicles, and the traffic volume is very little, from 8:00 to 10:00 the traffic volume increases to 200 vehicles at 10:00, between 10:00 and 15:00 the traffic volume decreases to 70 vehicles, then from 15:00 to 20:00 the traffic volume increases to 240 vehicles., after which the traffic volume decreases.

Figure 2.71 shows the time distribution of traffic volume for stake number K3743, which is located near Lhasa. From the figure, the traffic volume fluctuates, the traffic

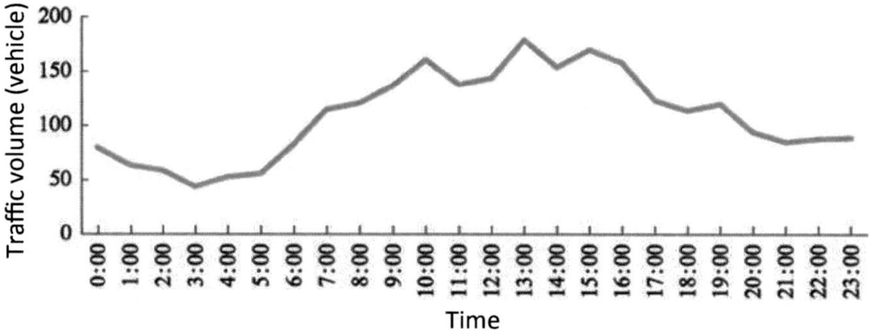

Fig. 2.68 Time distribution of traffic volume at stake K2285

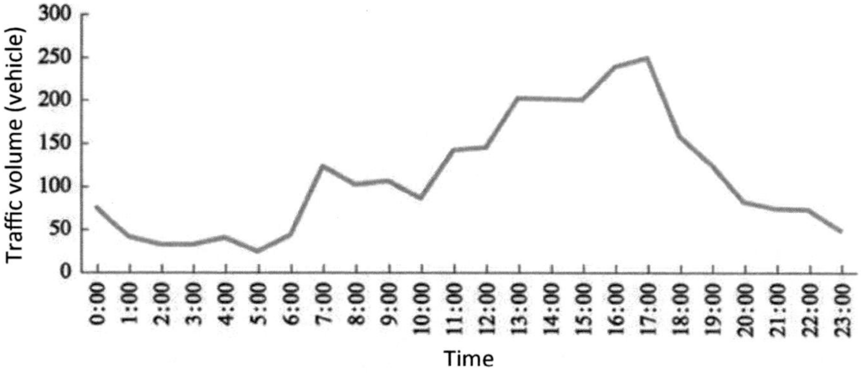

Fig. 2.69 Time distribution of traffic volume at stake K3168

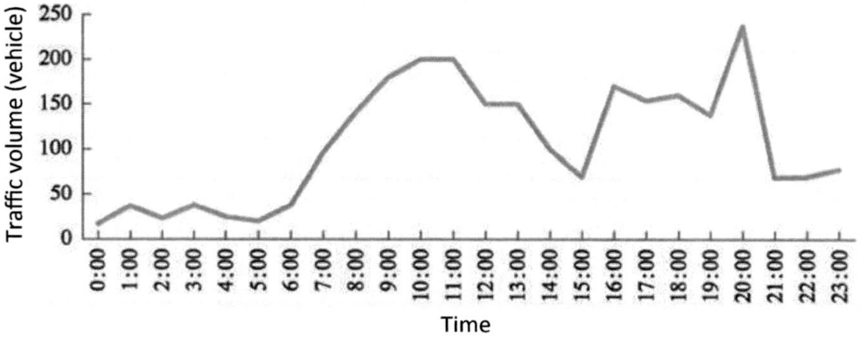

Fig. 2.70 Time distribution of traffic volume at stake K3342

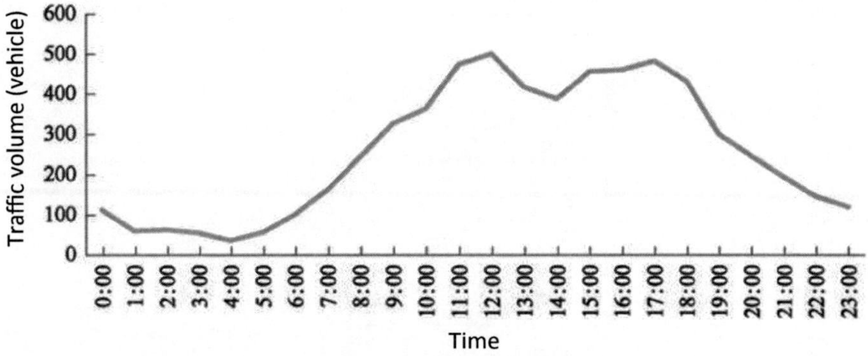

Fig. 2.71 Time distribution of traffic volume at stake K3743

volume from 0:00 to 6:00 is below 100 vehicles, and the traffic volume is very small, from 6:00 to 10:00 the traffic volume increases and reaches a maximum of 500 vehicles, from 10:00 to 18:00 the traffic volume changes smoothly and then decrease.

2.2.3 *Analysis of Traffic Flow Features*

Road construction indicators at high altitude area are generally higher, small traffic volume will easily make the driver to relax, and driver's workload increases with the elevation rises, the reaction time and action agility is reduced. Affected by cold weather and low pressure and oxygen deficient environment, vehicle fuel is not completely burned and vehicle power performance decreases. Considering the low oxygen content, high altitude, road alignment and other factors in the high altitude area, according to the collected operating speed information of typical road sections, the operating speed features and change pattern are studied for multi-type vehicles on freeways in various alignments and terrains in high altitude and oxygen deficient environment, and provide the basis for establishing the distribution model of operating speed for multiple types of vehicles.

(1) **Relationship between Altitude and Operating Speed**

The results of the speed survey of passenger car and trucks in the straight section of the Qinghai-Tibet highway between Golmud and Lhasa are as follows: the average speed of passenger car in the direction of Golmud-Lhasa is 75 km/h, and that of for trucks is 40 km/h; the average speed of passenger car in the direction of Lhasa-Golmud is 90 km/h, and that of for trucks is 70 km/h. The operating speed of each type of vehicle shows a general decreasing trend with the increase of altitude.

According to Fig. 2.72, the difference of speed distribution between two directions of Golmud-Lhasa of Qinghai-Tibet highway is large, the operating speed in the

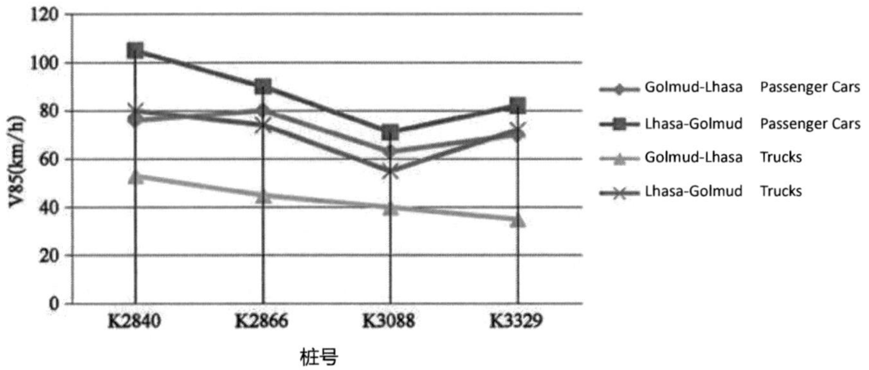

Fig. 2.72 Operating speed distribution in straight section of qinghai-tibet highway between Golmud and Lhasa

direction of Golmud-Lhasa is relatively low, and the speed difference between large and passenger car in the same direction is over 20 km/h (the speed difference between large and passenger car in Lhasa-Golmud is 35 km/h), the speed difference is large. This is mainly because the trucks in the direction of Golmud-Lhasa are basically fully loaded and have lower speed, while most of the trucks in the direction of Lhasa-Golmud are unloaded and have higher speed, thus the operating speed of the two types of vehicles in this direction is significantly higher than that of in Golmud-Lhasa direction.

(2) Relationship between Alignment and Operating speed

Figure 2.73 shows a velocity probability density distribution of two directions of stake K168, space between Sects. 2.1 and 2.2 of K168 is 527 m, with longitudinal slopes of 0.5% and 3.1% respectively. It can be seen from the figure that the velocity probability density distribution of K168 in Gonghe-Yushu direction is more discrete. Gonghe-Yushu is a uphill direction, and the central axis of the velocity distribution curve from Sects. 2.1 to 2.2 is shifted to the left; while Yushu-Gonghe is a downhill direction, and the central axis of the velocity distribution curve from Sects. 2.2 to 2.1 is shifted to the right.

Figure 2.74 shows the velocity probability density distribution of two directions of stake number K2285, the spacing for Sect. 2.1 of K2285, Sect. 2.2 of K2285 and Sect. 2.3 of K2285 3 are 340 m and 260 m respectively, and the longitudinal slopes are 1.8%, 1.5% and 2.5% respectively. It can be seen from the figure that the dispersion of the velocity probability density distribution in these two directions is similar.

(3) Relationship between Road Grade and Operating Speed

In order to study the influence of different road grade on traffic speed, cross-sectional speed of three different grades of highways was selected as the research objects: K1904 was selected as the survey point for freeway, with four lanes in both directions;

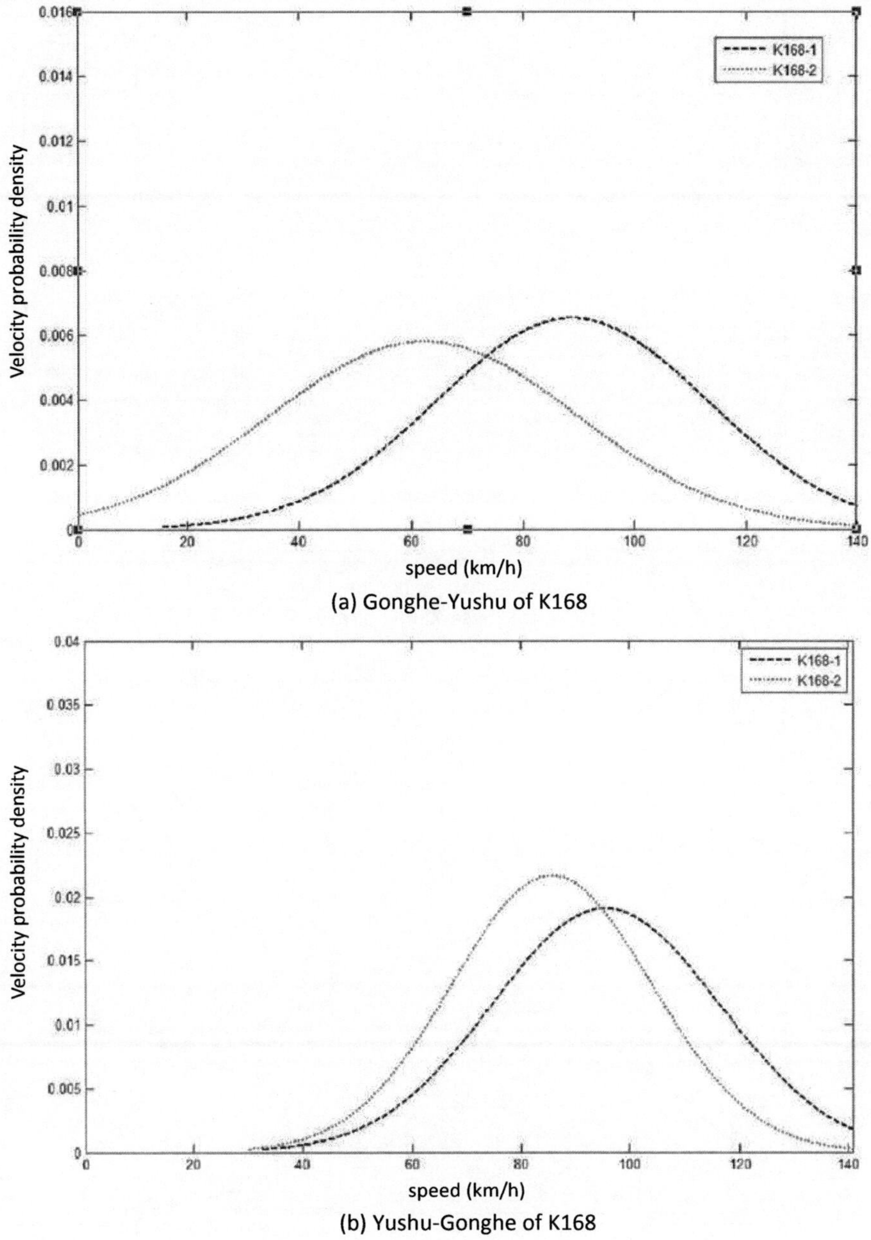

Fig. 2.73 Distribution of velocity probability density in two directions of stake K168

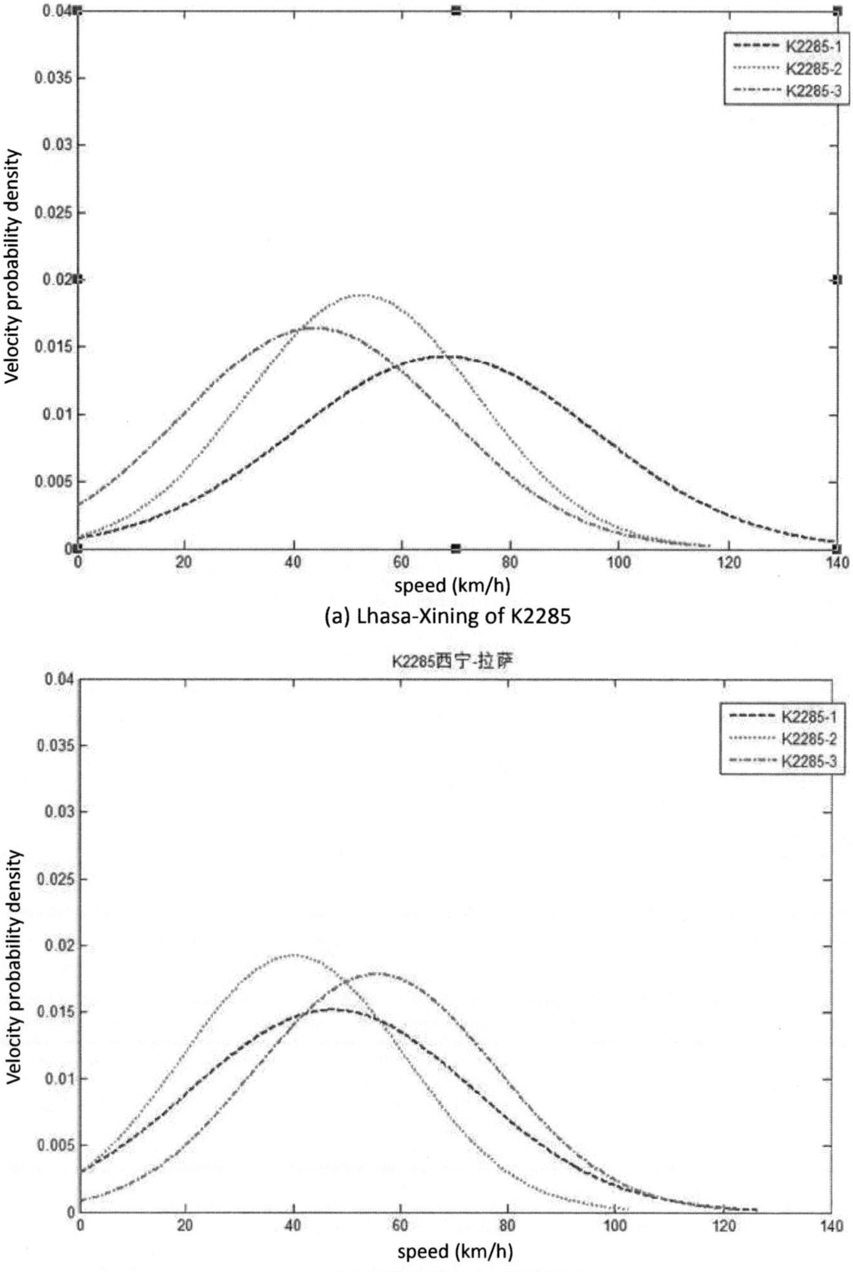

Fig. 2.74 Distribution of velocity probability density in two directions of stake K2285

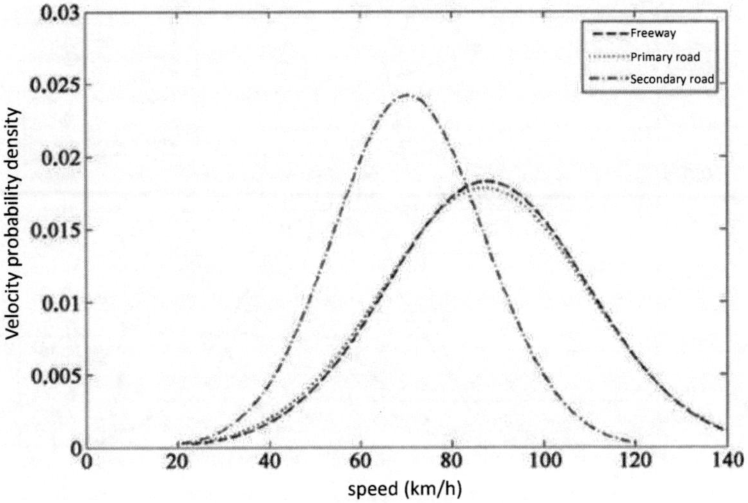

Fig. 2.75 Distribution of velocity probability density for different road grades (I)

K157 was selected as the survey point for primary roads, with two lanes in both directions; K2160 was selected as the survey point for secondary roads, with two lanes in both directions, and the alignment of these three survey points is approximately the same.

Figure 2.75 is a schematic diagram of the effect of different road grades on vehicle speed. It can be seen from the figure that, comparing to secondary roads, the center axis of the curve for freeways and primary roads are shifted to the right, i.e., the average speed of freeways and primary roads is greater than that of secondary roads.

Figure 2.76 is a diagram of effect of different road grades on the speed of truck and passenger car. It can be seen from the figure that the average speed of trucks is lower than that of the passenger car on each grade of road; the central axis of speed distribution curves of trucks and passenger car on freeways and primary roads is approximately the same; the central axis of speed distribution curves of trucks and passenger car on secondary roads is shifted to the left comparing to the freeways and primary roads.

In general, based on the traffic volume and its composition change pattern of the high altitude highway represented by G109, as well as the distribution pattern of traffic flow speed under different road environment conditions, it can be seen that: in the time distribution of traffic volume, the dominant vehicles on Qinghai-Tibet highway are small and medium-sized buses and trucks, and only these two types of vehicles show a continuous increase, which reflecting that the vehicles on Qinghai-Tibet highway are developing bipolar. In the spatial distribution of traffic volume, the distribution of traffic volume along the Qinghai-Tibet highway shows the features of high at both ends and low in the middle, regarding the distribution of each type of vehicle along the route, there are mainly trucks at Golmud section, while at Lhasa

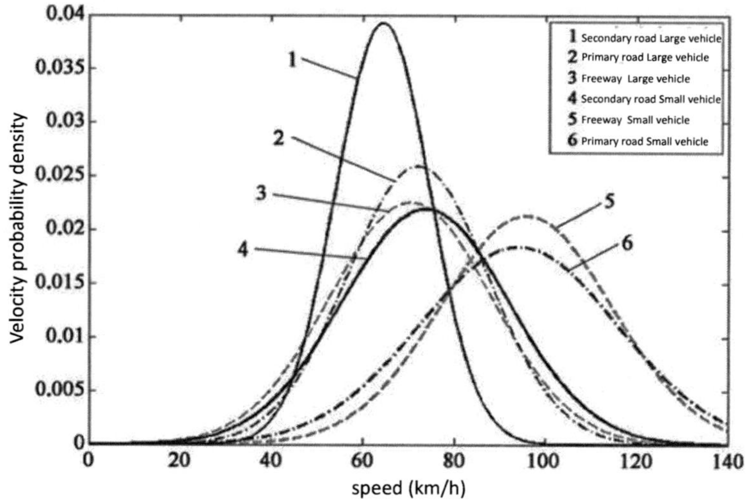

Fig. 2.76 Distribution of velocity probability density for different road grades (II)

section is mainly small and medium-sized buses, reflecting that Golmud along the Qinghai-Tibet highway mainly undertakes the function of freight input and output, while Lhasa mainly undertakes the function of passenger input and output.

By analyzing the relationship of altitude, alignment, road grade between operating speed distribution, the overall operating speed distribution of different vehicles in high altitude area is as follows: the overall operating speed decreases when altitude increases; the operating speed of traffic flow with heavy load vehicles is lower than that of with light load vehicles, and the overall speed difference between large and passenger car with heavy load is more than 20 km/h, which is due to the power performance of heavy-duty vehicles decreases significantly in high altitude areas, while passenger car are less affected by altitude.

2.3 Traffic Safety Service Level in High Altitude Areas

Traffic accident frequency prediction refers to describing the interrelationship between accident occurrence and influencing factors from the quantitative aspect, revealing the features and relationships of quantitative for accidents under certain conditions, carry out statistical and analytical studies on accidents, and its causes, so as to reasonably analyze the accident formation mechanism, and accordingly studying the regularity of accidents, evaluating road safety, predicting the development trend of accidents, and formulating accident prevention and safety protection systems.

A generalized linear model (GLM) and an event count model are used to predict the frequency of traffic accidents in high altitude areas, and the models are fitted with Poisson regression, negative binomial regression, zero-stack Poisson regression and zero-stack negative binomial regression, respectively.

2.3.1 Determination of Road Segmentation Method

The first step to establish accident frequency prediction model is to divide road segment length, which directly affects the accuracy of model fitting and the application efficiency of the model. This tested section is mostly distributed in G109 and G214, building as Class-II highway standard, with two-way two-lane carriageway and a design speed of 60 km/h. In China, a characteristic of two-lane highways is that there are many villages and towns on both sides of the road, with mixed traffic, and the accident distribution is more different than that of non-village sections. Many existing studies divide a highway into three forms, i.e., ordinary road section, village road section and intersection road section, and they model upon this these different forms respectively, to improve the accuracy and pertinence of the model. While in this book, in addition to the road sections crossing villages and towns, the road sections studied are largely located in mountainous areas, Gobi and other no man's land areas. These areas are sparsely populated and there are few intersections. In order to make full use of the data obtained, the road sections in this study are divided into non-urban road sections and village and town road sections.

(1) **Road Segmentation Method Based on Ordered Sample Clustering**

Mr. Zhong Liande of Beijing University of Technology used the method of ordered sample clustering to divide the road sections in his doctoral thesis, and then studied the freeway accident prediction model. He analyzed the advantages and disadvantages of traditional road segmentation methods (fixed length and variable length) and ordered sample clustering methods in detail. Through verification, he found that when the accident data is highly discrete, the model established by using ordered clustering analysis to divide road segments is better than the accident prediction model established by traditional road segmentation method.

The study in this book also applies the advantages of the orderly clustering road segmentation method to establish the accident prediction model for the non-urban road sections of two-lane highways in high altitude areas. This can well solve the problem of accident data dispersion caused by the non-urban road sections of highways in high altitude areas crossing a large area of no man's land. There are few villages and towns crossed in the survey section, and fewer villages and towns can be studied compared with non-urban sections, however, a large number of accidents are concentrated, so a fixed length segmentation method with 1 km as the statistical unit is adopted.

(2) **Calculation Steps and Formulas of Optimal Segmentation**

In generally, "cluster analysis" refers to that statistical objects (element individuals) have no order and cluster according to certain indicators, such as clustering schools according to their graduation rate and classifying factories according to the number of defective products. There is no order between these schools and factories, and anyone and everyone can be grouped together. This is the "cluster analysis" in general statistical books. Ordered samples are samples arranged in a certain order, and the second order cannot be disturbed. The difference between the ordered sample clustering method and the general clustering method is that the "status" of each sample is different. The Fisher optimal segmentation method is the most used. The essence of this method is to divide the ordered samples into *n* segments, so that the sum of squares of deviations within the segment is the minimum and the sum of squares of deviations outside the segment is the maximum. The spatial distribution of the dependent variables of the accident prediction model—accident indicators, i.e., accident frequency, number of injuries, number of fatal accidents, etc.—is an ordered sample, and the order of the accident indicators distributed by stake number cannot be disarranged.

The calculation of optimal segmentation has three steps, namely, range transformation, segment diameter matrix, the sum of squares of intra group deviations (or the sum of segment diameters) of all segments, and the optimal segmentation of various segments.

(1) Range Transformation

If the original data matrix is

$$X = \begin{matrix} x_{11} & x_{12} & \cdots & x_{1p} \\ x_{21} & x_{22} & \cdots & x_{2p} \\ \vdots & \vdots & & \vdots \\ x_{n1} & x_{n2} & \cdots & x_{np} \end{matrix}$$

Transform the element x_{il} in X into $z_{il} = (x_{il} = \min\{x_{il}\})/(\max\{x_{il}\}\text{-}\min\{x_{il}\})$ to form a new standardized Z matrix.

(2) Calculation Segment Diameter Matrix D

$$d_{ij} = \sum_{\alpha=i}^{j} \sum_{\beta=1}^{p} [\overline{z}_{\alpha\beta} - \overline{z}_{\beta 1} \underset{1 \le i \le j \le n}{(i, j)}]^2 \tag{2.7}$$

Among

$$\overline{z}_{\beta}(i, j) = \frac{1}{j - i + 1} \sum_{\alpha=i}^{j} z_{\alpha\beta} \tag{2.8}$$

So

$$d_{ij} = \sum_{k=i}^{j} \sum_{i=1}^{m} x_{k1}^{2} - \frac{1}{j - i + 1} \sum_{i=1}^{m} (\sum_{k=i}^{j} x_{k1})^{2} \tag{2.9}$$

Where,

m Number of variables.

(3) Calculate the sum of squared deviations in the group (or the sum of segment diameters) of all segments and the optimal segmentation of various segments

According to the segmentation calculation results of matrix D and the optimal segment k–1, for each $m = n, n-1, \ldots, k$, separately calculate the sum of squared deviations in the group corresponding segment k:

$$S_m(k; a_1(j), a_2(j), \cdots, a_{k-2}, j) = S_j(k - 1; a_1(j), a_2(j), \cdots, a_{k-2} + d(j + 1, n))$$
$$(j = k - 1, k, \cdots, m - 1, m; n, n - 1, \cdots, k)$$

Find the minimum value and determine the corresponding optimal segmentation point, namely

$$S_m(k; a_1(m), a_2(m), \ldots, a_{k-1}(m)) = \min\{S_m(k - 1; a_1(j), a_2(j), \ldots, a_{k-2}(j), j)\}$$

Thus, the optimal segment k of n samples ($m = n$) is

$$(x_1, \ldots, x_{a1(n)})(x_{a1(n)}, \ldots, x_{a2}), \ldots, (x_{a(k-2)(n)+1}, \ldots, x_{a(k+1)(n)})(x_{a(k-1)(n)+1}, \ldots, x_n)$$

Where $a_{1(\mathrm{n})}, a_{2(\mathrm{n})}, \ldots, a_{\mathrm{k}-1(\mathrm{n})}$ are the optimal segmentation points of segment k.

(3) **Statistics of Segmentation Results**

Traffic accidents on a highway should not be disarranged and should conform to the ordered sample. More accidents occurred in a road section have their own reasons, the same as with the less accidents. The idea of clustering road segmentation for orderly samples is that "the frequency of traffic accidents must have related influencing factors". At present, there is no mature software that can realize the processing of ordered clustering analysis. In this book the R language clustering software is used to realize the program of ordered clustering analysis with reference to relevant materials.

The number of road sections is mainly considered in two aspects: first, based on the principle of practical application, the average length of road sections should be about 3 km; second, the objective function is no longer significantly reduced. Among

Table 2.5 Descriptive statistics of non-urban sections

Counting	Maximum	Minimum	Average value	Standard deviation
218	30	1	2.73	3.45

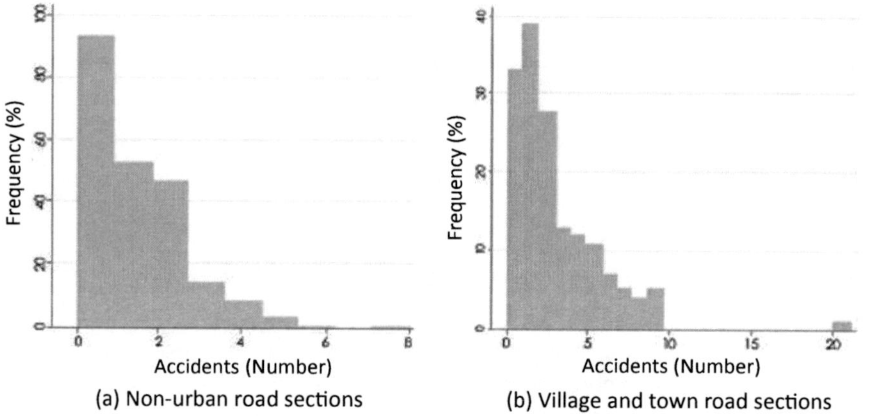

Fig. 2.77 Accident frequency distribution of non-urban sections and village and town road sections

the non-urban sections, a long distance is in Gobi no man's land, some sections are longer. A total of 218 non-urban road is segmented, see Table 2.5 for descriptive statistics.

The following analysis on the influencing factors of traffic accidents in non-urban road sections and the establishment of highway accident prediction model in high altitude areas take the 218 road sections divided by ordered sample clustering method as research samples.

Due to the concentration of accidents in village and town sections and the small length of sample sections, the fixed length segmentation method with smaller statistical units is adopted. In this book, a total of 158 section units are obtained by taking 1 km as the statistical unit.

It can be seen from Fig. 2.77 that the accident frequency distribution of non-urban road sections and village and town road sections is inclined to discrete Poisson and Negative Binominal distribution, which passes the X^2 test with a confidence level of 95%.

2.3.2 *Analysis of Traffic Accident Influencing Factors*

The purpose of analyzing traffic accident impact factor is to find out road alignment, traffic flow, environment and other factors related to accidents, so as to formulate prevention and control measures to reduce accident frequency and severity. At the

Table 2.6 Variables of traffic accident influencing factor analysis

Variables	English code	Remarks
Number of accidents	NC	–
Exposure variable	Expo	Million vehicle kilometers per year
Daily traffic volume	AADT	–
Proportion of trucks	PT	–
Direction coefficient	D_{ir}	–
Roadside danger level	L_{rs}	Classification according to the roadside safety net area
Density of access	D_{ac}	Number of access ports within the study section
Flat curve density	D_{hc}	Number of horizontal curves with radius less than 600 m
Altitude	E	Average elevation within the study section
Oxygen content	O^2	Ratio to oxygen content under standard pressure in plain
Relative altitude	E_{dif}	Absolute value of elevation difference of 1 km section

same time, it is the basis for establishing highway accident prediction model in high altitude areas, and the relevant factors affecting traffic accidents are also the basis for bringing the accident prediction model as independent variables. This chapter selects some variables of road alignment, traffic flow and environment (Table 2.6), analyzes the correlation between each variable and traffic accidents, and determines the independent variables used in the traffic accident frequency prediction model.

Analyzing the correlation between various factors and accidents can determine the impact of various factors on the occurrence of accidents, and preliminarily screen the indicators to be used for highway accident prediction models in high altitude areas.

Spearman correlation coefficient is a nonparametric form of Pearson correlation coefficient, which is calculated according to the rank of data rather than the actual value. It is suitable for ordered data or equispaced data with abnormal distribution assumption. The range of correlation coefficient values is $-1\sim1$. The larger the absolute value is, the stronger the correlation is. The sign of correlation coefficient also indicates the direction of correlation. The calculation formula of Spearman correlation coefficient is

$$r = 1 - \frac{6\sum_{i=1}^{n} D^2}{n(n^2 - 1)} \tag{2.10}$$

Where

$$\sum_{i=1}^{n} D^2 = \sum_{i=1}^{n} (U_i - V_i)^2 \tag{2.11}$$

Here (U_i, V_i) is the rank of two variables.

Table 2.7 shows the correlation coefficient analysis of non-urban road sections, in which crash Expo is the number of accidents under the unit exposure variable to help analyze variables unrelated to the length of the road section. In non-urban sections, the number of accidents has a positive correlation with the traffic volume, the proportion of trucks, the roadside grade, the number of flat curves, and the number of access ports, and weak with the last one. The relative altitude and the length of the road section are negatively correlated with the number of accidents. Because the base number of gentle sections at relative altitude is large, the absolute number of accidents is relatively larger, resulting in a negative correlation with the number of accidents. The section length is that in the study sample. Altitude and oxygen content have a negative correlation and strong, but both them have a weak correlation with the number of accidents, so choose one to enter the model. See Table 2.8 for the correlation coefficient analysis of rural road sections. Compared with non-urban sections, the number of access ports has a stronger positive correlation, the roadside danger level has a lower impact, and the proportion of trucks show a negative correlation.

2.3.3 Regional Highway Accident Prediction Model

(1) **Model Form Identification and Model Evaluation IndicatorIndicator**

(1) Model form Recognition

Generally, only when various traffic safety influencing factors are fixed to a specific value, the number of traffic accidents or accident rate variables will obey the above distribution, and a certain characteristic value (such as the mean value) of this distribution will have a specific correlation with the traffic safety influencing factors, which can be obtained by statistical regression method.

Assume the data matrix x_{ij} of influencing factors is an independent variable, and the vector i of specific influencing factors corresponding accident number Y_i is a dependent variable, then the conceptual model of its generalized estimation model is

$$Y_i \cap \text{Prob}(\mu_i) \tag{2.12a}$$

$$\mu_i = g(x_{i1}, x_{i2}, \cdots, x_{ij}, \cdots, x_{in}) \tag{2.12b}$$

$$\mu_i = f(\eta_i) \tag{2.12c}$$

Where,

μ_i – Characteristic value of distribution Prob of accident variable (usually mean value, see above);

Table 2.7 Correlation analysis results of various influencing factors in non-urban sections

Affecting factors		L	Crash	Crash_ expo	E	E_{dif}	0_2%	Truck%	Vol	D_{ir}	Expo	L_{rs}	D_{ac}	D_{hc}
L	Significance of correlation coefficient (two-tailed)	1.000	−0.296 0.000	−0.570 0.000	−0.227 0.001	0.547 0.000	0.207 0.002	0.005 0.945	−0.212 0.002	−0.223 0.001	0.930 0.000	−0.149 0.028	−0.019 0.776	−0.163 0.016
Crash	Significance of correlation coefficient (two-tailed)	−0.296 0.000	1.000	0.875 0.000	−0.040 0.552	−0.134 0.047	0.019 0.775	0.057 0.401	0.266 0.014	0.017 0.806	−0.283 0.000	0.193 0.004	0.100 0.142	0.134 0.049
Crash_ expo	Significance of correlation coefficient (two-tailed)	−0.570 0.000	0.875 0.000	1.000	−0.049 0.475	−0.315 0.000	−0.013 0.854	0.170 0.012	0.353 0.007	0.026 0.705	−0.610 0.000	0.152 0.024	0.027 0.039	0.227 0.001
E	Significance of correlation coefficient (two-tailed)	−0.227 0.001	−0.040 0.552	−0.049 0.475	1.000	0.290 0.000	−0.900 0.000	−0.070 0.304	0.494 0.000	0.652 0.000	−0.060 0.377	0.320 0.000	0.007 0.918	0.270 0.000
E_{dif}	Significance of correlation coefficient (two-tailed)	0.547 0.000	−0.134 0.047	−0.315 0.000	0.290 0.000	1.000	−0.371 0.000	0.061 0.371	0.001 0.987	0.148 0.029	0.553 0.000	0.182 0.007	−0.159 0.019	0.177 0.009

(continued)

Table 2.7 (continued)

Affecting factors		L	Crash	Crash_expo	E	E_{dif}	O_2%	Truck%	Vol	D_{ir}	Expo	L_{rs}	D_{ac}	D_{hc}
L	Significance of correlation coefficient (two-tailed)	1.000	−0.296 0.000	−0.570 0.000	−0.227 0.001	0.547 0.000	0.207 0.002	0.005 0.945	−0.212 0.002	−0.223 0.001	0.930 0.000	−0.149 0.028	−0.019 0.776	−0.163 0.016
O_2%	Significance of correlation coefficient (two-tailed)	0.207 0.002	0.019 0.775	−0.013 0.854	−0.900 0.000	−0.371 0.000	1.000	−0.123 0.071	−0.227 0.001	−0.603 0.000	0.135 0.047	−0.327 0.000	0.131 0.054	−0.267 0.000
Truck%	Significance of correlation coefficient (two-tailed)	0.005 0.945	0.057 0.401	0.170 0.012	−0.070 0.304	0.061 0.371	−0.123 0.071	1.000	−0.628 0.000	0.357 0.000	−0.225 0.001	0.105 0.123	−0.454 0.000	0.090 0.186
Vol	Significance of correlation coefficient (two-tailed)	−0.212 0.002	0.266 0.014	0.353 0.007	0.494 0.000	0.001 0.987	−0.227 0.001	−0.628 0.000	1.000	0.340 0.000	0.098 0.150	0.248 0.000	0.564 0.000	0.111 0.103

(continued)

Table 2.7 (continued)

Affecting factors		L	Crash	Crash_ expo	E	E_{dif}	O_2%	Truck%	Vol	D_{ir}	Expo	L_{rs}	D_{ac}	D_{hc}
L	Significance of correlation coefficient (two-tailed)	1.000	−0.296 0.000	−0.570 0.000	−0.227 0.001	0.547 0.000	0.207 0.002	0.005 0.945	−0.212 0.002	−0.223 0.001	0.930 0.000	−0.149 0.028	−0.019 0.776	−0.163 0.016
%	Significance of correlation coefficient (two-tailed)	−0.223 0.001	0.017 0.806	0.026 0.705	0.652 0.000	0.148 0.029	−0.603 0.000	0.357 0.000	0.340 0.000	1.000	−0.126 0.062	0.453 0.000	0.055 0.416	0.110 0.105
expo	Significance of correlation coefficient (two-tailed)	0.930 0.000	−0.283 0.000	−0.610 0.000	−0.060 0.377	0.553 0.000	0.135 0.047	−0.225 0.001	0.098 0.150	−0.126 0.062	1.000	−0.079 0.246	0.182 0.007	−0.163 0.016
L_{rs}	Significance of correlation coefficient (two-tailed)	−0.149 0.028	0.193 0.004	0.152 0.024	0.320 0.000	0.182 0.007	−0.327 0.000	0.105 0.123	0.248 0.000	0.453 0.000	−0.079 0.246	1.000	0.187 0.006	0.247 0.000
D_{ac}	Significance of correlation coefficient (two-tailed)	−0.019 0.776	0.100 0.142	−0.027 0.689	0.007 0.918	−0.159 0.019	0.131 0.054	−0.454 0.000	0.564 0.000	0.055 0.416	0.182 0.007	0.187 0.006	1.000	−0.131 0.053

(continued)

Table 2.7 (continued)

Affecting factors		L	Crash	Crash_ expo	E	E_{dif}	0_2%	Truck%	Vol	D_{ir}	Expo	L_{rs}	D_{ac}	D_{hc}
L	Significance of correlation coefficient (two-tailed)	1.000	−0.296 0.000	−0.570 0.000	−0.227 0.001	0.547 0.000	0.207 0.002	0.005 0.945	−0.212 0.002	−0.223 0.001	0.930 0.000	−0.149 0.028	−0.019 0.776	−0.163 0.016
D_{hc}	Significance of correlation coefficient (two-tailed)	−0.163 0.016	0.134 0.049	0.227 0.001	0.270 0.000	0.177 0.009	−0.267 0.000	0.090 0.186	0.111 0.103	0.110 0.105	−0.163 0.016	0.247 0.000	−0.131 0.053	1.000

Table 2.8 Correlation analysis results of various influencing factors in village and town sections

Affecting factors		Crash	Crash_expo	E	Edit	O_2%	Vol	Truck%	D_{ir}	Expo	L_{rs}	D_{ac}	D_{hc}
Crash	Significance of correlation coefficient (two-tailed)	1.000	0.935 0.000	−0.188 0.018	0.008 0.920	0.212 0.007	0.566 0.000	−0.339 0.000	−0.065 0.418	0.524 0.000	0.121 0.000	0.367 0.000	0.033 0.683
Crash_expo	Significance of correlation coefficient (two-tailed)	0.935 0.000	1.000	−0.033 0.680	0.014 0.862	0.028 0.727	0.285 0.000	−0.243 0.002	−0.073 0.362	0.234 0.003	0.183 0.021	0.369 0.000	−0.023 0.778
E	Significance of correlation coefficient (two-tailed)	−0.188 0.018	−0.033 0.680	1.000	−0.074 0.357	−0.963 0.000	−0.560 0.000	0.726 0.000	0.510 0.000	−0.565 0.000	−0.399 0.000	−0.216 0.006	−0.439 0.000
E_{dif}	Significance of correlation coefficient (two-tailed)	0.008 0.920	0.014 0.862	−0.074 0.357	1.000	0.059 0.459	−0.014 0.865	0.059 0.463	0.015 0.847	−0.018 0.823	0.029 0.720	−0.001 0.993	−0.069 0.392
O_2%	Significance of correlation coefficient (two-tailed)	0.212 0.007	0.028 0.727	−0.963 0.000	0.059 0.459	1.000	0.632 0.000	−0.728 0.000	−0.426 0.000	0.638 0.000	0.470 0.000	0.267 0.001	0.458 0.000
Vol	Significance of correlation coefficient (two-tailed)	0.566 0.000	0.285 0.000	−0.560 0.000	−0.014 0.865	0.632 0.000	1.000	−0.797 0.000	−0.279 0.000	0.994 0.000	0.470 0.000	0.487 0.000	0.249 0.002

(continued)

Table 2.8 (continued)

Affecting factors		Crash	Crash_expo	E	Edit	O_2%	Vol	Truck%	D_{ir}	Expo	L_{rs}	D_{ac}	D_{hc}
Crash	Significance of correlation coefficient (two-tailed)	1.000	0.935 0.000	−0.188 0.018	0.008 0.920	0.212 0.007	0.566 0.000	−0.339 0.000	−0.065 0.418	0.524 0.000	0.121 0.000	0.367 0.000	0.033 0.683
Truck%	Significance of correlation coefficient (two-tailed)	−0.339 0.000	−0.243 0.002	0.726 0.000	0.059 0.463	−0.728 0.000	−0.797 0.000	1.000	0.237 0.003	−0.789 0.000	−0.436 0.000	−0.404 0.000	−0.209 0.009
D_{ir}	Significance of correlation coefficient (two-tailed)	−0.065 0.418	−0.073 0.362	0.510 0.000	0.015 0.847	−0.426 0.000	−0.279 0.000	0.237 0.003	1.000	−0.281 0.000	−0.003 0.965	0.038 0.635	−0.317 0.000
Expo	Significance of correlation coefficient (two-tailed)	0.524 0.000	0.234 0.003	−0.565 0.000	−0.018 0.823	0.638 0.000	0.994 0.000	−0.789 0.000	−0.281 0.000	1.000	0.465 0.000	0.463 0.000	0.270 0.001
L_{rs}	Significance of correlation coefficient (two-tailed)	0.121 0.000	0.183 0.021	−0.399 0.000	0.029 0.720	0.470 0.000	0.470 0.000	−0.436 0.000	−0.003 0.965	0.465 0.000	1.000	0.337 0.000	0.321 0.000
D_{ac}	Significance of correlation coefficient (two-tailed)	0.367 0.000	0.368924 0.000	−0.216 0.006	−0.001 0.993	0.267 0.001	0.487 0.000	−0.404 0.000	0.038 0.635	0.463 0.000	0.337 0.000	1.000	0.130 0.102

(continued)

Table 2.8 (continued)

Affecting factors		Crash	Crash_expo	E	Edit	O_2%	Vol	Truck%	D_{ir}	Expo	L_{rs}	D_{ac}	D_{hc}
Crash	Significance of correlation coefficient (two-tailed)	1.000	0.935 0.000	−0.188 0.018	0.008 0.920	0.212 0.007	0.566 0.000	−0.339 0.000	−0.065 0.418	0.524 0.000	0.121 0.000	0.367 0.000	0.033 0.683
D_{hc}	Significance of correlation coefficient (two-tailed)	0.033 0.683	−0.023 0.778	−0.439 0.000	−0.069 0.392	0.458 0.000	0.249 0.002	−0.209 0.009	−0.317 0.000	0.270 0.001	0.321 0.000	0.130 0.102	1.000

(g) – Correlation model established between transition variable η_i and independent variable x_{ij};
(f) – Correlation function, which is used to describe the specific mathematical relationship between η_i and μ_i.

Equation (2.12a) is called "probability distribution part".

Equation (2.12c) is called "correlation part". It is mainly to establish the mathematical relationship between the statistical characteristic value of the accident and the transition variable, whether it is necessary to exist, and in what mathematical form, depending on Eq. (2.12b). That is, after the study of Eq. (2.12b) is completed, we can determine whether Eq. (2.12c) is required and what form of Eq. (2.12c) is.

In this way, the research on Eq. (2.12b) becomes the focus and difficulty.

Equation (2.12b) is called the "relevant model part". It is to establish the relevant model between the accident statistical characteristic value (such as mean value) or transition variable (when required) and independent variable. The previous research shows that the accident frequency distribution does not obey the normal distribution, but tends to obey the discrete Poisson distribution and negative binomial distribution and other probability forms. Therefore, this study uses the generalized linear regression method to establish the accident prediction of two-lane highways, and the probability distribution uses the distribution forms that have been studied more recently such as Poisson, Negative and Binomial. Non-urban sections contain many zero accident sections, so the zero accumulation Poisson regression and negative binomial regression models with logit/probit discrete selection model are also selected. The mixed model first uses logit/probit model to identify zero accidents, and then uses Poisson/negative binomial regression for non-zero sections.

In the study, descriptive statistics and correlation analysis were carried out on the independent variables, and finally irrelevant independent variables were determined for modeling. Road section length and traffic volume are absolute variables that affect the number of accidents. Many studies have separated them from other variables and used the form of Exposure product in the model, rather than treating them equally with other variables.

By this, the Eq. (2.12b, 2.12c) can be combined into the following formula:

$$\mathrm{NC}_i = \mathrm{expo} \times \exp\left(\beta_0 + \sum_{j=1}^{n} \beta_j x_{ij}\right) \tag{2.13}$$

Where,

X_{ij} – Road environmental traffic attribute value of section i;
β – Model parameter coefficient independent of accident location i;

Expo – Road utilization, also known as traffic risk exposure indicator, with the unit of million vehicle kilometers per year.

(2) Model evaluation indicator

i. Maximum likelihood estimation. The maximum likelihood estimation method is widely used to evaluate the regression models of Poisson model, Negative Binomial model, ZIP, ZINB and other discrete distribution data. If the population X belongs to discrete type, its form of distribution rule $P\{X = x\} = p(x;\theta)$, $\theta \in \Theta$ is known, θ is the parameter to be estimated. Θ is the possible values range of θ. If $X_1, X_2, \ldots,$ X_n are samples from X, then the joint distribution law of $X_1, X_2, \ldots, X_n$ is

$$\prod_{i=1}^{n} p(x_i;\theta) \tag{2.14}$$

Suppose $x_1, x_2, \ldots, x_n$ is a sample value corresponding to sample $X_1, X_2, \ldots, X_n$. It is known that the probability to take observation value$x_1, x_2, \ldots, x_n$ for sample X_1, $X_2, \ldots, X_n$, that is, probability of event $\{X_1 = x_1, X_2 = x_2, \ldots, X_n = x_n\}$ occurrence is

$$L(\theta) = L(x_1, x_2, \cdots, x_n;\theta) = \prod_{i=1}^{n} p(x_i;\theta), \theta \in \Theta \tag{2.15}$$

This probability varies with the value of θ, and is a function of θ. $L(\theta)$ is called sample likelihood function. The maximum likelihood estimation method is to fix sample observation value $x_1, x_2, \ldots, x_n$, and select the likelihood function $L(x_1, x_2,$ $\ldots, x_n; \theta)$ within the possible range Θ of value θ, to reach the maximum parameter value $\hat{\theta}$ as estimation value of θ,, that is, get $\hat{\theta}$ to make

$$L(x_1, x_2, \cdots, x_n; \hat{\theta}) = \max_{\theta \in \Theta} L(x_1, x_2, \cdots, x_n;\theta) \tag{2.16}$$

The $\hat{\theta}$ thus obtained is related to the sample value $x_1, x_2, \ldots, x_n$, always recorded as $\hat{\theta}$ $(x_1, x_2, \ldots, x_n)$, called maximum likelihood estimation value of parameter θ. While the corresponding statistic $\hat{\theta}$ $(X_1, X_2, \ldots, X_n)$ is called the maximum likelihood estimation amount of parameter θ.

According to the definition of maximum likelihood estimation method, the estimation parameter is the best when obtaining maximum likelihood value. The same is true for maximum log-likelihood. Because the data is the same, it can be compared between different models.

ii. Akaike Information Criterion (AIC). It was proposed by Japanese scholar Akaike in 1973 and is widely used in the determination of autoregressive order in time series analysis, the screening of independent variables in multiple regression and generalized linear regression, and the comparison and optimization of models in nonlinear regression. AIC is defined as follows.

When the model or equation is estimated by the least square method, see the following formula:

$$\text{AIC} = n\ln\left(\frac{n-k}{n} \times MS\right) + 2k \tag{2.17}$$

When the model or equation is estimated by maximum likelihood method, see the following formula:

$$\text{AIC} = -2\ln L + 2k \tag{2.18}$$

Where,

k Number of independent variables in the model;
L Maximum likelihood function of the model;
N Number of samples.

MS calculation is shown in the following formula:

$$MS = \sum\left(y - \hat{y}\right)^2 / (n - k - 1) \tag{2.18}$$

AIC consists of two parts. The first part reflects the fitting accuracy of the regression equation, and the smaller the value, the better; the other part reflects the number of variables in the regression, i.e., the complexity of the model, the smaller the k is, the better, this ""punishment" for the number of independent variables or the number of parameters in the model. Therefore, the smaller the AIC, the better. And its basic principle is "less but better". AIC is not only applicable to the observation within the sample, but also to predict the performance of a regression model outside the sample.

In addition to the two model evaluation indicators described above, there are BIC, Vuong test statistics, likelihood law (LR), PseudoR^2 and other statistics.

2.3.4 General Distribution Characteristics of Traffic Accidents

According to the accident statistics, the number of accidents near stake number G214 is counted, including Xining and Jinghong. The initial stake number ranges from K107 to K992, 885 km in total. 188 accidents totally occurred, with an accident rate of 0.21 per kilometer. The specific stake number distribution is shown in the Fig. 2.78.

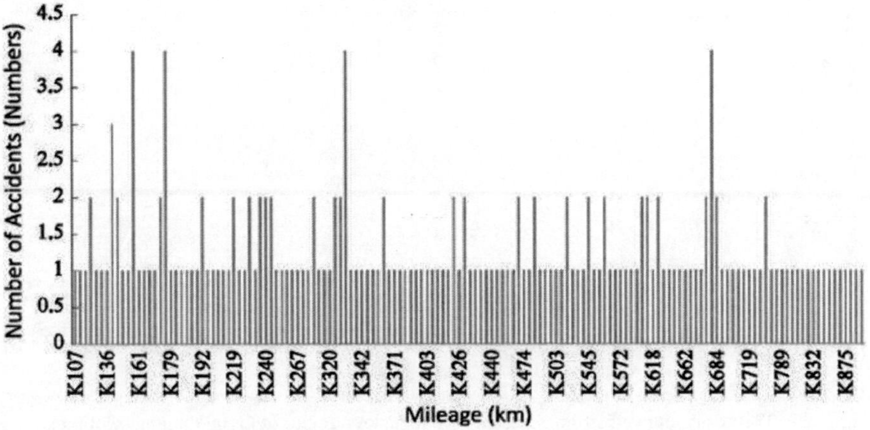

Fig. 2.78 Traffic accidents distributed according to stake number

It can be seen from Fig. 2.79 that the accident rate of stake number G214 ranges from K100 to K200 is high, 0.38Nos./100 km, and the overall trend decreases with the increase of stake number. From the perspective of traffic accident risk coefficient, the accident rate is the highest from stake number number K200 to K300, 1.77Nos./ person, and then gradually decreases. In general, the accident rate is high and the accident risk is serious in the range of K100 to K300.

According to the accident data of G109 Qinghai section, the statistics of the accidents occurred near the stake number number show that the initial stake number number range is K1844~K2736, totally 892 km, 388 accidents occurred, and the

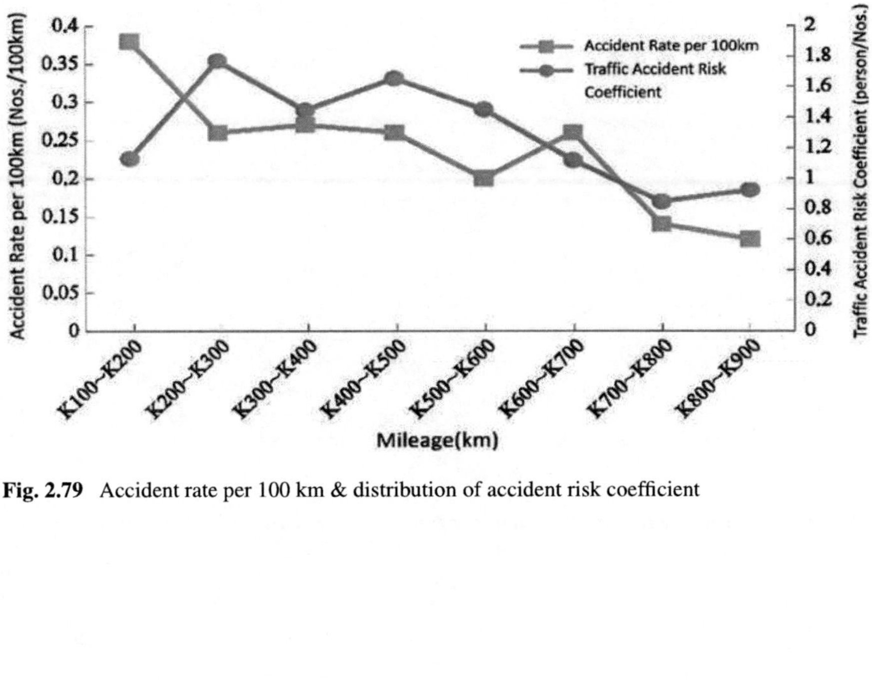

Fig. 2.79 Accident rate per 100 km & distribution of accident risk coefficient

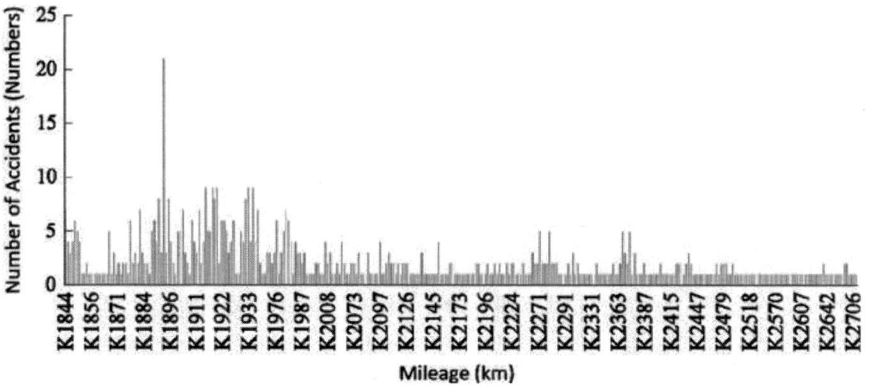

Fig. 2.80 Traffic accidents distributed according to stake number of G109 Qinghai section

accident rate per kilometer is 0.43. The specific stake number distribution is as Fig. 2.80.

In general, the accident rate of 100 km from K1900 to K2000 is the highest, reaching 0.6. From the perspective of accident risk, the severity of traffic accidents between K2000 and K2100 is relatively high. Combining these two aspects (Fig. 2.81), it can be concluded that the accident rate and severity are relatively high within the range of Kl900 to K2100. From the statistical data, 94 accidents have occurred in this range in the past five years, causing 144 deaths, these two parts accounted for 24.85% of the total accidents and 26.52% of the total deaths.

According to the accident statistics, the number of accidents occurred near the stake number of G109 Tibet section was counted, including Golmud to Lhasa. The stake number starting from K2773 to K3876, totally 1103 km, 113Nos. of accidents

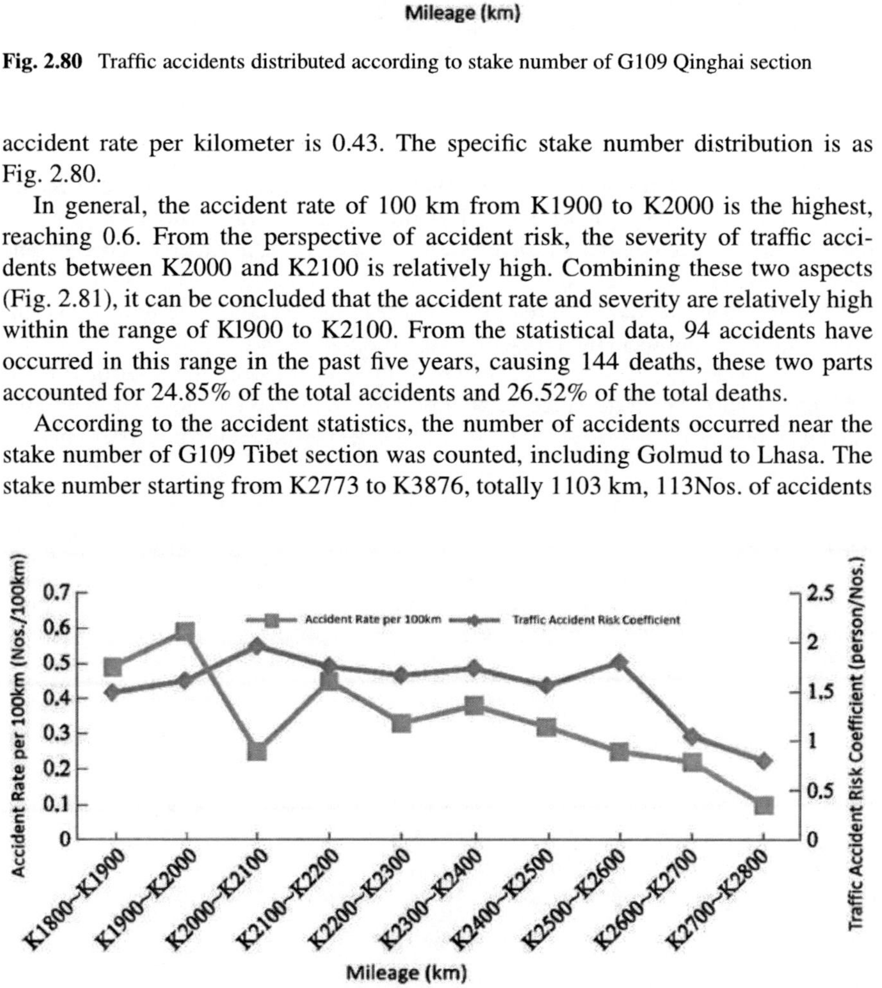

Fig. 2.81 Accident rate per 100 km & traffic accident risk coefficient of G109 Qinghai section

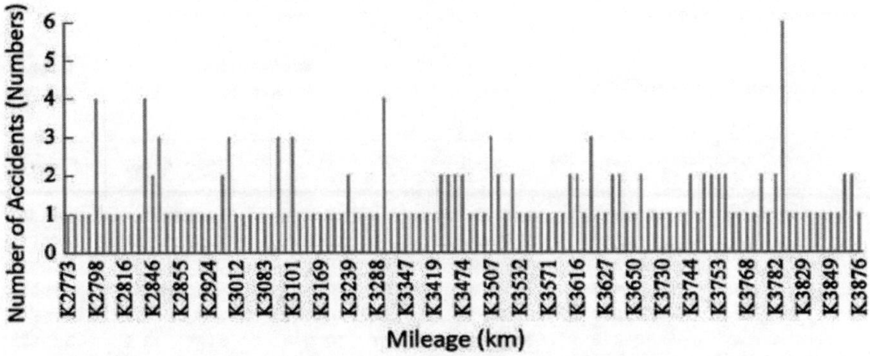

Fig. 2.82 Traffic accidents distributed according to stake number of G109 Tibet section

occurred, with an accident rate of 0.1 per kilometer. The specific stake number distribution is shown in Fig. 2.82.

(2) **Parameter Estimation & Test of Model**

The regression analysis method of backward elimination is adopted in this study: first, establish a full model, and then, remove one variable that most does not conform to the model (for example, the minimum P value) each time according to the judgment of relevant indicators in the output results, until there is no independent variable that does not conform to the criteria in the regression equation. Therefore, the inconsistent independent variables are removed in turn and regressed again. After several steps, it is finally concluded that the variables in the model contribute to the model, and each system is significant. The following models are established for the two sections.

(1) Accident Prediction Model for Non-rural Road Sections

Through the regression analysis method of backward elimination, Poisson Regression, Negative Binomial Regression, Zero-Inflated Poisson Regression and Zero Inflated Negative Binomial Regression are respectively used to model the non-rural road sections. The parameter estimation results are shown in Figs. 2.83, 2.84, 2.85 and 2.86.

It shows in Fig. 2.83 that the estimation results of Poisson regression parameters. The P value of the chi-square test of the model is 0.0000, which means that the model is statistically significant. Relative altitude, proportion of trucks, roadside grade, access port density and horizontal curve density coefficient are significant, contributing to the model. Figure 2.85 shows that the 95% confidence interval of alpha is (0.50,1.25) and chibar2 = 49.21, so the original hypothesis of over dispersion parameter alpha = 0 can be rejected at the 5% significance level (corresponding to Poisson regression). Table 2.9 shows that log likelihood, AIC and BIC of negative binomial regression are slightly better than Poisson regression. Figure 2.84 shows that the Vuong test statistic of zero inflated Poisson regression is 3.04, greater than 1.96, indicating that this model is better than Poisson regression. The log likelihood,

Poisson regression | Number of obs = 218
LR chi2(5) = 165.44
Prob > chi2 = 0.0000
Log likelihood = -350.63025 | Pseudo R2 = 0.1909

NC	Coef.	Std. Err.	z	P>|z|	[95% Conf.	Interval]
E_dif	-.0198143	.0027621	-7.17	0.000	-.0252279	-.0144007
PT	5.003214	1.469923	3.40	0.001	2.122219	7.88421
L_rs	.9355271	.2464758	3.80	0.000	.4524435	1.418611
D_ac	.2564455	.10212	2.51	0.012	.056294	.4565971
D_hc	.4459533	.0788897	5.65	0.000	.2913322	.6005743
_cons	-3.923272	.6754544	-5.81	0.000	-5.247138	-2.599406
ln(expo)	1	(exposure)				

Fig. 2.83 Estimation results of poisson regression parameters

Zero-inflated Poisson regression | Number of obs = 218
Nonzero obs = 125
Zero obs = 93

Inflation model = logit | LR chi2(4) = 68.23
Log likelihood = -320.4008 | Prob > chi2 = 0.0000

NC	Coef.	Std. Err.	z	P>|z|	[95% Conf.	Interval]
NC						
E_dif	-.0154057	.0031183	-4.94	0.000	-.0215174	-.009294
PT	4.851567	1.484666	3.27	0.001	1.941675	7.761458
L_rs	.4868504	.2506315	1.94	0.002	-.0043783	.9780791
D_hc	.3126266	.0779182	4.01	0.000	.1599096	.4653435
_cons	-2.702025	.6867791	-3.93	0.000	-4.048087	-1.355963
ln(expo)	1	(exposure)				
inflate						
_cons	-.7214386	.1845822	-3.91	0.000	-1.083213	-.3596642

Vuong test of zip vs. standard Poisson: z = 3.04 Pr>z = 0.0012

Fig. 2.84 Parameter estimation results of zero-inflated poisson regression

AIC and BIC test statistics of this model are better than negative binomial regression. In combination with the average accident value 1.13 of non-rural road sections, it is close to its variance value 1.33, and many sample road sections with zero accidents are more suitable for the establishment of zero-inflated Poisson Regression Model. The zero-inflated negative binomial regression Vuong < 1.96, and the model with low chibar value is not better than the negative binomial regression. From the above results, the zero-inflated Poisson Regression model is selected to predict the accidents in non-rural road sections.

```
Negative binomial regression                    Number of obs     =        218
                                                LR chi2(4)        =      59.81
Dispersion     = mean                           Prob > chi2       =     0.0000
Log likelihood = -328.97285                     Pseudo R2         =     0.0833
```

NC	Coef.	Std. Err.	z	P>|z|	[95% Conf.	Interval]
E_dif	-.0219955	.0034069	-6.46	0.000	-.028673	-.015318
PT	3.521492	1.824605	1.93	0.004	-.0546685	7.097653
L_rs	1.189369	.3624337	3.28	0.001	.479012	1.899726
D_hc	.3914371	.128844	3.04	0.002	.1389076	.6439667
_cons	-3.226751	.8459334	-3.81	0.000	-4.88475	-1.568752
ln(expo)	1	(exposure)				
/lnalpha	-.2334143	.2326553			-.6894103	.2225817
alpha	.7918254	.1842224			.5018719	1.249298

```
Likelihood-ratio test of alpha=0: chibar2(01) =   49.21 Prob>=chibar2 = 0.000
```

Fig. 2.85 Negative binomial regression parameter estimation results

```
Zero-inflated negative binomial regression      Number of obs     =        218
                                                Nonzero obs       =        125
                                                Zero obs          =         93

Inflation model = logit                         LR chi2(4)        =      48.02
Log likelihood  = -319.8415                     Prob > chi2       =     0.0000
```

NC	Coef.	Std. Err.	z	P>|z|	[95% Conf.	Interval]
NC						
E_dif	-.0158967	.0033185	-4.79	0.000	-.022401	-.0093925
PT	4.633897	1.549772	2.99	0.003	1.596399	7.671394
L_rs	.6248625	.2943191	2.12	0.034	.0480078	1.201717
D_hc	.2956929	.0863247	3.43	0.001	.1264996	.4648861
_cons	-2.749477	.7066448	-3.89	0.000	-4.134475	-1.364478
ln(expo)	1	(exposure)				
inflate						
_cons	-.7729176	.2046841	-3.78	0.000	-1.174091	-.3717441
/lnalpha	-2.855588	1.112757	-2.57	0.010	-5.036552	-.6746238
alpha	.057522	.064008			.0064961	.509348

```
Likelihood-ratio test of alpha=0: chibar2(01) =   1.12 Pr>=chibar2 =  0.1451
Vuong test of zinb vs. standard negative binomial: z =    1.84  Pr>z = 0.0326
```

Fig. 2.86 Parameter estimation results of zero-inflated negative binomial regression

Table 2.9 Comparison of statistical indicators of four distribution models for non-rural road sections

Model	Log Likelihood	AIC	BIC	Vuong	Chibar2
Poisson	−0.63	3.27	−750.53		
NB	−328.97	3.07	−938.97		49.21
ZIP	−320.40	3.01	−1003.57	3.04	
ZINB	−319.84	3.01	−1005.33	1.84	1.12

Table 2.10 Comparison of statistical indicators of four distribution models for rural road sections

Model	Log Likelihood	AIC	BIC	Vuong	Chibar2
Poisson	−297.43	3.83	−532.82		
NB	−288.22	3.68	−684.69		18.42
ZIP	−296.61			0.48	
ZINB	−319.84			−0.00	1.12

The non-zero part of the zero-inflated Poisson regression model is expressed as follows:

$$\begin{aligned} NC_i = \text{expo} \times \text{EXP}(-2.7020 + 4.8516PT + 0.4869L_{rs} + \\ 0.3126D_{HC} - 0.0154E_{dif})NC_i \sim \text{Poisson} \end{aligned} \tag{2.20}$$

(2) Accident Prediction Model for Rural Road Sections

Using the same method to model and analyze the rural road sections. The comparison of statistical indicators of the four distribution models shows in Table 2.10. There are few samples of zero accident road sections in rural road sections. Zero-inflated Poisson and negative binomial regression model test statistics Vuong are used in modeling. The z test value is less than 1.96, and the P value is greater than 0.05. The model is not significant. All statistical indicators of the negative binomial regression model are slightly better than Poisson regression, the 95% confidence interval of alpha is $(0.09, 0.35)$ and chibar2 $= 18.42$, so the original assumption that the over dispersion parameter alpha $= 0$ can be rejected at the 5% significance level (corresponding to Poisson regression). From the above, the negative binomial regression model is selected to predict the accidents of rural and town sections. The cumulative standard residuals of the model prediction result for expo range from −7.68 to 2.26. The sum of the cumulative standard residuals is 1.7237, which is far less than 12.57 ($\sqrt{158}$). The model is stable, and the inspection of other variables is also stable.

The expression of negative binomial regression model is as follows:

$$\begin{aligned} NC_{ti} = \text{expo} \times \text{EXP}(-18.4388 - 6.3571PT + 0.1957D_{AC} - \\ 10.0928O_2 - 10.3742D_{ir})NC_{ti} \sim \text{NB} \end{aligned} \tag{2.21}$$

(3) Safety Impact Factor Analysis

Based on the above modeling results, the accident reduction factors of each factor can be calculated to reflect the impact of each influencing factor on the accidents of rural sections and non-rural sections in high altitude areas.

Accident reduction factor refers to the percentage of accident reduction when one independent variable increases by one unit and other independent variables remain unchanged according to the accident prediction model. A negative accident reduction factor means that when the independent variable increases by one unit, the accident increases, while a positive one means that it decreases. The analysis results are shown in Table 2.11 (1% is taken as a unit for variables in the form of ratios such as oxygen content and truck proportion). In non-rural and town sections, the proportion of truckss, roadside hazard level and density of horizontal curves all have a positive impact on the occurrence of accidents. For each 1% increase in the proportion of truckss, the accident will increase by 4.97%, and the roadside hazard level will increase by one level. The accident will increase by 62.73%. For each 1 increase in the density of horizontal curves in a unit section, the accident will increase by 36.70%, which has little impact on the relative altitude. In rural and town sections, only the density of access ports has a strong positive impact. For each additional access port, the number of accidents increases by 21.62%. Other factors have a negative impact. Among them, the impact of oxygen content is greatly affected by the size of the rural and town in the sample section, and the impact is limited in microscopic accident analysis (the impact of oxygen content is not significant in non-rural and town sections). For every 1% increase in the proportion of trucks, the accidents will decrease by 6.16%. The investigated road section is a national road with the main function of distribution, and the number of freight vehicles is relatively stable. The increase in the proportion of trucks is due to the decrease in the mixing rate of passenger car, which is reflected in the negative impact on the accidents. G109 Qinghai section is mainly designed for two-way two-lane traffic. The increase of direction coefficient is reflected in the increase of vehicles in the same direction and the decrease of vehicles in the opposite direction, which is conducive to reducing the risk of overtaking.

Table 2.11 Statistics of accident reduction factors

Influencing Factors	Non-rural road	Rural road
O_2		9.60%
E_{dif}	1.53%	
PT	−4.97%	6.16%
D_{ir}	/	9.85%
L_{rs}	−62.73%	
D_{hc}	−36.70%	
D_{ac}		−21.62%

2.3.5 *Traffic Safety Service Level of Secondary Highway*

(1) Accident Measurement Equation

The level of service of safety (LOSS) is used to measure the quality of traffic safety services provided by traffic facilities to traffic participants. Since it is used to measure the quality of service, it is inevitable that such quality of service can be quantified. Therefore, the safety service level is defined as a quality indicator used to describe the traffic safety status of traffic facilities and provide traffic safety services for traffic participants.

Many studies have shown that the level of accidents is most closely related to the length of road sections and traffic volume (AADT). In order to facilitate the practical application of safety service level, the average accident measurement equation established in this chapter only considers the two variables that have the greatest impact on the number of accidents: road section length and traffic volume. So far, a consensus has been reached on the nonlinear and non-Gaussian relationship between traffic volume and road safety. This relationship can be expressed by different models, and with the increase of the length of the road section, the number of accidents will inevitably increase. The research and analysis prove that the road sections are divided into rural road sections and non-rural highway sections, so that the accident distribution is easy to follow the Poisson distribution form. Therefore, the probability form of Poisson distribution is used in this section to establish the mean value measurement equation of accidents. The equation form is a generalized linear regression model in which the distribution family follows the Poisson distribution and the connection function is an identity function:

$$\mathrm{E(crash)} = \mathrm{L}(\beta_0 + \beta_1 X + \beta_2 \mathrm{X}^2 + \cdots) \tag{2.22}$$

Where:

E(crash)	—Expected number of accidents per unit time of a section;
L	—Length of road section (km);
X	—Annual Average Daily Traffic (AADT);
β_0	—Parameters to be estimated.

The sample tree data selected in this study are: 158 rural roads and 218 non-rural roads.

According to the characteristics of the Poisson distribution of accidents, the maximum likelihood method is used to estimate the parameters to be estimated in the accident mean measurement equation β_i. The accident mean value measurement equation provides the expected accident rate prediction value of a certain length of road facilities relative to a certain traffic volume in a certain period. Through regression analysis, the measurement equations shown in Formula (2.23) and Formula (2.24) are obtained, where Formula (2.23) is the accident mean value measurement equation for non-rural road sections, and Formula (2.24) is the accident mean value measurement equation for rural road sections

$$\begin{gathered}\mathrm{E(crash)} = \mathrm{L} \times (-26.2918 + 0.0200\mathrm{X} - 4.72 \times 10^{-10} X^3) \\ \mathrm{E(crash)} \sim \text{poisson}\end{gathered} \tag{2.23}$$

$$\begin{gathered}\mathrm{E(crash)} = \mathrm{L} \times (5.1842 - 0.025\mathrm{X} - 4.4 \times 10^{-7}\mathrm{X}^2 - 1.89 \times 10^{-11}\mathrm{X}^3) \\ \mathrm{E(crash)} \sim \text{poisson}\end{gathered} \tag{2.24}$$

(2) Quantitative Classification of Highway Traffic Safety Service Level in High Altitude Areas

The main basis of the current study is the σ-theoretical safety service level division method, takes the predicted value of the above average accident measurement equation as the center line, based on the principle of practical application, according to ±σ the safety service level of the rural and non-rural sections of the secondary highway in high altitude areas is divided into four levels according to the classification criteria. The classification and quantification results are shown in Figs. 2.87 and 2.88.

The qualitative description of security service level shows in Table 2.12. Level I and II safety service levels are better than expected. Level II safety service level has the potential to reduce accidents, and certain safety improvement measures can be taken, while Level I have little potential to reduce accidents, so no safety improvement measures can be taken; Level III and IV are lower than the expected safety level. It is necessary to continue to explore the factors that affect the safety level and improve the safety level, especially in the case of Level IV, which is in poor safety conditions.

The accident rate of roads in rural areas at high altitude increases with the continuous increase of traffic volume. When the traffic volume increases to a certain extent, the change of accident rate is no longer obvious, and the research conclusion is

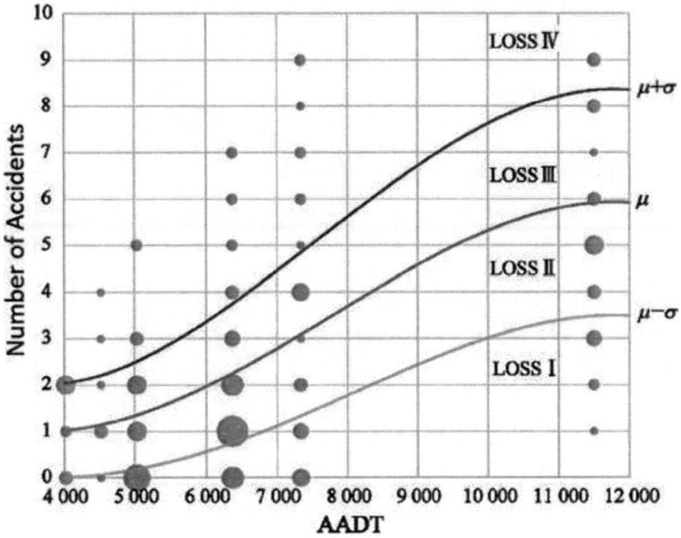

Fig. 2.87 Classification of safety service level for rural sections of high altitude highways

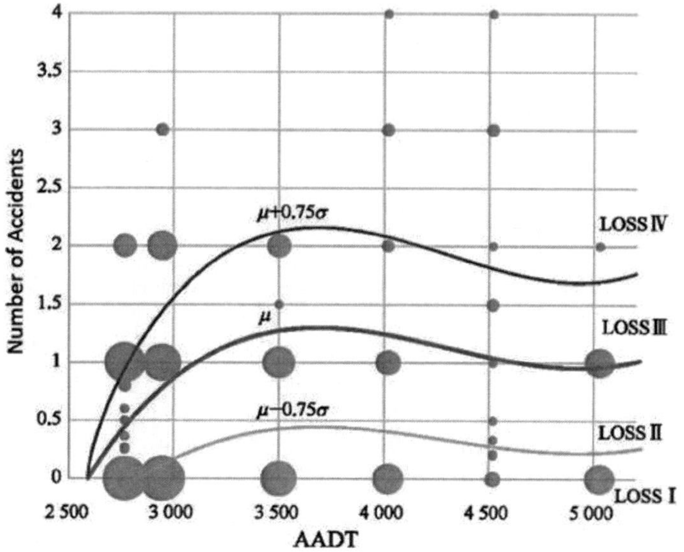

Fig. 2.88 Classification of safety service level for non-rural sections of high altitude highways

Table 2.12 Qualitative description of highway traffic safety service level in high altitude areas

Security Service Level	Qualitative Description of Traffic Facility Safety Service Level
Level I	The quality of safety service provided to traffic participants is very high, the safety condition is good, and the potential for further reduction of accidents is small, so the existing level can be maintained
Level II	The quality of safety service provided to traffic participants is higher than expected. Based on maintaining the existing level, appropriate measures can be taken to further improve the safety level
Level III	The quality of safety service provided to traffic participants is lower than expected, and the potential for further reduction of accidents is large, so improvement measures must be taken
Level IV	The quality of safety service provided to traffic participants is very low, the safety situation is very poor, the potential for further reduction of the accident rate is great, there are great potential safety hazards, and it is urgent to take positive improvement measures

completely consistent with the actual traffic operation. As the traffic volume of high altitude non-rural roads is generally at a low level, the road service level is high, and vehicle overspeed is common. When the traffic volume increases to a certain extent, the speed can be effectively controlled, and the accident rate shows a downward trend. The tail of the curve in Fig. 2.88 shows an upward trend. If the traffic volume continues to increase, the accident rate will rise again. When the traffic volume increases to a certain extent, the change of the accident rate will not be obvious.

2.3.6 *Preliminary Study on Freeway Traffic Safety Service Level*

(1) **Freeway at High Altitude**

In May 2011, the commencement ceremony of Gonghe-Yushu Freeway (Gong-Yu Freeway) was held in Jiegu Town, Yushu Prefecture, Qinghai Province, which is the first freeway built in the high altitude permafrost region of the Qinghai-Tibet Plateau. Recently, the first phase of Gong-Yu Freeway was basically open to traffic, but the traffic safety facilities, mechanical and electrical equipment and other later projects are still not perfect. There are many forks on the way, and vehicles are driving alternately and in reverse. There are hidden dangers in traffic safety.

The Beijing-Tibet Freeway is a radial freeway in the capital, which does not meet the operation standard of the freeway. It starts from Beijing and ends at Lhasa, Tibet Autonomous Region, passing through Beijing, Hebei, Inner Mongolia, Ningxia, Gansu, Qinghai, and Tibet, with a total length of 3 km. The freeway is an important part of the national freeway network. At present, Cha'ge section has not been completed, and some parts of Golmud-Lhasa section have not been started. The open road section in Qinghai is an open freeway with main line toll collection. There are many level crossings across rural areas, and there are potential traffic safety hazards.

(2) **Study on Freeway Traffic Safety in High altitude Area**

The freeway is a fully closed, multi-lane, dedicated automobile highway with a median, comprehensive control of the entrance and exit, and a variety of safety service facilities. It has good alignment indicators, less roadside interference, and is completely separated from the opposite traffic flow. Compared with the Qinghai-Tibet Freeway, it has many advantages. In view of the current situation of freeway construction in Qinghai-Tibet region, most sections have not yet reached the freeway operation standard, and the operation time is relatively short. There is a lack of accident data on the freeway, which is difficult to support the construction of the model. Therefore, the model form is determined according to the research results of accident prediction on Qinghai-Tibet Highway, the characteristics of high-speed highway operation and the historical data of highway accident prediction model research.

In previous studies, regression models subject to negative binomial distribution were mostly used in freeway accident prediction models, and the main influencing factors involved whether rural sections, bridge sections, average corner of horizontal curve, proportion of trucks, etc. In this chapter, considering the distribution of rural areas and population in high altitude areas, the accident prediction model is divided into two models, namely, rural road section and non-rural road section. The main influencing factors involved include oxygen content, relative elevation, proportion of trucks, direction coefficient, roadside grade, horizontal curve density and contact population density.

In combination with the characteristics of high altitude highways and the inherent characteristics of highways, the above main influencing factors are screened out. The alignment of the freeway is excellent and the design data is detailed, so the curve corners among the influencing factors are included, the relative altitude is replaced by the average gradient of the vertical curve, and the direction coefficient is eliminated. The road section in the bridge area is affected by the access port, so the density of the access port is included in the model. Considering the unique environment of high altitude, the oxygen content and roadside grade are retained. The model form is as follows:

$$NC = \text{expo} \times \text{EXP}(\beta_0 + \beta_1 PT + \beta_2 D_{AC} + \beta_3 O_2 + \beta_4 A_b + \beta_5 S_V) \tag{2.25}$$

Among them, the rural sections are subject to the negative binomial distribution, and the non-rural sections are subject to the zero inflated Poisson distribution.

Chapter 3
Dynamic Features of Typical Vehicles Under Low Pressure and Oxygen Deficient Environment in High Altitude Areas

3.1 Vehicle Types and Methods for Testing

3.1.1 Vehicle Types for Testing

3.1.1.1 Traffic Volume and Composition Distribution in Qinghai-Tibet Plateau Area

Combining the changes in the traffic volume and its composition of G109 from 2012 to 2014 with the observation results in 2015 (Table 3.1) in Sect. 2.2.2 of this book, it can be derived the change pattern in time distribution of traffic volume and its composition of G109 in these four years.

According to the analysis results, passenger cars and trucks running on Qinghai–Tibet Highway grow year-by-year, with an average annual growth rate of about 7%, while the agricultural vehicles such as tractors decrease yearly, reflecting that the leading vehicle type on Highway is developing more rapid and motorized; from analysis of the number of each vehicle type, it can be seen that the leading types on Qinghai–Tibet Highway are passenger cars and trucks (i.e. articulated combination of vehicle), and only these two types of vehicles shows continuous growth, reflecting a polarized development of vehicles on Qinghai–Tibet Highway. According to the statistics of each vehicle equivalent, the proportion of articulated combination of vehicle equivalents on Qinghai–Tibet Highway from 2012 to 2015 is about 50%, which is the most typical vehicle type other than minibuses, therefore, attention should be paid for this vehicle type when selecting indicators such as longitudinal grade design, sight distance check and superelevation setting.

J. Liu, *Technical Indicators and Safety Design of Freeway in High Altitude Area*,
https://doi.org/10.1007/978-981-99-0620-8_3

Table 3.1 Annual average daily traffic volume of each observation station in 2015 (vehicle)

Survey sites	Minibus	Bus	Minivan	Medium truck	Large truck	Combination of vehicle	Total of natural numbers	Total of equivalent minibus
Naijigou observation station	858	32	157	94	472	605	2,219	4,805
Anduo traffic observation station	1,068	21	143	149	546	726	2,653	5,737
	1,214	52	169	186	689	740	3,050	6,423
	1,326	16	523	111	722	818	3,517	7,117
Qingqu traffic observation station of G317	1,136	42	231	392	317	206	2,323	3,632
Heilong Deqing traffic observation station	1,463	164	192	86	640	642	3,187	6,198
	3,019	287	573	205	385	696	5,166	8,078
	2,542	252	1,154	323	434	510	5,215	7,683

According to the *Technical Standards for Highway Engineering* (JTG B01-2014) and *Road Vehicle Outline Dimensions, Axle Load and Mass Limits* (GB 1589–2016), the design vehicle outline dimensions used for road design are specified in Table 3.2. This project focuses on the selection of heavy vehicles and articulated combination of vehicle for testing.

Table 3.2 Designing vehicle dimensions (m)

Types	Total length	Total width	Total height	Front suspension	Wheelbase	Rear suspension
Minibus	6	1.8	2	0.8	3.8	1.4
Bus	13.7	2.55	4	2.6	6.5 + 1.5	3.1
Articulated bus	18	2.5	4	1.7	5.8 + 6.7	3.8
Loaded bus	12	2.5	4	1.5	6.5	4
Articulated combination of vehicle	18.1	2.55	4	1.5	3.3 + 11	2.3

Note The wheelbase of the articulated combination of vehicle (3.3 + 11) m: 3.3 m is the distance from the first axis to the articulation point, and 11 m is the distance from the articulation point to the last axis

3.1.1.2 Typical Vehicle Types for Testing

(1) **Dongfeng DFL4251A10 Heavy-Duty Semi-trailer Tractor (Six-Axle Trucks)**

Driving features of six-axle truck are tested, in which the tractor adopts Dongfeng DFL4251A10 heavy-duty semi-trailer tractor and the trailer adopts Chusheng CSC9400CXY warehouse grill type transport semi-trailer. The test vehicle is shown in Fig. 3.1.

(1) Test Vehicle Parameters

Tables 3.3, 3.4, and 3.5 shows basic structure and performance parameters of Dongfeng DFL4251A10 heavy-duty semi-trailer tractor and trailer.

(2) Measurement Parameters of Test Vehicle

Actual gross vehicle weight: m = 47,600 kg (including 3 person accompanying the vehicle, 120 kg of instruments and equipment).

(3) Rotating Mass Conversion Factor

By reviewing relevant literature and considering the actual load of the vehicle, the transmission ratio of each gear, the main gear ratio and the range of the rotating mass conversion coefficient, the rotating mass conversion coefficients for each gear were obtained after fitting, as shown in Table 3.6.

Fig. 3.1 Articulated combination of vehicle for testing

Table 3.3 Basic structure and performance parameters of heavy-duty semi-tractor

Curb weight	8,700 kg	Gross vehicle weight	25,000 kg	Engine model	Dongfeng dCi375-30
Tire model	12R22.5	Max. engine torque	1,700 N m	Min. stable engine speed	700 r/min
Wheel radius	0.526 m	Max. output power	276 kW	Max. stable engine speed	2,100 r/min
Max. speed	95 km/h	Max. torque speed	1,300 r/min	Rated speed	1,900 r/min
Axle speed ratio	3.42	Transmission efficiency	89%	Max. design traction mass	40,000 kg
Length	6.96 m	Width	2.5 m	Height	3.7 m

Table 3.4 Transmission ratio of each gear of heavy-duty semi-trailer tractor

Gear	1	2	3	4	5	6	7	8
Transmission ratio	14.03	11.64	9.6	7.97	6.62	5.49	4.55	3.78
Gear	9	10	11	12	13	14	15	16
Transmission ratio	3.08	2.56	2.11	1.75	1.45	1.21	1	0.83

Table 3.5 Basic structure and performance parameters of trailer

Curb weight	8,300 kg	Gross vehicle weight	39,600 kg
Tire model	12R22.5	Wheel radius	0.526 m

Table 3.6 Rotating mass conversion coefficients for each gear (I)

Gear	Rotating mass conversion factor δ	Gear	Rotating mass conversion factor δ
1	1.5953867	9	1.077696566
2	1.425863349	10	1.069592776
3	1.306136807	11	1.063785999
4	1.227002441	12	1.059946314
5	1.172577895	13	1.057293682
6	1.134766008	14	1.055529682
7	1.108688426	15	1.0542473
8	1.090965252	16	1.053387682

(2) **Dongfeng DFL5311CCYAX9A Warehouse Grill Truck (Four-Axle Truck)**

The driving features of the four-axle truck were studied during the project research, and the test vehicle was a Dongfeng DFL5311CCYAX9A warehouse truck, as shown in Fig. 3.2.

Fig. 3.2 Heavy-load truck for testing

Table 3.7 Basic structure and performance parameters of test vehicle

Curb weight	10,070 kg	Gross vehicle weight	31,000 kg	Engine model	ISL95-340E40A
Tire model	12R22.5	Max. engine torque	1,500 N m	Min. stable engine speed	700 r/min
Wheel radius	0.526 m	Max. output power	245 kW	Max. stable engine speed	1,850 r/min
Max. speed	90 km/h	Max. torque speed	1,500 r/min	Rated speed	1,850 r/min
Axle speed ratio	4.44	Transmission efficiency	89%	Max. design traction mass	31,000 kg
Length	11.98 m	Width	2.3 m	Height	3.9 m

(1) Test Vehicle Parameters

Tables 3.7 to 3.8 shows basic structure and performance parameters of Dongfeng DFL5311CCYAX9A warehouse grill truck.

(2) Measurement Parameters of Test Vehicle

Actual gross vehicle weight: m = 30,990 kg (including 3 person accompanying the vehicle, 120 kg of instruments and equipment).

Table 3.8 Transmission ratio of each gear of test vehicle

Gear	1	2	3	4	5	6
Transmission ratio	12.10	9.41	7.31	5.71	4.46	3.48
Gear	7	8	9	10	11	12
Transmission ratio	2.71	2.11	1.64	1.28	1	0.78

Table 3.9 Rotating mass conversion coefficients for each gear (II)

Gear	Rotating mass conversion factor δ	Gear	Rotating mass conversion factor δ
1	1.584737522	7	1.105822895
2	1.385472927	8	1.095863436
3	1.264554352	9	1.089793745
4	1.192813189	10	1.086173627
5	1.149033929	11	1.083975107
6	1.122237061	12	1.082626517

(3) Rotating Mass Conversion Factor

By reviewing relevant literature and considering the actual load of the vehicle, the transmission ratio of each gear, the main gear ratio and the range of the rotating mass conversion coefficient, the rotating mass conversion coefficients for each gear were obtained after fitting, as shown in Table 3.9.

3.1.2 *Test Method*

In order to accurately obtain the engine operating external features curve and engine braking features curve under different altitudes, a road with a longitudinal grade of $\leq 1.5\%$ was selected as the test section, and the transmission gears were selected as 4th, 5th, 6th, 7th, 8th and 9th gears, and round-trip tests were conducted on the test section to reduce the influence of road slope, wind direction and wind speed. During the test, vehicle speed was collected by the RLVB3iSL RTK speed sensor from Racelogic, UK; the road slope was collected by the RT3100 inertial navigation system from Oxford Technology, UK; the accelerator pedal stroke was collected by the SENST2 displacement sensor from ISAAC, Canada; In order to improve the acquisition accuracy of vehicle speed and road slope, RLVBBS4RG differential base station was used to improve the satellite positioning accuracy. The specific altitude is shown in Table 3.10, and the specific instrumentation information is shown in Table 3.11.

Table 3.10 Altitude of the experimental site

Site	Xi'ning	Gonghe	Xinghai	Yushu	Maduo
Nominal altitude (m)	2,300	2,890	3,638	4,188	4,545

Table 3.11 Equipment Information

No.	Equipment	Equipment information	Quantity	Accuracy	Origin and company
1.	Vehicle speed sensor	RLVB3iSL-RTK	1	0.1 km/h	Racelogic UK
2.	Inertial navigation systems	TR3100	1	0.04°	Oxford Tech UK
3.	Displacement sensors	SENST2	1	0.25%F.S	ISAAC Canada
4.	Differential base stations	RLVBBS4RG	1	2em	Racelogic UK

Table 3.12 Weather and temperature conditions of the test sites

No.	Sites	Weather	Temperature (°C)
1	Xi'ning	Clear skies	24
2	Gonghe	Cloudy	20
3	Xinghai	Cloudy	13
4	Maduo	Cloudy	10
5	Yushu	Partly cloudy	8

3.1.3 Test Site and Environmental Conditions

Weather and temperature conditions of the test sites, are listed in Table 3.12.

3.2 Principles of Vehicle Dynamics in Plateau Areas

3.2.1 Principles of Vehicle Driving Dynamics

The driving force of a car is the force generated by the car engine by burning fuel to produce energy to overcome the resistance and make the car move forward. The torque generated by the car engine and transmitted to the driving wheels through the transmission system produces a circumferential force on the ground, and the ground reaction force on the wheels is the driving force that drives the car (Fig. 3.3). The formula for calculating the driving force of a car is as follows:

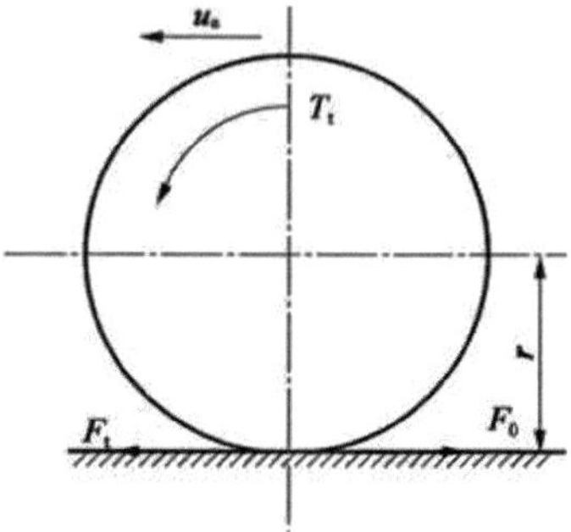

Fig. 3.3 Schematic diagram of mechanical analysis

$$F_t = \frac{T_t}{r} = \frac{T_{\rm tq} i_{\rm g} i_0 \eta_{\rm T}}{r} \tag{3.1}$$

where

T_t Torque acting on the driving wheels (N m);
T_{tq} Effective engine torque (N m);
i_0 Transmission ratio of main reducer;
i_g Transmission ratio of the gearbox;
η_T Mechanical efficiency of the driveline, for large trucks generally are 0.82–0.85;
r Rolling radius of tires (generally approximately equal to the radius of the tire at rest, unit: m).

The external features of the engine determine the relationship between the engine output torque and speed, after the driveline reaches the wheels, it can be expressed as the relationship between the driving force and the speed of the vehicles, such a relationship can be expressed in the traction features of the vehicles (driving force diagram) (Fig. 3.4).

According to mechanical analysis, the driving force equals to the sum of driving resistance:

$$F_t = \sum F = F_{\rm f} + F_{\rm w} + F_{\rm i} + F_{\rm j} \tag{3.2}$$

where

F_t driving resistance (N);
ΣF Sum of driving resistance (N);
F_f Rolling resistance(N);
F_w Air resistance(N);
F_i Slope resistance(N);
F_j Acceleration resistance(N).

In which Rolling resistance

$$F_f = Gf \tag{3.3}$$

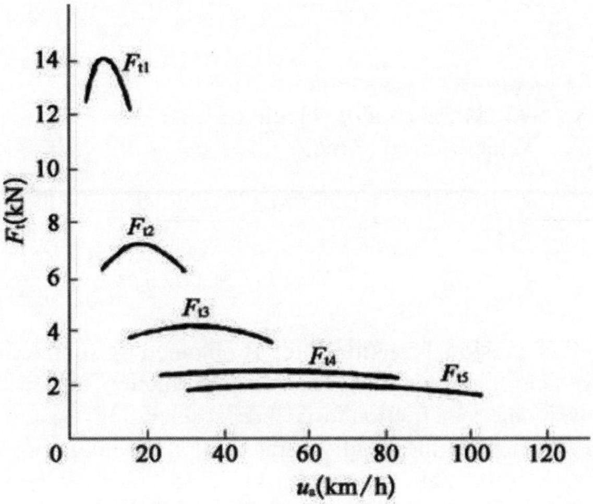

Fig. 3.4 Diagram of truck traction features of Model NKR552/555

Air resistance

$$F_{\mathrm{w}} = \frac{C_{\mathrm{D}}A}{21.15G}u_{\mathrm{a}}^2 \tag{3.4}$$

Slope resistance $F_i = Gi$ (3.5)

$$F_i = Gi \tag{3.5}$$

Acceleration resistance

$$F_{\mathrm{j}} = \delta m \frac{du}{dt} \tag{3.6}$$

where

G Self-weight of vehicle (kg).
F Friction coefficient.
i Gradient.
δ Rotating mass conversion factor of vehicle.

In formula

$$\delta = 1 + \frac{1}{m}\frac{\sum I_{\mathrm{W}}}{r^2} + \frac{1}{m}\frac{I_{\mathrm{f}}i_{\mathrm{g}}^2 i_0^2 \eta_{\mathrm{T}}}{r^2}$$

I is rotational inertia (rotational inertia of flywheel (If) and wheel (IW)); δ can be estimated according to empirical formula for estimation as follow:

$$\delta = 1 + \delta_1 + \delta_2 i_g^2, \delta_1 \approx \delta_2 = 0.03 \sim 0.05$$

CD Air resistance coefficient.
A Windward area of vehicle (m^2).
U_a Vehicle speed (Km/h).

Therefore, the traction force is calculated as follows

$$F_t = Gf + Gi + \delta m \frac{du}{dt} + \frac{C_D A}{21.15} u_a^2 \tag{3.7}$$

The residual traction force is obtained by subtracting the air resistance from the vehicle traction force. The residual traction force represents the force needed to overcome road friction resistance, acceleration resistance and grade resistance, which is divided by the dead weight to be the vehicle power factor. The vehicle power factor is the main indicator of vehicle traction performance, larger the value, greater the ability of the vehicle to accelerate, climb and overcome road resistance. The expression of vehicle power factor D is as follows.

$$D = \frac{Gf + Gi + \delta m \frac{du}{dt}}{G} = \frac{T_{tq} i_g i_0 \eta_T}{rG} - \frac{C_D A}{21.15G} u_a^2 \tag{3.8}$$

By calculating the traction force and air resistance of the vehicle, dynamic features of the vehicles can be obtained (Fig. 3.5). In order to calculate the maximum speed, the drag force can be considered to consist of rolling resistance and air resistance only. The above formula is used to calculate the speed of the vehicle in a state of force balance. If the maximum speed for a given slope must be calculated, the resistance is composed of rolling resistance, air resistance and slope resistance (for a given slope i), and the maximum speed is obtained by the equation.

3.2.2 Principle of Vehicle Driving Resistance

While driving, vehicle is always subject to air resistance and rolling resistance, in which the air resistance is affected by the driving speed far more, so when the speed is low, the proportion of rolling resistance is larger; when the speed is high, the proportion of air resistance is larger. There are two main methods to obtain air resistance and rolling resistance: the first is test method, and the second is theoretical calculation method. Considering the test is mainly carried out in the higher altitude areas with large gradient, while the test method is only suitable at test site for it requires a long length of flat road and the test results will greatly be influenced by external factors, therefore the test method is not applicable for testing air resistance and rolling resistance on the actual road. In the following, the air resistance and

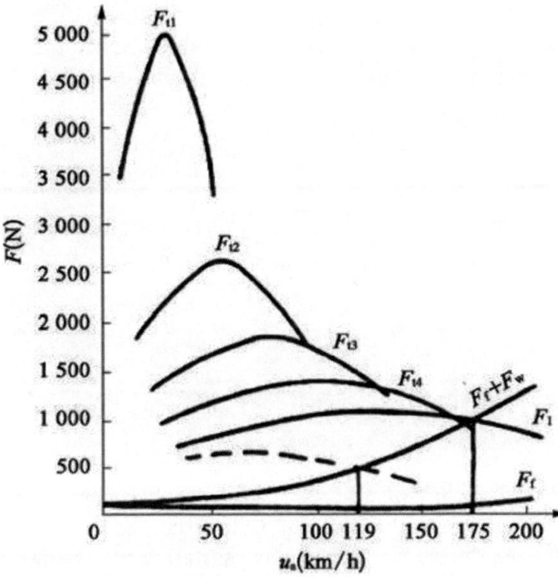

Fig. 3.5 Diagram of vehicle traction versus resistance

rolling resistance models are obtained by theoretical calculation method with the automobile dynamics as the theoretical guide.

(1) **Air Resistance and Rolling Resistance**

It is known from automobile dynamics that the value of air resistance is usually proportional to the dynamic pressure of the relative velocity of the airflow 1/2pu^2 in the running speed range of the vehicle, i.e.

$$F_{\mathrm{w}} = \frac{1}{2} C_{\mathrm{D}} A \rho \mu_{\mathrm{r}}^2 \tag{3.9}$$

where

- C_D The air resistance coefficient, in general, it should be a function of the Reynolds number Re. At higher speeds, higher dynamic pressures and lower viscous friction of the corresponding gases, C_D will not vary with Re.
- ρ Air density, generally, p = 1.2258 kg/^{3};
- A Windward area, i.e. the projected area in the direction of car travel (m^2);
- u_r Relative speed, the speed of the car when there is no wind (m/s).

From the statistical data, the air resistance coefficient of the truck is $C_D = 0.6$–1.0, combined with the shape parameters of the test vehicle and experience to choose $C_D = 0.72$, the windward area of the test articulated combination of vehicle is A = 7.14 m^2 (truck A = 6.98 m^2); ρ is corrected according to the air density at different altitudes. During the test, the wind speed is small relative to the vehicle speed, and the test is conducted in both directions to reduce the influence of wind speed on the test results, and u_r is taken as the vehicle speed u_a.

Based on the test and statistical results in vehicle dynamics, the empirical formula of rolling resistance coefficient for trucks is

$$f = 0.0076 + 0.000056u_a \tag{3.10}$$

Obtained from the formula (3.9) and formula (3.10), the relationship between air resistance and rolling resistance with the speed of test vehicle in driving is

$$F_w + F_f = \frac{1}{2}C_D A\rho u_r^2 + mgf = \frac{1}{2}C_D A\rho u_a^2 + mgf \tag{3.11}$$

(2) **Determination of Air Resistance and Rolling Resistance at Different Altitudes**

With increase of altitude, the atmospheric pressure, air density and atmospheric temperature will all change to a certain extent, and there is a specific relationship between these parameters. By mastering the change pattern of these parameters by means of theory and test, its influence on the driving resistance of whole vehicle can be better analyzed.

It is known from atmospheric science principles and testing experience that atmospheric pressure is related to altitude, atmospheric temperature, and weather conditions. The automobile-related road tests need to be conducted when the weather is clear, so the influence of weather conditions can be ignored. Within the range of the actual temperature change of the atmosphere, the effect of temperature change on atmospheric pressure is small. The atmospheric pressure measured at the same altitude becomes slightly larger with the increase of temperature, but the change is small, so the effect of temperature on pressure can be ignored.

The relationship between atmospheric pressure and altitude is

$$P_H = 1013.25 \times \left(1 - \frac{0.0065H}{288.15}\right)^{5.255} \tag{3.12}$$

where

P_H Atmospheric pressure at altitude H (hP_a);
H altitude (m).

The density of air is related to both atmospheric pressure and temperature, and can be derived from the gas equation of state

$$\rho = \frac{P_H M}{R(273.15 + T)} = \frac{28.98 P_H}{8.314 \times 10 \times (273.15 + T)} \tag{3.13}$$

where

ρ Air density (g/L);
M Air molar mass, M = 28.98 g/mol;
R Proportional constants, R = 8.314 J/(mol K);
T Air temperature (°C).

Atmospheric pressure and air density at different altitudes are shown in Table 3.13.

According to the formula (3.13) and Table 3.13, relationship between air resistance, rolling resistance and vehicle speed at different altitudes can be calculated as follows, and the coefficient values are shown in Tables 3.14 and 3.15.

$$F_{\mathrm{f}} + F_{\mathrm{w}} = a_{FfFw}u_{\mathrm{a}}^2 + b_{FfFw}u_{\mathrm{a}} + c_{FfFw} \quad (3.14)$$

3.2.3 Altitude-Based Power and Resistance Discounting Principle

According to the research results, the engine torque and air resistance of vehicles at high altitude are reduced by the decrease of oxygen content and air density. In this section, the discount pattern of engine torque and air resistance with the increase of

Table 3.13 Atmospheric pressure and air density at different altitudes

No. 1	Altitude (m)	Atmospheric pressure ($\times 10^2$ Pa)	Air density (g/L)
1	0	1013.25	1.225706
2	2,300	765.8202	0.898337
3	2,890	710.9901	0.8454
4	3,638	646.0908	0.787024
5	4,188	601.5018	0.740472
6	4,545	573.9179	0.711541

Table 3.14 Coefficient values of air resistance and rolling resistance for combination vehicle at different altitudes

Site	Altitude (m)	a_{FfFw}	b_{FfFw}	c_{FfFw}
Xi'ning	2,300	0.78170172	26.12288	3545.248
Gonghe	2,800	0.167671	26.12288	3545.248
Xinghai	3,500	0.156093093	26.12288	3545.248
Maduo	4,200	0.141122298	26.12288	3545.248
Yushu	4,600	0.14686028	26.12288	3545.248

Table 3.15 Coefficient values of air resistance and rolling resistance for heavy-load vehicle at different altitudes

Site	Altitude (m)	a_{FfFw}	b_{FfFw}	c_{FfFw}
Xi'ning	2,300	0.174177563	17.00731	2308.135
Gonghe	2,890	0.163913667	17.00731	2308.135
Xinghai	3,638	0.152595209	17.00731	2308.135
Maduo	4,188	0.137959894	17.00731	2308.135
Yushu	4,600	0.143569293	17.00731	2308.135

altitude are analyzed and to establish the corresponding model to lay the foundation for the study of the relationship between driving force and driving resistance of vehicles in high altitude areas.

(1) **Altitude-based Air Resistance Discount Analysis**

(1) Air Resistance Formula for Vehicles

According to the relevant information, the theoretical formula of wind resistance is:

$$F_{\mathrm{w}} = \frac{1}{2} A C_{\mathrm{D}} \rho_{\mathrm{a}} u^2 \tag{3.15}$$

Where

- A Windward area (m^2);
- C_D Coefficient of air resistance, related to streamline of large truck;
- ρ_a Air density, (kg/m^3);
- u Relative velocity of large truck and air, approximated as large truck travel speed (m/s).

In the *Study on the Design Method and Key Indicators of Freeway Slope*, the test vehicle is Dongfeng Tianlong DFL4251A9, and the relevant parameters are taken as follows:

$$A = 7\,\mathrm{m}^2;$$

$C_D = 0.9$ (Note: in *Road Survey and Design* (Zhang Yuhua), the air resistance coefficient for heavy vehicles is 0.6–1.0);

$$P_{\mathrm{a}} = 1.2258\,\mathrm{kg/m^3}.$$

When converting the speed unit to km/h, the formula is:

$$F_{\mathrm{w}} = \frac{A C_{\mathrm{D}} \rho_{\mathrm{a}} v^2}{2 \times 3.6^2} = \frac{A C_{\mathrm{D}} \rho_{\mathrm{a}} v^2}{25.92}$$

From which, the acceleration due to the relevant wind resistance can be calculated:

$$a_{\rm w} = \frac{F_{\rm w}}{m} = \frac{AC_{\rm D}\rho_{\rm a}v^2}{25.92m} = \frac{7 \times 0.9 \times 1.2258 \times v}{25.92 \times 56000} = 5.3 \times 10^{-6}v^2 \quad (3.16)$$

where

v Relative velocity of large trucks and air, approximation the traveling speed for large trucks (km/h).

(2) Air Density Versus Altitude

Under standard conditions, the atmospheric density is $\rho_0 = 1.293$ kg/m^3. The relationship between air density and altitude is shown in Table 3.16. The fitting is shown in Fig. 3.6.

The atmospheric relative density is calculated as:

$$k_p = 1 - 1.0028 \times 10^{-4}H + 3.3795 \times 10^{-9}H^2 \quad (3.17)$$

where

H Altitude (m).

Table 3.16 Different air density and altitude

Altitude (m)	0	1,000	2,000	2,500	3,000	4,000	5,000
Relative atmospheric pressure	1	0.881	0.774	0.724	0.677	0.591	0.514
Relative air density	1	0.903	0.813	0.77	0.73	0.653	0.583

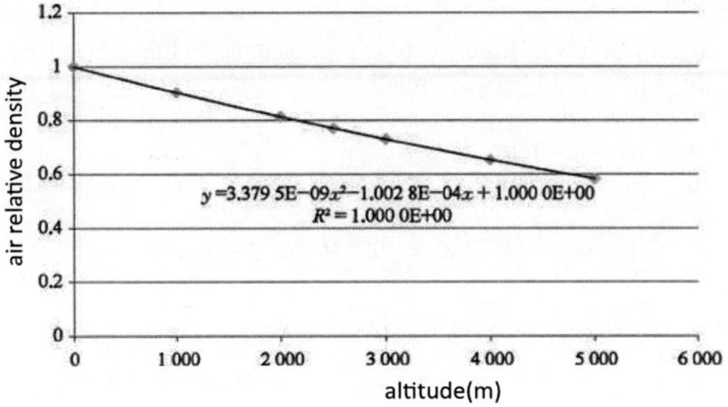

Fig. 3.6 Air relative density and altitude

For example, in the altitude of 5,000 m, the air relative density is

$$k_\rho = 1 - 1.0028 \times 10^{-4} \times 5000 + 3.3795 \times 10^{-9} \times 5000^2 = 0.583$$

The air density is calculated as follows:

$$\rho_a = k_\rho \rho_0 = 1.293 \times (1 - 1.0028 \times 10^{-4} H + 3.3795 \times 10^{-9} H^2) \tag{3.18}$$

According to this formula, corresponding altitude for $P_a = 1.2258$ kg/m^3 is 527.65 m

(3) Altitude-Based Air Resistance Discount Factor

The relative density of the atmosphere with respect to the altitude of 0 m is

$$k_\rho = 1 - 1.0028 \times 10^{-4} \times 5000 + 3.3795 \times 10^{-9} \times 5000^2 = 0.583$$

In the actual calculation, the coefficient relative to the usual altitude (527.65 m) is more meaningful. The formula for calculating the coefficient of relative air density (relative to the altitude of 527.65 m) is

$$\begin{aligned} k_a &= \frac{1 - 1.0028 \times 10^{-4}\, H + 3.3795 \times 10^{-9}\, H^2}{1 - 1.0028 \times 10^{-4} \times 527.65 + 3.3795 \times 10^{-9} \times 527.65^2} \\ &= \frac{1 - 1.0028 \times 10^{-4}\, H + 3.3795 \times 10^{-9}\, H^2}{0.948} \end{aligned} \tag{3.19}$$

where

H Altitude (m).

(2) **Altitude-Based Engine Torque Discounting Analysis**

(1) Air Content Versus Altitude

The oxygen content in the atmosphere is proportional to the atmospheric pressure; therefore, the relationship between relative atmospheric pressure and altitude is the relationship between relative atmospheric oxygen content and altitude. According to Table 3.16, the fit is shown in Fig. 3.7.

The relative atmospheric pressure (relative oxygen content) is calculated as

$$k_{O_2} = 1 - 1.2360 \times 10^{-4} H + 5.2960 \times 10^{-9} H^2$$

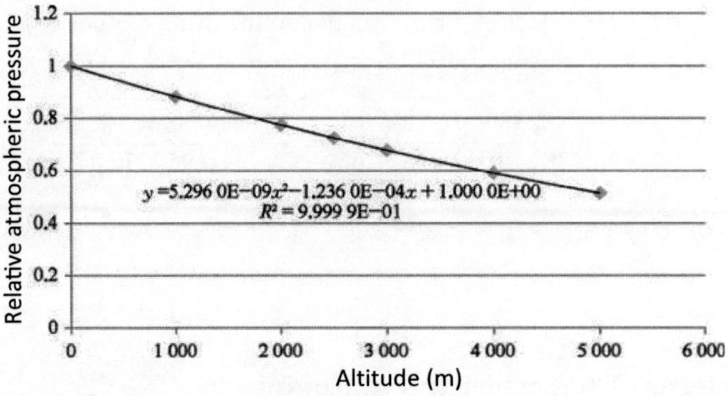

Fig. 3.7 Relative atmospheric pressure and altitude

where

H Altitude (m).

(2) Altitude-Based Vehicle Torque Reduction

i. Natural inhalation. The relative atmospheric pressure (relative oxygen content) relative to the altitude of 0 m is

$$k_{O_2} = 1 - 1.2360 \times 10^{-4} H + 5.2960 \times 10^{-9} H^2$$

While in practical calculations, the coefficient relative to the usual altitude (527.65 m) is more meaningful. The formula for calculating the coefficient of relative atmospheric pressure (relative oxygen content) (with respect to the altitude of 527.65 m) is

$$\begin{aligned} k_p &= \frac{1 - 1.2360 \times 10^{-4} \times H + 5.2960 \times 10^{-9} \times H^2}{1 - 1.2360 \times 10^{-4} \times 527.65 + 5.2960 \times 10^{-9} \times 527.65^2} \\ &= \frac{1 - 1.236 \times 10^{-4} \times H + 5.2960 \times 10^{-9} \times H^2}{0.93626} \end{aligned} \tag{3.20}$$

where

H Altitude (m).

ii. Intercooled boost. The intercooled supercharged engine torque reduction factor (relative to 0 m above sea level) is

$$k_p = 1 - 7.6238 \times 10^{-5} H$$

The engine torque reduction factor relative to the usual altitude (527.65 m) can be calculated by the following formula

$$k_{\mathrm{p}} = \frac{1 - 7.6238 \times 10^{-5} H}{1 - 7.6238 \times 10^{-5} \times 527.65} = \frac{1 - 7.6238 \times 10^{-5} H}{0.95977} \tag{3.21}$$

Where

H Altitude(m).

3.3 External Characteristics of Engine

3.3.1 Test and Calculation Process of Engine External Characteristics

The calculation steps of the external characteristics of the engine are as follows:

First, conducting a full throttle acceleration test on the test vehicle in each gear (4th, 5th, 6th, 7th, 8th, 9th gear). During the test, recording the data of the vehicle speed, grade, and other changes along with time, according to the original data based on the vehicle speed u_a (km/h) and time t (s), the vehicle speed-time relational graph is drawn and the abnormal points are eliminated. At the same time, since the test site is a high altitude areas, and there are slopes in roads, in order to ensure the smoothness of the external characteristic curve of the engine, the gradient change of the test section during the whole test process is ignored in the process of test data processing, and the average gradient during the test process is substituted into the calculation of slope resistance, that is:

$$F_{\mathrm{i}} = mg\frac{i}{\sqrt{i^2 + 1}} \tag{3.22}$$

Second, fitting the vehicle speed u (m/s) and time t (s) with cubic polynomial function to obtain the formula between vehicle speed u (m/s) and time t (s), that is:

$$u = at^3 + bt^2 + ct + d \tag{3.23}$$

Third, Deriving the vehicle speed u, the formula of acceleration du/dt (m/s^2) and time t (s) is obtained, that is:

$$\frac{\mathrm{d}u}{\mathrm{d}t} = 3at^2 + 2bt + c \tag{3.24}$$

Fourth, according to the relationship between acceleration du/dt (m/s^2) and time t (s), speed u_a (km/h) and time t (s), the relational graph between acceleration du/dt

(m/s^2) and speed u_a (km/h) can be obtained. Then the quadratic polynomial fitting is performed on acceleration du/dt (m/s^2) and speed u_a (km/h) to obtain the relationship between acceleration du/dt (m/s^2) and vehicle speed u_a (km/h). According to $F_j = \delta m$ du/dt, can get the functional relationship between acceleration resistance and vehicle speed, that is:

$$F_j = \delta m \frac{du}{dt} = \delta m (a_{Fj} u_a^2 + b_{Fj} u_a + c_{Fj}) \tag{3.25}$$

Fifth, Adding the acceleration resistance F_j, rolling resistance, air resistance $F_f + F_w$, and slope resistance F_i together to get the driving force F_t, and then get the relationship between driving force F_t (N) and vehicle speed u_a (km/h), that is:

$$\begin{aligned} F_t &= F_j + F_f + F_w + F_i = \delta m \frac{du}{dt} + (F_f + F_w) + mg\frac{i}{\sqrt{i^2+1}} \\ &= \delta m (a_{Fj} u_a^2 + b_{Fj} u_a + c_{Fj}) + (a_{FfFw} u_a^2 + b_{FfFw} u_a + c_{FfFw}) + \left(mg\frac{i}{\sqrt{i^2+1}} \right) \end{aligned} \tag{3.26}$$

Sixth, Calculating the relationship between engine torque and speed external characteristics.

Converting the vehicle linear speed to the revolution speed, and the conversion formula is as follows:

$$u_a = \frac{0.377rn}{i_g i_0} \tag{3.27}$$

Converting the driving force to the engine driving torque, and the conversion formula is as follows:

$$T = \frac{F_t r}{i_g i_0 \eta_T} \tag{3.28}$$

Bring formula (3.26) into formula (3.28) to get the formula between engine torque and revolution speed:

$$\begin{aligned} T &= \frac{\delta m \left(a_{Fj} u_a^2 + b_{Fj} u_a + c_{Fj}\right) + \left(a_{FjFw} u_a^2 + b_{FjFw} u_a + c_{FjFw}\right) + \left(mg\frac{i}{\sqrt{i^2+1}}\right)}{i_g i_0 \eta_T} r \\ &= \frac{\delta m \left[a_{Fj}\left(\frac{0.377rn}{i_g i_0}\right)^2 + b_{Fj}\left(\frac{0.377rn}{i_g i_0}\right) + c_{Fj}\right] + \left[a_{FjFw}\left(\frac{0.377rn}{i_g i_0}\right)^2 + b_{FjFw}\left(\frac{0.377rn}{i_g i_0}\right) + c_{FjFw}\right] + \left(mg\frac{i}{\sqrt{i^2+1}}\right)}{i_g i_0 \eta_T} \end{aligned}$$

$$= \left[\left(\delta m a_{Fj} + a_{FfFw}\right)\left(\frac{0.377^2 r^3}{i_g^3 i_0^3 \eta_T}\right)\right] n^2 + \left[\left(\delta m a_{Fj} + a_{FfFw}\right)\left(\frac{0.377 r^2}{i_g^2 i_0^2 \eta_T}\right)\right] n$$
$$+ \left(\frac{\delta m c_{Fj} + c_{FfFw} + mg\frac{i}{\sqrt{i^2+1}} r}{i_g i_0 \eta_T}\right) d \tag{3.29}$$

Seventh, Averaging the external characteristic curve of engine torque and revolution speed at the same altitude.

In the same altitude areas, first, obtaining multiple groups of data in the given gear, second, deleting the abnormal data from the data curve in the same gear and average them to obtain the engine torque and revolution speed characteristic curve in that gear, third, averaging the different gear curves at the same altitude to obtain the engine torque and revolution speed external characteristic curve at a certain altitude. Then the test and analysis results of engine torque and speed external characteristic curves at five altitudes are obtained.

3.3.2 Comparison of Engine External Characteristics at Different Altitudes

(1) **Six Axle Articulated Train**

The external characteristic curve of six-axle articulated train engine at different altitudes is shown in Fig. 3.8, and the relationship is shown in Formula (3.30) to Formula (3.34).

The engine external characteristics of Xi'ning (Altitude: 2300 m) are as follows:

$$T = -8.07974 \times 10^{-4} n^2 + 3.180009 \times 10^0 n - 1.425276 \times 10^3 \tag{3.30}$$

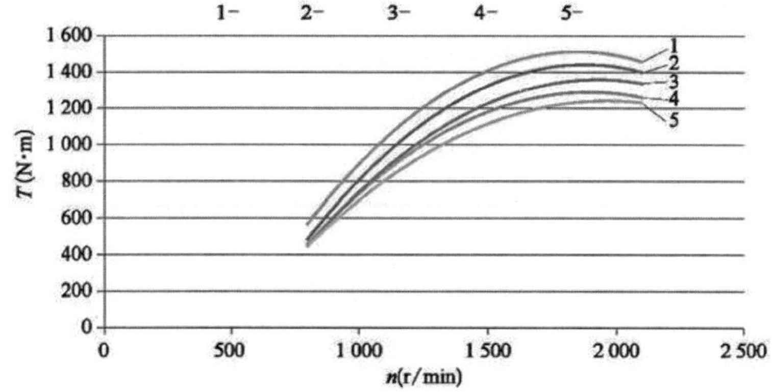

Fig. 3.8 Comparison of engine characteristic curves of articulated train at different altitudes

The engine external characteristics of Gonghe (Altitude: 2890 m) are as follows:

$$T = -8.271719 \times 10^{-4} n^2 + 3.103933 \times 10^0 n - 1.470650 \times 10^3 \quad (3.31)$$

The engine external characteristics of Xinghai (Altitude: 3638 m) are as follows:

$$T = -7.170240 \times 10^{-4} n^2 + 2.765362 \times 10^0 n - 1.307850 \times 10^3 \quad (3.32)$$

The engine external characteristics of Yushu (Altitude: 4188 m) are as follows:

$$T = -6.930665 \times 10^{-4} n^2 + 2.627628 \times 10^0 n - 1.201803 \times 10^3 \quad (3.33)$$

The engine external characteristics of Ma'duo (Altitude: 4545 m) are as follows:

$$T = -5.858343 \times 10^{-4} n^2 + 2.303183 \times 10^0 n - 1.021282 \times 10^3 \quad (3.34)$$

Table 3.17 shows the comparison of maximum torque of articulated train engine at different altitudes. The relationship between the maximum driving torque and the altitude is shown in Fig. 3.9.

(2) **Four-Axle Truck**

At five level of altitudes, 2,300 m, 2,890 m, 3,638 m, 4,188 m and 4,545 m, the external characteristic curve of truck engine is shown in Fig. 3.10, and the relationship is shown in Formula (3.35) to Formula (3.39).

The engine external characteristics of Xi'ning (Altitude: 2300 m) are as follows:

$$T = -1.446873 \times 10^{-3} n^2 + 4.541418 \times 10^0 n - 2.286501 \times 10^3 \quad (3.35)$$

The engine external characteristics of Gonghe (Altitude: 2890 m) are as follows:

$$T = -1.598071 \times 10^{-3} n^2 + 5.076007 \times 10^0 n - 2.830965 \times 10^3 \quad (3.36)$$

The engine external characteristics of Xinghai (Altitude: 3638 m) are as follows:

Table 3.17 Comparison of maximum engine torque ratio of articulated trains at different altitudes

Altitude (m)	Maximum torque (N m)	Maximum torque ratio (%)
0	1,700	100
2,300	1,511.659	88.92157
2,890	1,441.199	84.7761
3,638	1,358.457	79.90911
4,188	1,288.712	75.80626
4,545	1,242.429	73.08267

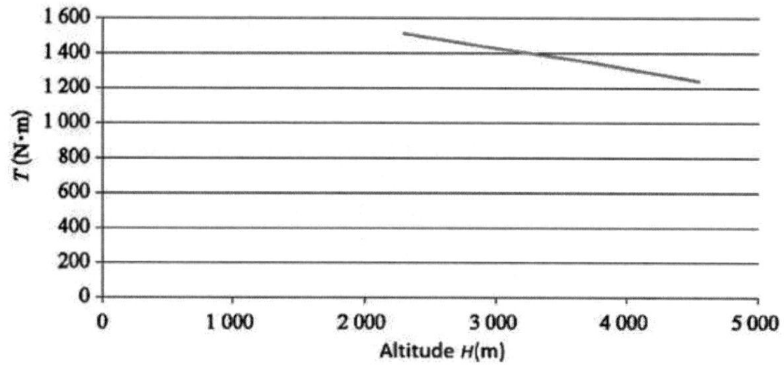

Fig. 3.9 Relation curve of engine torque of articulated train changing with altitude

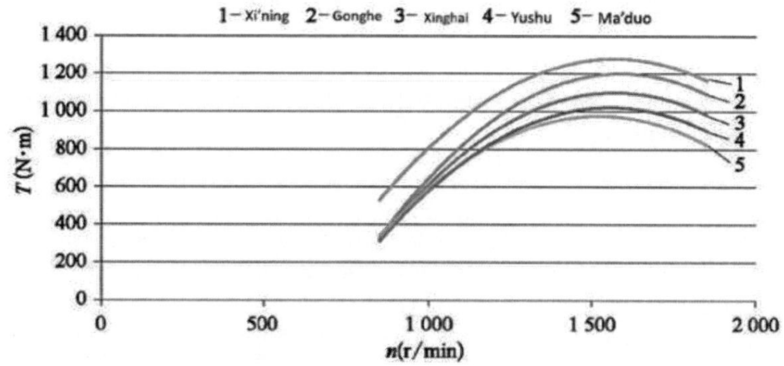

Fig. 3.10 Comparison of external characteristic curves of truck engines at different altitude

$$T = -1.488408 \times 10^{-3}n^2 + 4.674565 \times 10^0 n - 2.571938 \times 10^3 \tag{3.37}$$

The engine external characteristics of Yushu (Altitude: 4188 m) are as follows:

$$T = -1.428507 \times 10^{-3}n^2 + 4.445911 \times 10^0 n - 2.437557 \times 10^3 \tag{3.38}$$

The engine external characteristics of Ma'duo (Altitude: 4545 m) are as follows:

$$T = -1.412456 \times 10^{-3}n^2 + 4.300680 \times 10^0 n - 2.297679 \times 10^3 \tag{3.39}$$

Table 3.18 shows the comparison of the maximum torque of the truck engine at different altitudes. The relationship between maximum torque and altitude is shown in Fig. 3.11.

Table 3.18 Comparison of maximum torque of truck engine under different altitudes

Altitude (m)	Maximum torque (N m)	Maximum torque ratio (%)
0	1,500	100
2,300	1,277.121	85.1396
2,890	1,199.807	79.98602
3,638	1,098.338	73.222
4,188	1,021.669	68.11045
4,545	975.9971	65.06812

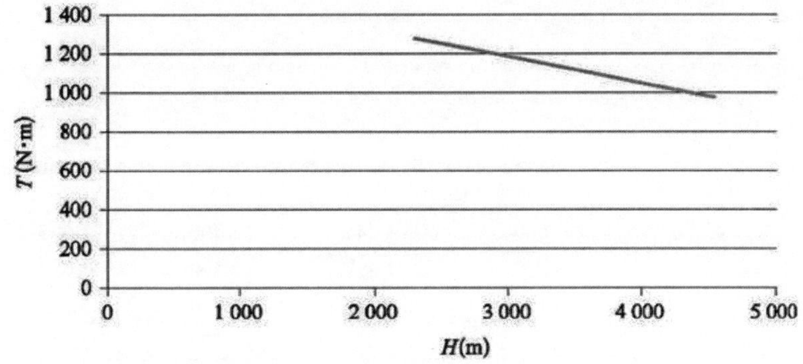

Fig. 3.11 Relational curve of maximum torque of truck engine changing with altitude

3.3.3 Analysis of Engine Torque Reduction Coefficient

(1) **Six-Axle Articulated Train**

In order to study the torque reduction coefficient of six-axle articulated train engine at specified speed at different altitudes, considering the limitations of 0 m altitude and test section, the external characteristic curve of the engine at 0 m altitude is provided by the manufacturer. The external characteristics of the engine provided by the engine manufacturer are shown in Fig. 3.12, and the relationship between revolution speed and torque is shown in Table 3.19.

It can be seen from Table 3.19 that when the altitude is 0 m, the revolution speed is 1,500 r/min, the engine torque of articulated train is 1,670 N m.

When the revolution speed of articulated train is 1,500 r/min, the torque comparison at different altitudes is shown in Table 3.20, and the relationship between torque and altitude is shown in Fig. 3.13.

(2) **Four-Axle Truck**

In order to study the torque reduction coefficient of four-axle truck engine at specified speed at different altitudes, considering the limitations of 0 m altitude and test section, the external characteristic curve of the engine at 0 m altitude is provided by

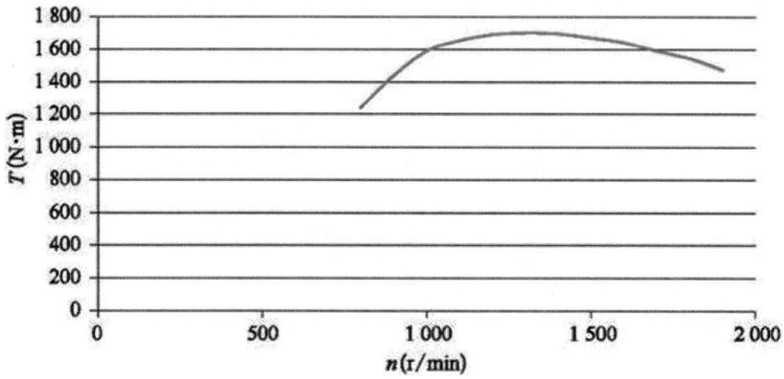

Fig. 3.12 External characteristic curve of articulated train engine

Table 3.19 Comparison of engine speed and torque of articulated train

Revolution speed (r/min)	Torque (N m)	Revolution speed (r/min)	Torque (N m)
800	1,241	1,400	1,695
900	1,437	1,500	1,670
1,000	1,589	1,600	1,641
1,100	1,653	1,700	1,591
1,200	1,689	1,800	1,544
1,300	1,700	1,900	1,475

Table 3.20 Torque reduction factor comparison of articulated train at 1500 r/min at different altitudes

Altitude (r/min)	Torque (N m)	Torque reduction factor (r/min) (%)
0	1,670	100.00
2,300	1,408.57	84.35
2,890	1,321.45	79.13
3,638	1,226.89	73.47
4,188	1,180.24	70.67
4,545	1,115.36	66.79

the manufacturer. The external characteristics of the engine provided by the engine manufacturer are shown in Fig. 3.14. The relationship between revolution speed and torque is shown in Table 3.21.

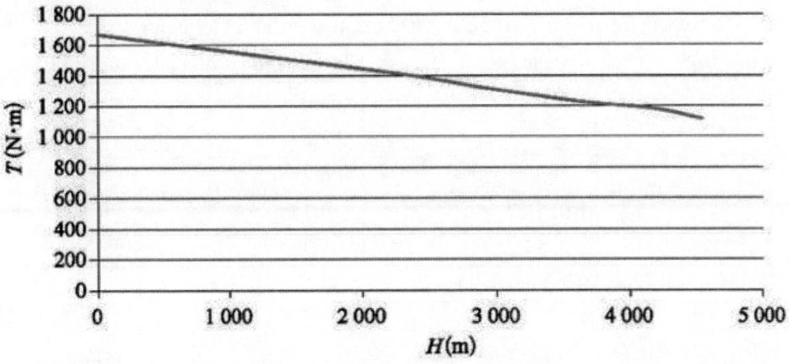

Fig. 3.13 Torque curve of articulated train with altitude at revolution speed 1500 r/min

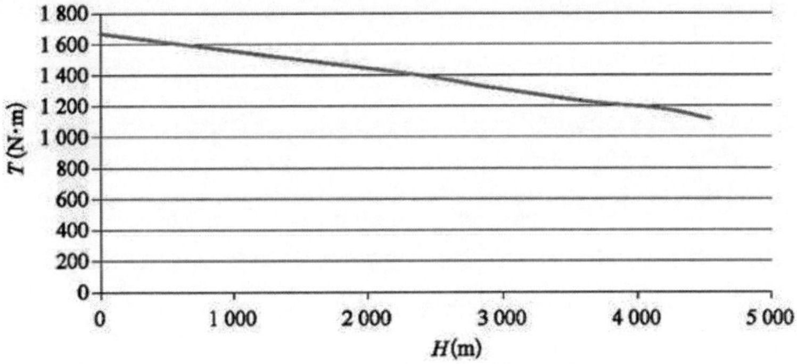

Fig. 3.14 Relational curve of external characteristics of truck engine

Table 3.21 Comparison of truck engine revolution speed and torque

Speed of revolution (r/min)	Torque (N m)	Speed of revolution (r/min)	Torque (N m)
800	1,100	1,400	1,500
900	1,270	1,500	1,500
1,000	1,400	1,600	1,460
1,100	1,500	1,700	1,405
1,200	1,500	1,800	1,327
1,300	1,500	1,900	1,258

It can be seen from Table 3.21 that the truck engine torque is 1,500 N m when the altitude is 0 m and the revolution speed is 1,400 r/min.

When the revolution speed of the truck is 1,400 r/min, the torque comparison at different altitudes is shown in Table 3.22, and the torque relational curve along with altitude is shown in Fig. 3.15.

Table 3.22 Torque comparison of truck at 1400 r/min at different altitudes

Altitude (r/min)	Torque (N m)	Torque reduction factor (r/min) (%)
0	1,500	100.00
2,300	1,235.61	82.37
2,890	1,143.23	76.22
3,638	1,055.17	70.34
4,188	986.84	65.79
4,545	954.86	63.66

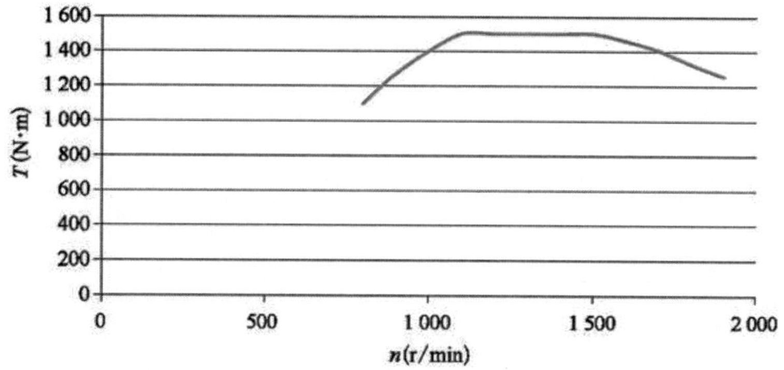

Fig. 3.15 Torque relational curve with altitude at 1400 r/min

(3) Result Analysis and Formula Fitting

The reduction factors of six-axle articulated train test vehicle and four-axle truck test vehicles are summarized in Table 3.23.

It can be seen from Table 3.23 that the corresponding reduction coefficients of the two test models have little difference, which supports the compose that supercharged and intercooled engines can be fitted and unified into an engine torque reduction formula. The difference in reduction rates has the following two reasons:

Table 3.23 Summary of engine torque reduction factors for two test models

Altitude (m)	Six-axle vehicle reduction factor (%)	Four-axle vehicle reduction factor (%)	Difference	Relative difference (%)
0	1.000	1.000	0.000	0.00
2,300	0.844	0.824	0.020	2.35
2,890	0.791	0.762	0.029	3.68
3,638	0.735	0.703	0.031	4.26
4,188	0.707	0.658	0.049	6.91
4,545	0.668	0.637	0.031	4.69

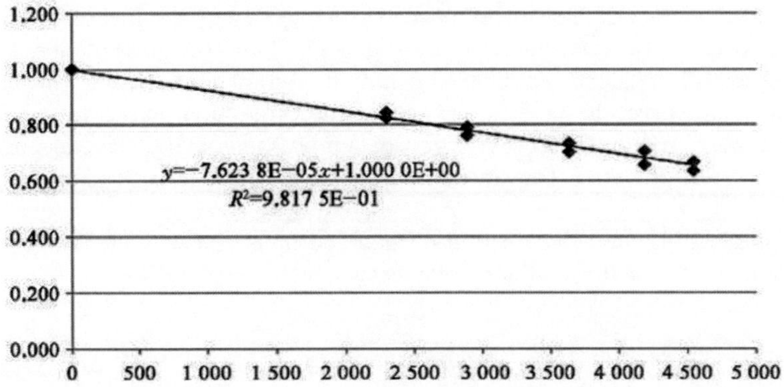

Fig. 3.16 Engine torque reduction coefficient of two test vehicles at different altitudes

first, certain errors cannot be avoided in all tests; Second, although they are all supercharged and intercooled engines (Generality), there are still some individual differences (Individuality).

The reduction coefficients of the two test models are used into fitting formula, and the fitting curve is shown in Fig. 3.16.

The formula is as follows:

$$k_p = 1 - 7.6238 \times 10^{-5} H \tag{3.40}$$

The engine torque reduction factor relative to the normal altitude (527.65 m) can be calculated as follows:

$$k_p = \frac{1 - 7.6238 \times 10^{-5} H}{1 - 7.6238 \times 10^{-5} \times 527.65} = \frac{1 - 7.6238 \times 10^{-5} H}{0.95977} \tag{3.41}$$

where,

H Altitude(m).

3.4 Engine Braking Characteristics

3.4.1 *Engine Braking Characteristic Test and Calculation Process*

The calculation steps of engine braking characteristics are as follows.

First, conducting engine braking road test on all gears of the test vehicle (4th, 5th, 6th, 7th, 8th and 9th gear). During the test, recording the data of speed and distance changes along with time, drawing the curve of speed time relationship according to the original data of speed u_a (km/h) and time t (s), and eliminate the abnormal points; At the same time, since the test site is a high altitude road section, and the general roads have slopes, in order to ensure the smoothness of the engine braking characteristic curve, the gradient change of the test section during the whole test process is ignored in the process of test data processing, and the average gradient during the test process is substituted into the slope resistance calculation Formula (3.5).

Second, fitting the vehicle speed u (m/s) and time t (s) with cubic polynomial function, and obtain the relationship between vehicle speed u (m/s) and time t (s) as follows:

$$u = a_1 t^3 + b_1 t^2 + c_1 t + d_1 \tag{3.42}$$

Third, derivatizing the vehicle speed u to obtain the relationship between deceleration a' (m/s^2) and time t (s) as follows:

$$a' = 3a_1 t^2 + 2b_1 t + c_1 \tag{3.43}$$

Fourth, multiplying the left and right sides of Formula (3.43) by −1 to get the functional relationship between the negative deceleration a = −a'(m/s^2) and time t (s). According to the relationship between the negative deceleration a' (m/s^2) and time t (s), the relationship between the speed u_a (km/h) and time t (s), the relationship curve between the negative deceleration a (m/s^2) and the speed u_a (km/h) can be obtained. Then, perform quadratic polynomial fitting between the negative deceleration a (m/ s ^{2}) and the speed u_a (km/h) to get the relationship formula as follows:

$$a = d_1 u_a^2 + e_1 u_a + f_1 \tag{3.44}$$

Fifth, multiplying the negative deceleration value a (m/s^2) by the total vehicle mass m (kg) and the rotating mass conversion coefficient δ, the total resistance F_{b_total}(N) is obtained, the total resistance is the sum of the braking force, rolling resistance, air resistance and slope resistance provided by the continuous braking system. In order to calculate the continuous braking force provided by the continuous braking system, the rolling resistance, air resistance and slope resistance must be subtracted to obtain the relationship between the continuous braking force F_{brake} (N) and the vehicle speed u_a (km/h) as follows:

$$\begin{aligned} F_{\mathrm{brake}} &= F_{\mathrm{b\_total}} - F_{\mathrm{w}} - F_{\mathrm{f}} - F_{\mathrm{i}} \\ &= (\delta m d_1 - a_{FfFw})\left(\frac{0.377nr}{i_g i_0}\right)^2 + (\delta m e_1 - b_{FfFw})\left(\frac{0.377nr}{i_g i_0}\right) + (\delta m f_1 - c_{FfFw}) \\ &\quad - mg\frac{i}{\sqrt{i^2+1}} \end{aligned} \tag{3.45}$$

Sixth, calculating the relationship between continuous braking torque and revolution speed.

Converting the vehicle speed to the revolution speed, and the formula is as follows:

$$u_a = \frac{0.377rn}{i_g i_0} \tag{3.46}$$

Converting the continuous braking force into the continuous braking torque provided by the continuous braking system. The formula is as follows:

$$T_b = \frac{F_{brake} r \eta_T}{i_g i_0} \tag{3.47}$$

By introducing formula (3.45) into formula (3.47), the relationship between continuous braking torque and speed is obtained as follows:

$$T_b = \frac{\left(\delta m d_1 - a_{FfFw}\right)\left(\frac{0.377nr}{i_g i_0}\right)^2 + \left(\delta m e_1 - b_{FfFw}\right)\left(\frac{0.377nr}{i_g i_0}\right) + \left(\delta m f_1 - c_{FfFw} - mg\frac{i}{\sqrt{i^2+1}}\right) r \eta_t}{i_g i_0} \tag{3.48}$$

3.4.2 *Comparison of Engine Braking Characteristics at Different Altitudes*

(1) **Six-Axle Articulated Train**

The engine braking characteristic curve at different altitudes is shown in Fig. 3.17. Formula (3.49) to Formula (3.53) shows the relationship.

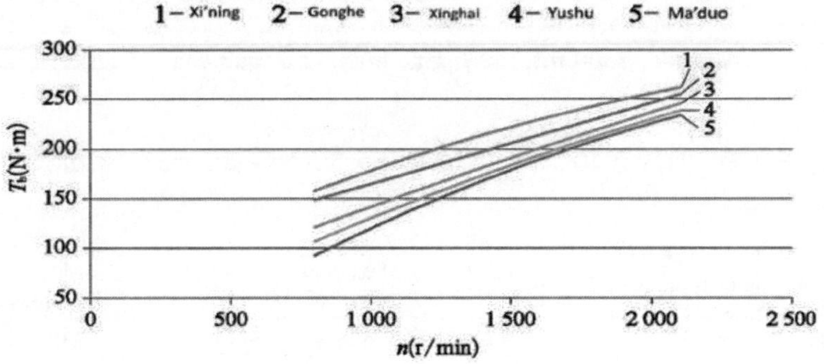

Fig. 3.17 Comparison curve of engine braking characteristic of articulated train at different altitude

Xi'ning (Altitude: 2300 m), the engine braking torque revolution speed relationship is as follows:

$$T_b = -1.909542 \times 10^{-5}n^2 + 1.348711 \times 10^{-1}n - 6.226588 \times 10^1 \quad (3.49)$$

Gonghe (Altitude: 2890 m), the engine braking torque-revolution speed relationship is as follows:

$$T_b = 8.375532 \times 10^{-7}n^2 + 7.912300 \times 10^{-2}n - 8.467711 \times 10^1 \quad (3.50)$$

Xinghai (Altitude: 3638 m), the engine braking torque-revolution speed relationship is as follows:

$$T_b = -6.257424 \times 10^{-6}n^2 + 1.138641 \times 10^{-1}n - 3.392294 \times 10^1 \quad (3.51)$$

Yushu (Altitude: 4188 m), the engine braking torque-revolution speed relationship is as follows:

$$T_b = -1.394100 \times 10^{-5}n^2 + 1.415179 \times 10^{-1}n + 2.449807 \times 10^0 \quad (3.52)$$

Ma'duo (Altitude: 4545 m), the engine braking torque-revolution speed relationship is as follows:

$$T_b = -2.414649 \times 10^{-5}n^2 + 1.784187 \times 10^{-1}n - 3.472807 \times 10^1 \quad (3.53)$$

Table 3.24 shows the comparison of maximum torque of articulated train engine at different altitudes. The relationship between the maximum braking torque and the altitude is shown in Fig. 3.18.

(2) **Four-Axle Truck**

The braking characteristic curve of four-axle truck engine at different altitudes is shown in Fig. 3.19. Formula (3.54) to Formula (3.58) shows the relationship between them.

Xi'ning (Altitude: 2300 m), the engine braking torque characteristic formula is as follows:

$$T_b = -4.439419 \times 10^{-5}n^2 + 2.028263 \times 10^{-1}n - 1.336312 \times 10^1 \quad (3.54)$$

Table 3.24 Comparison of maximum engine torque of articulated trains at different altitudes

Altitude (m)	Maximum torque (N m)	Altitude (m)	Maximum torque (N m)
2,300	261.2844	4,188	238.1575
2,890	254.529	4,545	233.4651
3,638	245.4423		

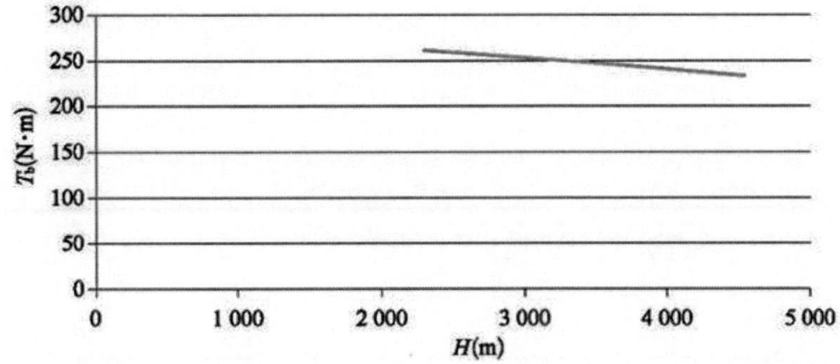

Fig. 3.18 Relational curve of braking torque of articulated train changing with altitude

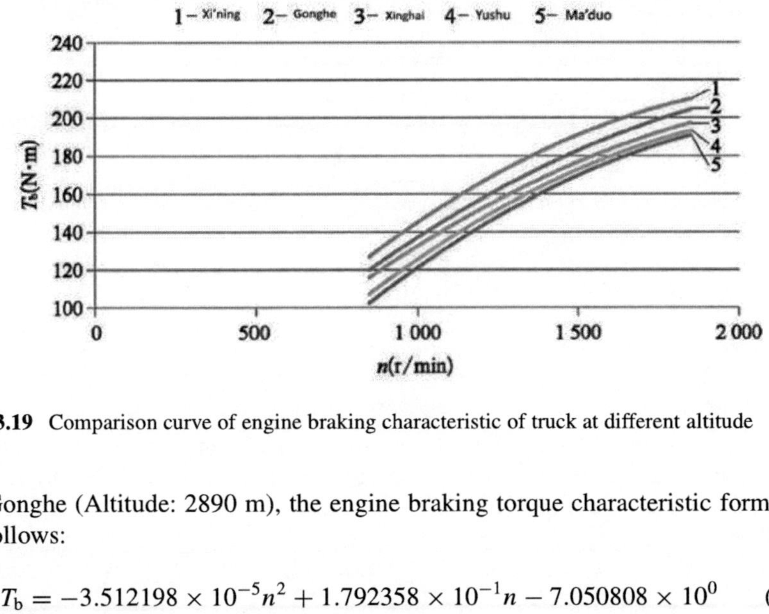

Fig. 3.19 Comparison curve of engine braking characteristic of truck at different altitude

Gonghe (Altitude: 2890 m), the engine braking torque characteristic formula is as follows:

$$T_b = -3.512198 \times 10^{-5} n^2 + 1.792358 \times 10^{-1} n - 7.050808 \times 10^0 \quad (3.55)$$

Xinhai (Altitude: 3638 m), the engine braking torque characteristic formula is as follows:

$$T_b = -3.452823 \times 10^{-5} n^2 + 1.745952 \times 10^{-1} n - 7.512564 \times 10^0 \quad (3.56)$$

Yushu (Altitude: 4188 m), the engine braking torque characteristic formula is as follows:

$$T_b = -4.255862 \times 10^{-5} n^2 + 2.007797 \times 10^{-1} n - 3.266302 \times 10^1 \quad (3.57)$$

Table 3.25 Comparison of maximum braking torque of truck engines downloaded at different altitude

Altitude (m)	Maximum torque (N m)	Altitude (m)	Maximum torque (N m)
2,300	209.9264	4,188	193.1226
2,890	204.3305	4,545	190.8735
3,638	197.3156		

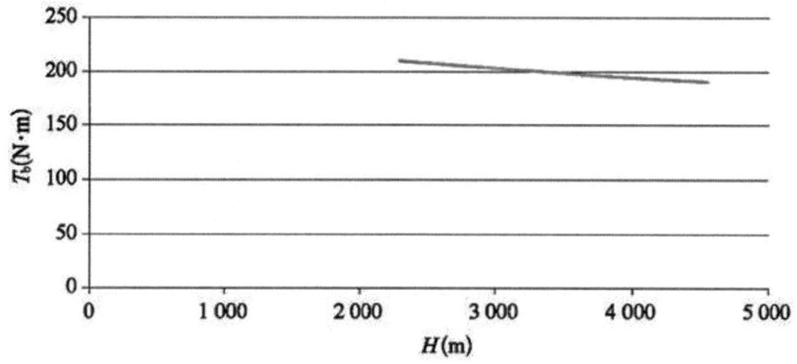

Fig. 3.20 Relational curve of maximum braking torque of truck changing with altitude

Ma'duo (Altitude: 4540 m), the engine braking torque characteristic formula is as follows:

$$T_b = -4.257117 \times 10^{-5} n^2 + 2.033151 \times 10^{-1} n - 3.955959 \times 10^1 \tag{3.58}$$

Table 3.25 shows the comparison of the maximum braking torque of truck engine at different altitudes. The relationship curve of the maximum braking torque with the altitude is shown in Fig. 3.20.

3.4.3 *Analysis of Engine Braking Torque Reduction Coefficient*

(1) **Six-Axle Articulated Train**

In order to study the reduction coefficient of braking torque of articulated train engine at specified speed at different altitudes, considering the limitations of 0 m altitude and test section, the project plans to calculate the engine braking torque at 0 m altitude according to the fitting formula of engine braking torque at 5 level of different altitudes, to calculate the reduction coefficient of engine braking torque (Table 3.26, Fig. 3.21, and Table 3.27).

Table 3.26 Torque comparison of articulated train at 1500 r/min at different altitudes

Altitude (m)	Torque (N m)	Altitude (m)	Torque (N m)
2,300	221.61	4,188	183.36
2,890	205.25	4,545	178.57
3,638	190.64		

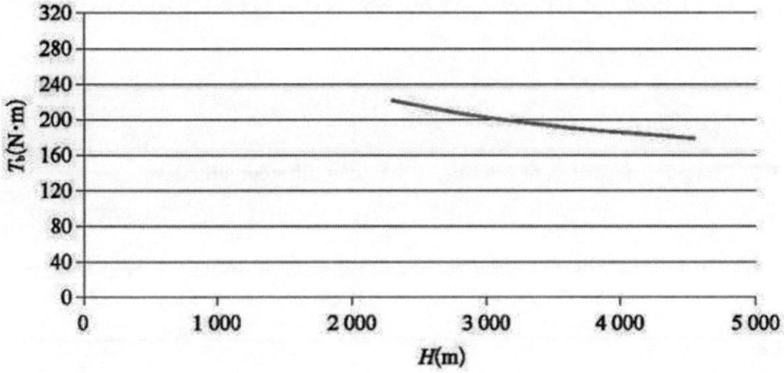

Fig. 3.21 Relational curve of braking torque with altitude at revolution speed 1500 r/min

Table 3.27 Torque reduction coefficient of articulated train at 1500 r/min at different altitudes

Altitude (m)	Torque (N m)	Reduction factor (%)
0	364.46	100.00
2,300	221.61	60.80
2,890	205.25	56.32
3,638	190.64	52.31
4,188	183.36	50.31
4,545	178.57	49.00

Fitting the torque corresponding to the revolution speed of 1500 r/min at different altitudes, the relationship between altitude and braking torque is as follows:

$$T_b = -1.425456 \times 10^{-9} H^3 + 1.920742 \times 10^{-5} H^2 - 9.873340 \times 10^{-2} H + 3.644617 \times 10^2 \tag{3.59}$$

According to the above formula, when the altitude is 0 m, the engine braking torque is 364.46 N m.

(2) **Four-Axle Articulated Truck**

In order to study the reduction coefficient of braking torque at constant speed of truck engine at different altitudes, considering the limitations of 0 m altitude area and test section, the project plans to calculate the dynamic torque of the engine at 0 m altitude according to the fitting formula of engine braking torque at five level of altitudes, to calculate the reduction coefficient of engine braking torque (Table 3.28, Fig. 3.22, and Table 3.29).

Fitting the torque corresponding to the truck speed of 1400 r/min at different altitudes, the relationship between braking torque and altitude is:

$$T_b = -2.185247 \times 10^{-9} H^3 + 2.401868 \times 10^{-5} H^2 - 9.465114 \times 10^{-2} H + 3.007888 \times 10^2 \quad (3.60)$$

Table 3.28 Torque comparison of truck at 1400/min at different altitudes

Altitude (m)	Torque (N m)	Altitude (m)	Torque (N m)
2,300	183.58	4,188	165.01
2,890	175.04	4,545	161.64
3,638	169.25		

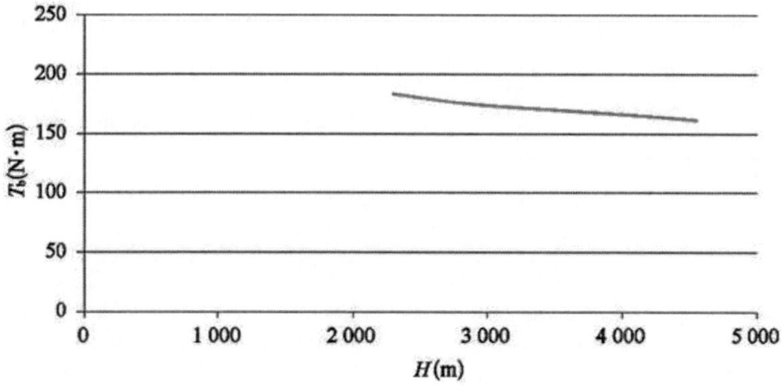

Fig. 3.22 Relational curve of braking torque with altitude at 1400/min

Table 3.29 Torque reduction coefficient of truck at 1400 r/min at different altitudes

Altitude (m)	Torque (N m)	Reduction factor (%)
0	300.78	100.00
2,300	183.58	61.03
2,890	175.04	58.20
3,638	169.25	56.27
4,188	165.01	54.86
4,545	161.64	53.74

It can be seen from the above formula that the engine braking torque is 300.78 N m when the altitude is 0 m and the revolution speed is 1400 r/min.

(3) Result Analysis and Formula Fitting

The reduction factors of six-axle test vehicle and four-axle test vehicles are summarized in Table 3.30.

The engine brake torque of the two test vehicles is similar to the engine torque, the corresponding reduction coefficient has little difference, which supports the compose of fitting a unified engine brake torque reduction formula.

The fitting of reduction coefficient of two test vehicle types is shown in Fig. 3.23. The formula is as follows:

$$k_p = 1 - 1.6667 \times 10^{-4} H + 1.2000 \times 10^{-8} H^2 \tag{3.61}$$

Relative to the normal altitude (527.65 m), the engine braking torque reduction factor can be calculated as follows:

Table 3.30 Summary of engine baking torque reduction factors for two test models

Altitude (m)	Six-axle vehicle reduction factor (%)	Four-axle vehicle reduction factor (%)	Difference	Relative difference (%)
0	1.000	1.000	0.000	0.00
2,300	0.608	0.610	−0.002	−0.38
2,890	0.563	0.582	−0.019	−3.34
3,638	0.523	0.563	−0.040	−7.57
4,188	0.503	0.549	−0.046	−9.04
4,545	0.490	0.537	−0.047	−9.67

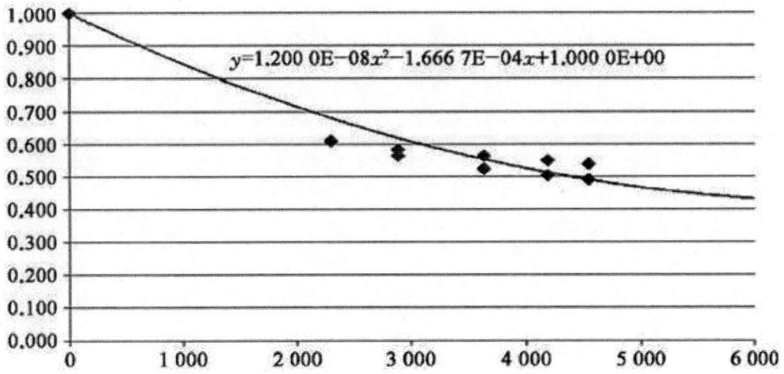

Fig. 3.23 Engine braking torque reduction coefficient of two test vehicles under different altitudes

$$
\begin{aligned}
k_{\mathrm{p}} &= \frac{1 - 1.6667 \times 10^{-4} H + 1.2000 \times 10^{-8} H^2}{1 - 1.6667 \times 10^{-4} \times 527.65 + 1.2000 \times 10^{-8} \times 527.65^2} \\
&= \frac{1 - 1.6667 \times 10^{-4} H + 1.2000 \times 10^{-8} H^2}{0.91540}
\end{aligned}
\tag{3.62}
$$

where,

H Altitude (m).

Through five altitudes of 2,300 m (Xi'ning), 2,890 m (Gonghe), 3,638 m (Xinghai), 4,188 m (Yushu), 4,545 m (Ma'duo) along the Qinghai–Tibet Highway, the acceleration driving test and engine braking deceleration driving test of six-axle articulated train (power weight ratio 5.6 kW/t) and four-axle truck (power weight ratio 8.3 kW/t), under full load were conducted, and the external characteristic curves of the engine at various altitudes were obtained. It can be seen from the curves that the engine torque reduced with the increase of altitude, the maximum torque reducing, and the altitude and engine torque are decreasing. With the increase of altitude, the engine braking torque characteristic decreases, the maximum braking torque decreases, and the altitude and engine braking torque show a decreasing trend.

Aiming at the driving characteristics of two test vehicles in high altitude areas, the engine external characteristics model, engine braking characteristics model, engine driving torque variation model and engine braking torque variation model with altitude under full load of two test vehicles at different altitudes are established. The test model established in the study can accurately reflect the uphill and downhill capacities of the 8.3 kW power weight ratio truck at different altitudes. The research results have practical application significance and research value, and provide the experimental and theoretical basis for the highway design and construction departments to select the slope when conducting line design at high altitudes. It also provides technical guidance for truck drivers to drive safely on continuous downhill roads in high altitude areas.

Chapter 4
Characteristics of Psychophysiological Change of Drivers in the Low Pressure and Oxygen Deficient Environment in High Altitude Area

According to the influence of the traffic operation environment on the drivers in high altitude areas, in this chapter we selected the motor vehicle drivers with different driving experience on the highways on Qinghai-Tibet Plateau, collected the heart rate, eye movement, electromyography, blood oxygen saturation and other indicators at different elevations (3,000–5,000 m) through driving simulation tests and real vehicle tests. After analysis and screening, indicators whose heart rate are easy to monitor during the test and less affected by individual differences are selected as the research objects, the change rules of drivers' perception and operation ability and fatigue characteristics in high altitude and low oxygen environment were analyzed, and the driver heart rate change model and driver fatigue model in low oxygen environment were established, which provides a basis for proposing that driver's physiological and physiological status affects the geometric design indicator, parameters and reasonable spacing of service facilities.

4.1 Drivers' Perception and Operation Ability in Low Oxygen Environment

4.1.1 Test Plan Design

4.1.1.1 Test Plan for Driver's Responsiveness

(1) **Mobile Driving Simulation Platform**

In the test the Mobile Driving Simulator (Fig. 4.1) was used which was developed by CCCC First Highway Consultants Company based on the latitude terrain roaming system. The built-in scene of the simulator is consistent with the G109 Qinghai-Tibet Highway section K2866–K3450, with a total length of 584 km.

J. Liu, *Technical Indicators and Safety Design of Freeway in High Altitude Area*,
https://doi.org/10.1007/978-981-99-0620-8_4

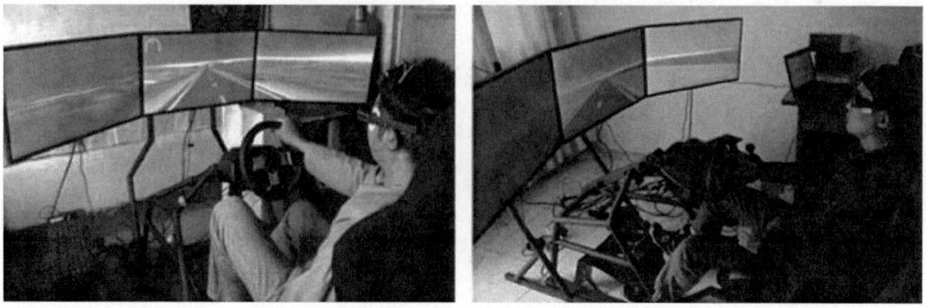

Fig. 4.1 Driving simulation test site

(2) **Reaction Speed Test Software**

Testing the driver's reaction time by using the "reaction speed test" function module of the mobile phone APP (Fig. 4.2). Starting by touching the middle of the square, and click again when the background color of the square changes, the software will display the reaction time spent by the test subject. The software measures the time from the change of the background color of the square to the time when the subject clicks the mobile phone screen, which belongs to the simple reaction time, but basically meets the test requirements.

(3) **Test Locations**

A total of three locations with different elevations were selected for the test, namely Nachitai (3,540 m altitude), Xidatan (4,150 m altitude) and Tuotuo River (4,533 m altitude).

(4) **Test Driver**

The test driver shall be a driver from other places with rich driving experience and a driving license of C or above. Considering the particularity and convenience of the test, the test driver shall be mainly male, aged 20–50 years, driving for more than 5 years, and free from cardiovascular distresses, so as to ensure that the driving behavior is closer to the actual situation.

(5) **Test Plan**

The test steps are as follows:

i. Preparations. The test-driving simulator operates normally. After the commissioning of all test instruments is completed, explain the basic process and principle of the test, the use of test instruments and precautions for the subjects, and make records.

Fig. 4.2 Reaction speed test software

ii. The subjects fill in the driver information registration form, test the reaction time before the test and record it. The test can be started after the current test subject gets the driving simulator in place and the instrument debugging and calibration is completed. The rest of the subjects sit aside and wait quietly.

iii. The test subject starts the test vehicle from the starting point, drives it for 2–2.5 h, stops and closes all test instruments, stores the data of all test instruments, ensures that the test data is not missing or abnormal due to instrument problems, tests the reaction time of the test subject again, and asks the test subject to fill in the test feedback form. After the first test is completed, the second test subject is replaced for the test until all test subjects complete the test.

(6) **Test Sample Size**

See Table 4.1 for the final number of samples collected from each test site.

Table 4.1 Number of people tested at each test site

Test location	Number of people
Nachitai (3,540 m altitude)	10
Xidatan (4,150 m altitude)	9
Tuotuo river (4,533 m altitude)	9

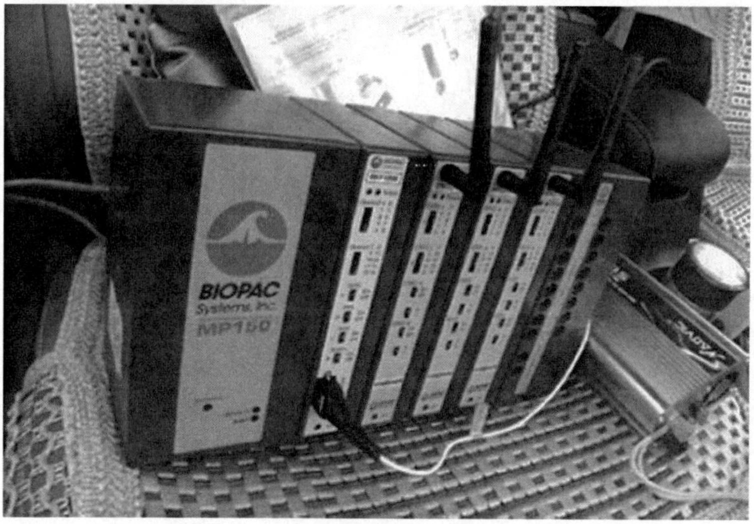

Fig. 4.3 M150 multi conductive physiological recorder

4.1.1.2 Drivers' Psychophysiological Data Collection Under Hypoxia Environment Based on Real Vehicle Test

The on-site vehicle test is adopted to record the drivers' psychophysiological parameters during the test, and the measurement indicators that can significantly and objectively reflect the drivers' psychophysiological state are proposed.

(1) **Test Instrument**

1. **Multi Conductive Physiological Recorder**

In this test the MP150 Multi Conductivity Physiological Recorder (16 channels) (Fig. 4.3) is used, which is made by BIOPAC Company in the United States. The system is composed of MP150 host, amplifier, transducer, bioelectricity wire and filter working in nuclear magnetic environment.

The amplifier selected for this real vehicle test is ECG100C ECG amplifier, which is used to collect the change of driver's heart rate.

2. **Dynamic GPS**

In this test it uses the Tianbao Trimble SPS351 beacon to collect the real-time three-dimensional position, altitude and speed of the test vehicle.

3. **Test Vehicle**

According to the analysis of accident data of the test section in recent five years and the actual vehicle type composition on site, Toyota RAV4 miniature multi-function off-road vehicle (4.265 m × 1.785 m × 1.705 m) was selected as the test vehicle.

(2) **Test Section**

This section (Fig. 4.4) starts from Nachitai, with an altitude of 3,540 m, and ends at Kunlun Mountain Pass, with an altitude of 4,776 m, chainage of K2828–K2893, and a total length of 64.7 km. It is a Class II highway. It runs in both directions without a median. The design speed is 60–80 km/h.

(3) **Test Driver**

The test driver shall be a driver from other places with rich driving experience and a driving license of C or above. Considering the particularity and convenience of the test, the test driver shall be mainly male, aged 20–50 years, driving for more than 5 years, and free from cardiovascular distresses, so as to ensure that the driving behavior is closer to the actual situation.

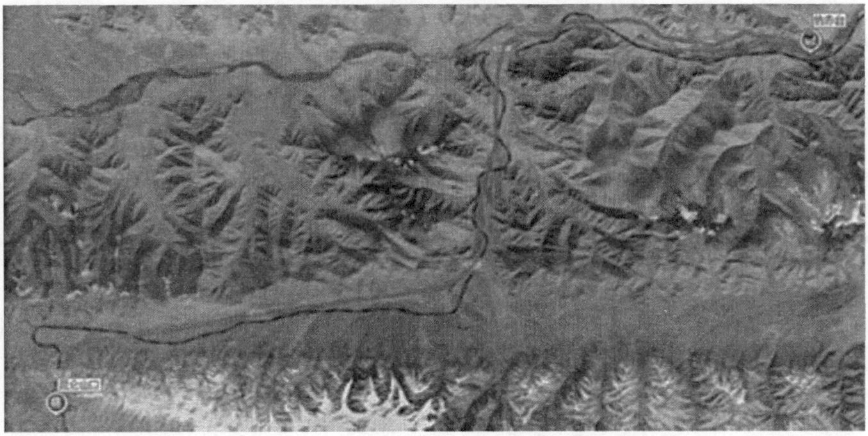

Fig. 4.4 Schematic diagram of section Nachitai-Kunlun mountain pass

4.1.2 Change Characteristics of Drivers' Reaction Ability

(1) Selection of Influencing Factors and Indicators

Based on the summary of domestic and foreign research literature, the main influencing factors of drivers' perception and operating ability in high altitude regions include altitude (unit: m), driver age and driving time (unit: h).

(2) Relationship between Altitude and Drivers' Response Time

The initial reaction time of drivers at the three altitude test sites before driving simulation test is drawn into a Box-plot (Fig. 4.5). Except for individual abnormal values, the driver's reaction time increases with that of altitude, from 3,540 to 4,150 m, and from 4,150 to 4,533 m, the reaction time increases by about 29%.

(3) Relationship between Drivers' Age and Reaction Time

The subjects were divided into four groups according to their ages: under 30 years old, 30–40 years old, 40–50 years old and over 50 years old. The relationship diagram with the initial response time of drivers (Fig. 4.6) was drawn. Through comparison, it was found that the response time of drivers of different ages at the same altitude was different. The initial response time of older subjects was slightly greater than that of younger subjects, but it was not obvious.

(4) Relationship between Driving Duration and Reaction Time Difference

Taking the difference valve between the reaction time measured before and after the driving simulation test as the reaction time difference Δt (ms). The driving time of drivers is divided into three groups: 1–1.5 h, 1.5–2 h and 2–2.5 h. The relationship between the driving time and the reaction time difference is drawn, as shown in Fig. 4.7.

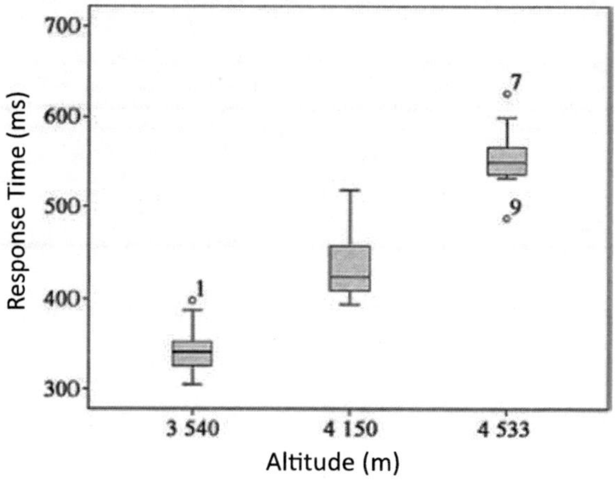

Fig. 4.5 Relationship between altitude and driver response time

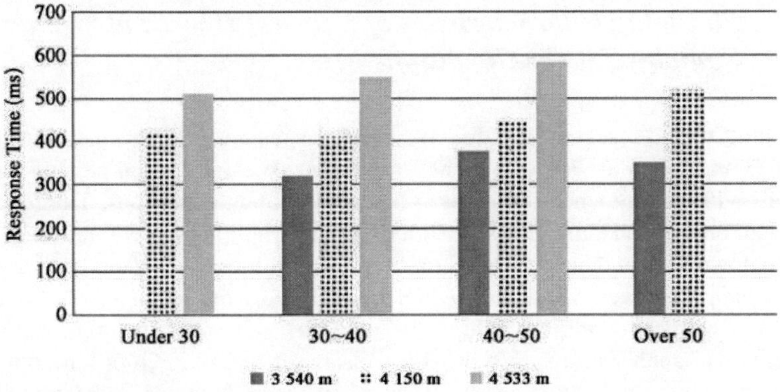

Fig. 4.6 Initial reaction time at different ages

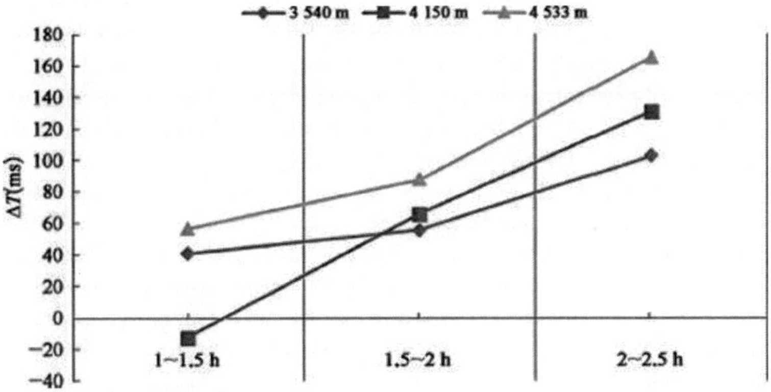

Fig. 4.7 Relationship between driving duration and reaction time

There are three conclusions can be drawn from Fig. 4.7: (i) When the driving time is 1–1.5 h, the drivers' reaction time is slightly affected by the driving time, and some subjects show a phenomenon of faster reaction. (ii) When the driving time is 1.5–2 h, the drivers' reaction time is extended by 70 ms on average; the driving time is 2–2.5 h, and the average reaction time is extended by 133 ms. (iii) When the driving time is more than 1.5 h, the drivers' reaction time interval at high altitude is greater than that at low altitude, indicating that the higher the altitude is, the greater the drivers' psychophysiological load is, and the longer driving time is more likely to cause driver fatigue.

4.1.3 *Changing Characteristics of Drivers' Heart Rate Under Complex Linear Conditions*

According to the research, the drivers' speed is generally lower than that in low altitude areas when driving in high altitude areas, especially when passing through sections with complex horizontal and vertical alignment conditions. The following will focus on the road alignment indicators in high altitude areas, combine the drivers' psychophysiological characteristics, altitude environmental factors and running speed change characteristics, study the drivers' heart rate change rule under the complex alignment conditions in high altitude areas, and establish the corresponding mathematical analysis model, to provide a basis for the research of horizontal and vertical alignment indicators that meet the drivers' perception and operating ability.

(1) **Selection of Influencing Factors and Indicators**

Based on the summary of domestic and foreign research literature, it can be concluded that the altitude, road alignment and vehicle speed will have an impact on the drivers' psychology and physiology when the vehicle is driving in the plateau area. Among them, the drivers' psychology and physiology indicators choose the heart rate growth rate for two reasons: (i) The heart rate change indicator itself can best reflect the drivers' tension during driving, and the indicator is relatively less subject to external interference during the collection process, the data accuracy is reliable, which is convenient for later model analysis; (ii) Since the initial heart rate of each person is different in calm state, and the maximum difference can reach more than 20 times/min, the direct use of the real-time heart rate value of the driver during operation will have an impact on the final research results, so the heart rate growth rate is used as the final indicator to evaluate the change of the driver's tension, and its calculation method is as follows:

$$N = \frac{n' - n}{n} \times 100\% \tag{4.1}$$

where,

N Instantaneous heart rate growth rate of the driver at a certain time during driving;
n' Instantaneous heart rate of the driver at a certain time during driving;
n Average heart rate of driver in static state.

(2) **Section Alignment Division**

With reference to the alignment division standards of road sections in the *Guidelines for Safety Audit of Highway* (2004 Edition), and in combination with the actual alignment indicators of roads in high altitude areas, the radius of horizontal curve of 1,000 m and the longitudinal slope of $\pm 2\%$ are taken as the critical values for alignment division, and the number of horizontal curve sections, longitudinal slope sections and curved slope combination sections of section K2828–K2893 is finally obtained, as shown in Table 4.2.

Table 4.2 Division of road sections

Section type	Horizontal curve section	Longitudinal slope section	Curved slope combination section
Linear indicators	R ≤ 1000 m, lil ≤ 2%	R > 1000 m, 1H > 2%	R ≤ 1000 m, lil > 2%
Quantity	46	22	49

(3) The Change Rule of Drivers' Heart Rate under Different Linear Combinations

1. Flat Curved Line Section

In the plateau environment, with the increase of the radius of the horizontal curve, the heart rate growth rate changes generally in a downward trend, but different from the plain area, the test section data shows two curves with obvious changes after being plotted as a scatter plot. At 3,567–3,957 m altitude and 4,100–4,709 m altitude, the heart rate growth rate and the radius of the flat curve are negatively correlated. Within the range of 3,567–3,957 m above sea level, the heart rate growth rate decreases from 22.31 to 17.38%, with a decrease of about 22.1% (Fig. 4.8); in the range of 4,100–4,709 m above sea level, the heart rate growth rate decreased from 35.47% to 29.23%, with a decrease of about 17.6% (Fig. 4.9).

From Figs. 4.8 and 4.9, it is also found that there are obvious differences in the corresponding ordinate values of some of the same radii, which is mainly due to the influence of altitude. We choose the heart rate change rate value when the radius of the horizontal curve with the most sample points is 300 m, as shown in Fig. 4.10.

It can be seen from Fig. 4.10 that under the same radius of the horizontal curve, altitude becomes the main factor affecting the heart rate growth rate. Due to the small number of sample points, it is difficult to fit the corresponding statistical equation,

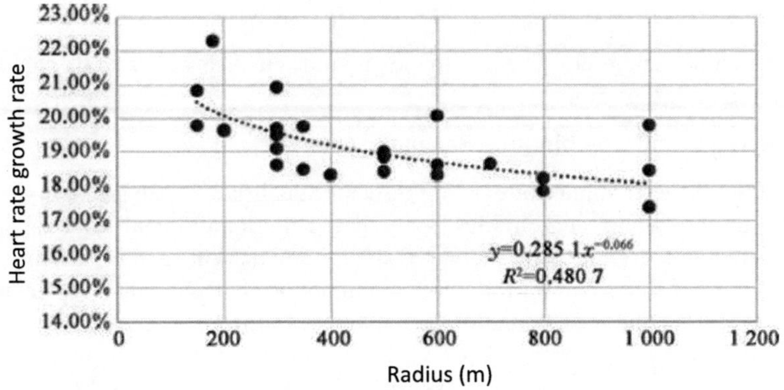

Fig. 4.8 Relationship between horizontal curve radius and heart rate growth rate at 3567–3957 m altitude

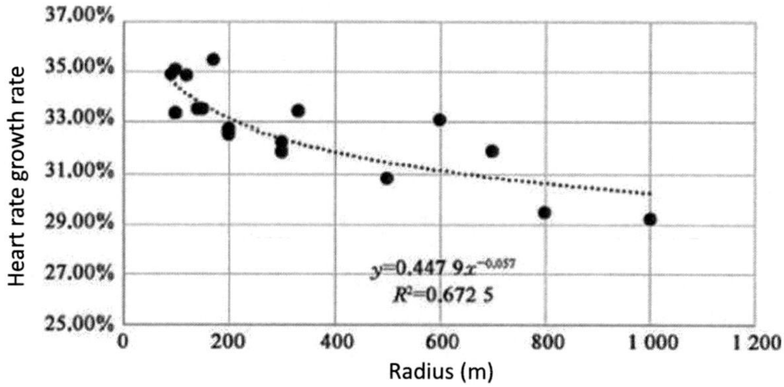

Fig. 4.9 Relationship between horizontal curve radius and heart rate growth rate at 4100–4709 m altitude

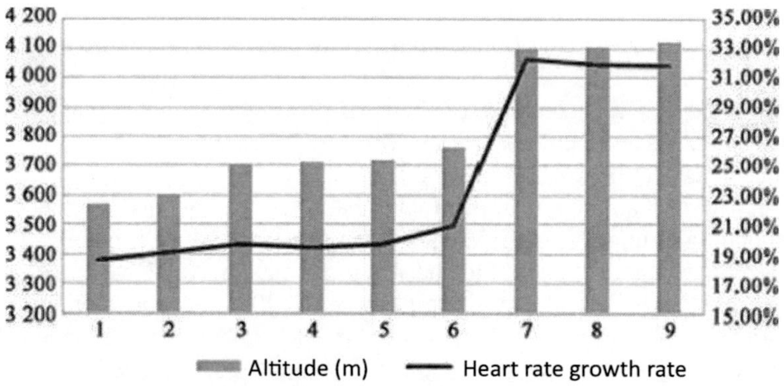

Fig. 4.10 Analysis on the change of heart rate growth rate with the radius of horizontal curve at 300 m

but the overall trend is consistent with the previous analysis of heart rate change rate and altitude relationship.

2. **Curved Slope Combination Section**

According to the summary of existing research, qualitative analysis shows that the smaller the horizontal curve radius is, the larger the longitudinal slope is, and the more dangerous driving is on the curve slope combination section. Therefore, the driving state of the driver on the linear combination road section is represented in the form of $|i/R|$, and the calculation formula is as follows:

$$W = |i/R| \tag{4.2}$$

where,

W Linear combination value;
i Longitudinal slope (%);
R Curve radius (km).

It can be seen from Figs. 4.11 and 4.12 that, with the increase of the linear combination value, the heart rate growth rate shows a slow growth trend. The fitting degrees of the two curves are 0.2216 and 0.4901, respectively. The fitting degree is relatively low. The reasons are as follows: (i) There are fewer curved slope combination sections in section K2828–K2893, which leads to too few sample points that can be collected, so the statistical law is not obvious; (ii) There is a sample point (421.33, 19.12%) in Fig. 4.11, due to its low altitude, although the linear combination value of it is large, the heart rate growth rate is relatively low.

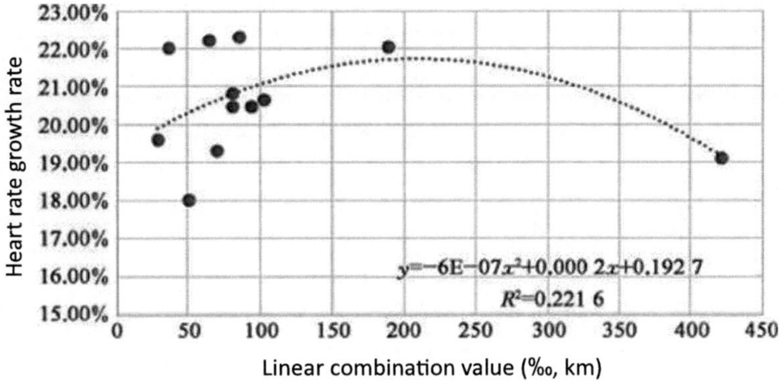

Fig. 4.11 Relationship between heart rate growth rate and linear combination value at 3616–3998 m altitude

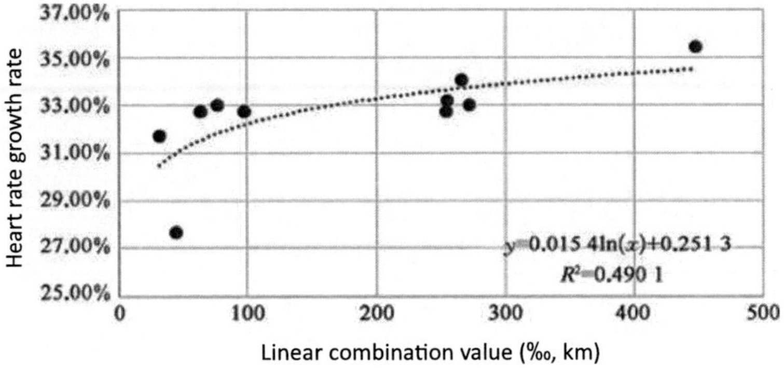

Fig. 4.12 Relationship between heart rate growth rate and linear combination value at 4127–4630 m altitude

3. **Longitudinal Slope Section**

It can be seen from Figs. 4.13 and 4.14 that the heart rate growth rate and longitudinal slope at 3,613–4,020 m and 4,036–4,727 m altitude are positively correlated. At 3,613–4,020 m altitude, with the increase of longitudinal slope, the heart rate growth rate increases from 17.9 to 27.1%, with an increase of 51.4%, while at 4,036–4,727 m, the heart rate growth rate increases from 24.7 to 37.36%, with an increase of 51.25%. Through linear fitting, it was found that the relationship between heart rate growth rate and the radius of the flat curve was basically quadratic parabola, and the fitting degrees were 0.5451 and 0.4139, respectively.

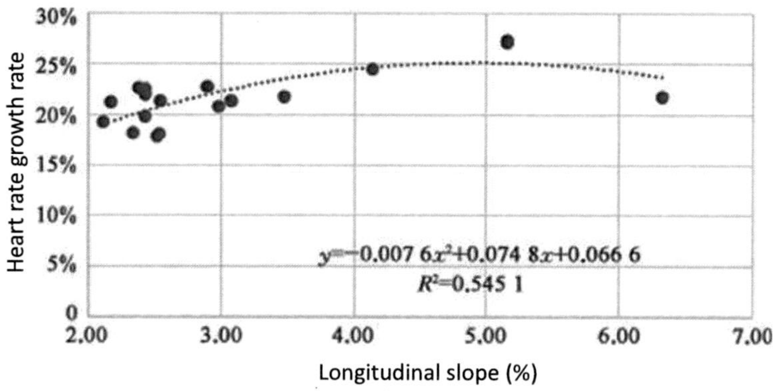

Fig. 4.13 Relationship between heart rate growth rate and longitudinal slope at 3613–4020 m altitude

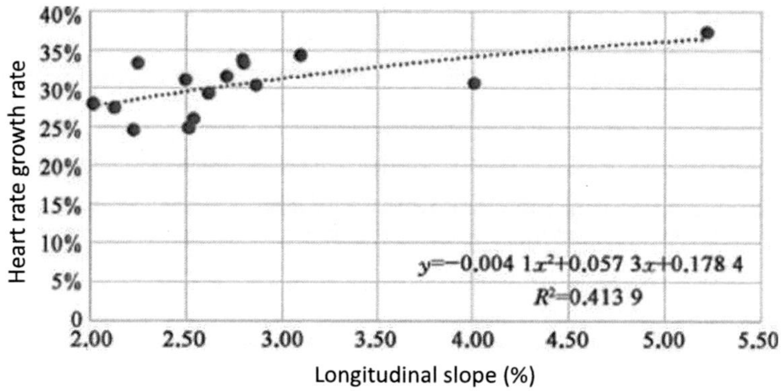

Fig. 4.14 Relationship between heart rate growth rate and longitudinal slope at 4036–4727 m altitude

4.1.4 Driver Heart Rate Change Model Under Complex Linear Conditions

Through the two factor and multi factor variable analysis, the change trend and partial correlation between the influencing heart rate growth rate and each influencing factor index are obtained. The next step is to conduct model analysis according to different linear sections and altitude intervals.

(1) **Flat Curved Line Section**

Through partial correlation analysis of heart rate growth rate N, altitude M, radius R and vehicle speed V, it is found that the linear relationship between heart rate growth rate and altitude, radius and vehicle speed is not obvious, but according to the results of two-factor variable analysis, M–N changes approximately in a positive power function, and V–N and R–N changes approximately in a negative power function. Considering that it is easy to convert into linear regression analysis and try to conform to the objective reality, it is assumed that the nonlinear relationship model between the four parameters is as follows:

$$N = b_0 \frac{M^{b_3}}{V^{b_1} R^{b_2}} \tag{4.3}$$

where,

N	Heart rate growth rate (%);
M	Altitude (m);
V	Vehicle speed (km/h);
R	Curve radius (m);
b_0, b_1, b_2, b_3	Undetermined coefficients.

Through calculation by SPSS software, $b_0 = 2.575 \times 10^{-5}$, $b_1 = 0.174$, $b_2 = 0.002$, $b_3 = 1.177$ can be obtained by substituting formula (4.3)

$$N = 2.575 \times 10^{-5} \frac{M^{1.177}}{V^{0.174} R^{0.002}} \quad M \in (3567, 3957) \tag{4.4}$$

The goodness of fit (R square) of the model is 0.701. After adjustment, the goodness of fit is 0.662. The Sig value of F test is less than 0.05, which is of good significance. The t test shows that the M is significant among the three variables M, V and R variables, which indicates that the altitude has a much greater impact on the rate of heart rate growth than the linear and vehicle speed.

Similarly, model calculation is carried out for the range of 4,100–4,709 m altitude, and $b_0 = 0.0382$, $b_1 = -0.012$, $b_2 = 0.052$, $b_3 = 0.284$ are obtained by substitution formula (4.3).

$$N = 0.0382 \frac{M^{0.284}}{V^{-0.012} R^{0.052}} \quad M \in (4100, 4709) \tag{4.5}$$

The goodness of fit (R square) of the model is 0.741. After adjustment, the goodness of fit is 0.690. The Sig value of F test is less than 0.05, which is of good significance. The *t* test shows that the *M* is significant among the three variables *M*, *V* and *R* variables, which indicating that the altitude has a much greater impact on the rate of heart rate growth than the linear and vehicle speed.

For more intuitive analysis, substitute $V = 80$ km/h into Eqs. (4.4) and (4.5), and use Matlab to draw *N–M–R* three-dimensional surface diagram, as shown in Figs. 4.15 and 4.16.

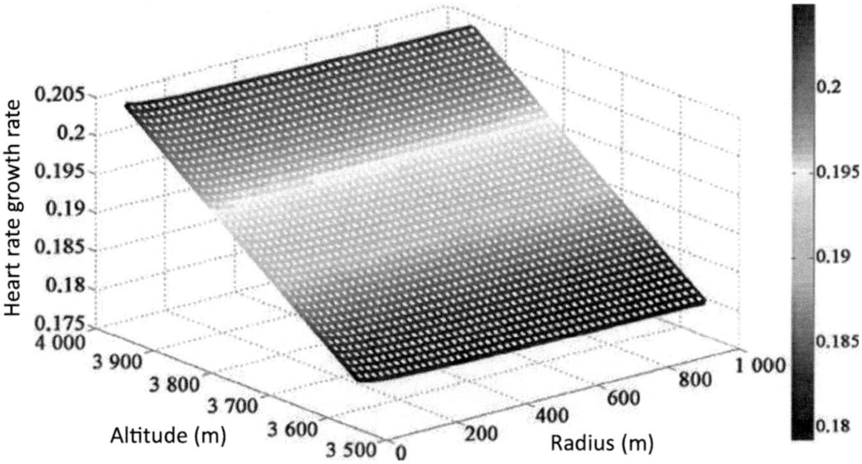

Fig. 4.15 *N–M–R* relationship at 3567–3957 m altitude

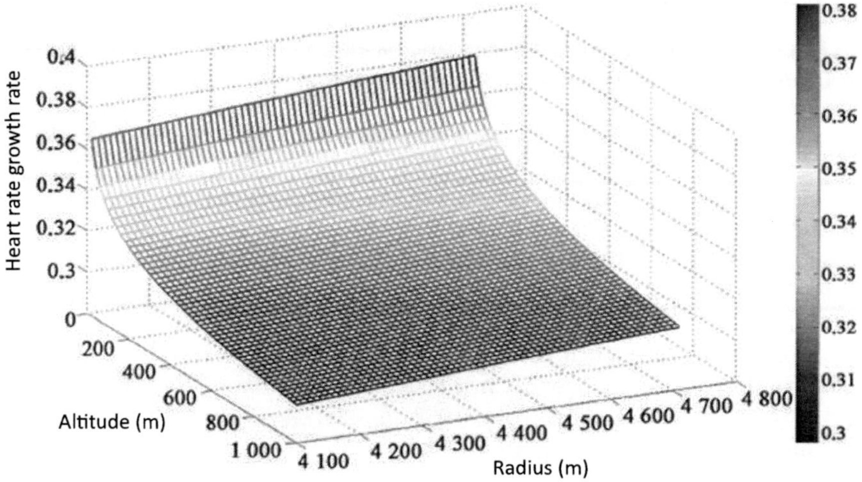

Fig. 4.16 *N–M–R* relationship at 4100–4709 m altitude

It can be clearly seen from Figs. 4.15 and 4.16 that when the driver is driving at the design speed in the flat curve section, the higher the altitude, the smaller the curve radius and the greater the heart rate growth rate are affected by both the altitude and the curve radius.

(2) **Curve Slope Combination Section**

Through the partial correlation analysis of heart rate growth rate W, altitude M, linear combination value 17 and vehicle speed V, it is found that the linear relationship between heart rate growth rate and linear combination value and vehicle speed is not obvious in the 3,616–3,998 m altitude, but according to the results of two-factor variable analysis, M–N changes approximately linearly, V–N and W–N changes approximately in a quadratic polynomial function, so it is assumed that the nonlinear relationship model between the four parameters is as follows:

$$N = aW^2 + bW + cV^2 + dV + e + fM \tag{4.6}$$

where,

N	Heart rate growth rate (%);
M	Altitude (km);
V	Vehicle speed (km/h);
W	Linear combination value (‰/km);
a, b, c, d, e, f	Undetermined coefficients.

Solve it with the nonlinear fitting tool in SPSS software, and get $a = -1.224 \times 10^{-5}$, $b = 0.008$, $c = -0.001$, $d = 0.166$, $e = -30.258, f = 11.472$, substitute into Eq. (4.6) to get the final model formula (4.7):

$$\begin{gathered} N = -1.224 \times W^2 + 0.008 \times W - 0.001V^2 + 0.166V + 11.472M - 30.258 \\ M \in (3616, 3998) \end{gathered} \tag{4.7}$$

The goodness of fit (R square) of the model is 0.847, which is good.

However, according to the two-factor variable analysis and multi factor variable analysis of 4,127–4,630 m above sea level, the correlation and partial correlation between heart rate growth rate and linear combination value, altitude and vehicle speed are high and have good linear correlation. Therefore, it is assumed that the linear relationship model between the four parameters is as follows:

$$N = aW + bV + cM + d \tag{4.8}$$

where,

N	Heart rate growth rate (%);
W	Combined value of alignment (‰/km);
V	Vehicle speed (km/h);

M Altitude (km);
a, b, c, d Undetermined coefficients.

Solve through multiple linear regression in SPSS software to obtain $a = 0.008$, $b = 0.025$, $c = 11.463$, $d = -22.317$, substitute into Eq. (4.8) to obtain model formula (4.9):

$$N = -22.317 + 0.008W + 11.643M + 0.025V \quad M \in (4127, 4630) \tag{4.9}$$

The goodness of fit (R square) of the model is 0.952. After adjustment, the goodness of fit is 0.928. The Sig value of F test is <0.05, which is of good significance. The t test shows that M is significant among M, V and W variables, which indicates that altitude has a greater impact on the rate of heart rate growth than linear and vehicle speed.

For more intuitive analysis, substitute $V = 80$ km/h into Eqs. (4.7) and (4.9), and use Matlab to draw N–M–W three-dimensional surface diagram, as shown in Figs. 4.17 and 4.18.

It can be clearly seen from Figs. 4.17 and 4.18 that when the driver drives at the design speed in the curve slope combination section, the higher the altitude, the greater the linear combination value and the greater the heart rate growth rate are affected by the altitude and linear combination value.

(3) **Longitudinal Slope Section**

Through the partial correlation analysis of heart rate growth rate N, altitude M, longitudinal slope i, slope length L and vehicle speed V, it is found that the relationship between slope length and heart rate growth rate is weak, so this factor is eliminated in the modeling process. According to the results of the two-factor variable analysis, M–N and i–N approximately changes in a quadratic parabola, and V–N approximately

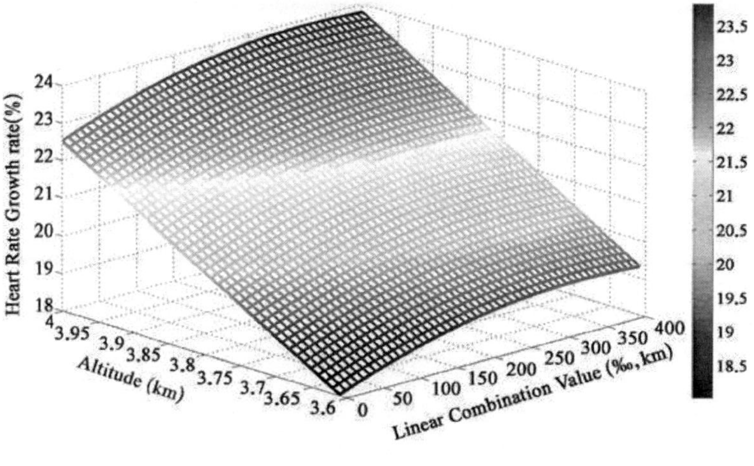

Fig. 4.17 N–M–W relationship at 3616–3998 m altitude

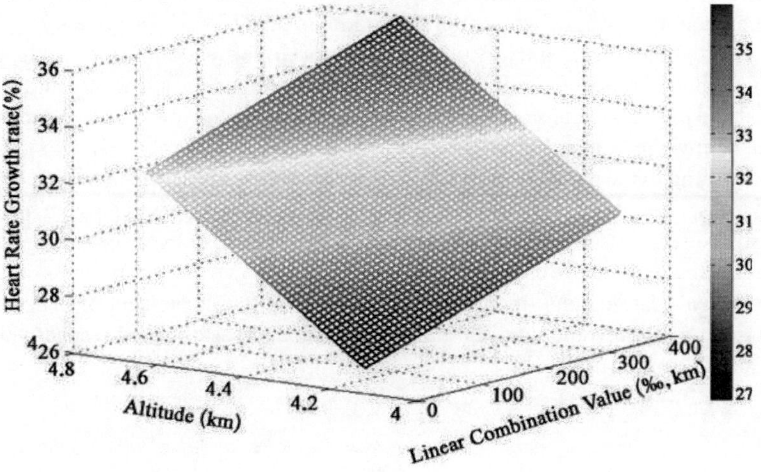

Fig. 4.18 *N–M–W* relationship at 4127–4630 m altitude

changes linearly. Therefore, assume that the nonlinear relationship model between the four parameters is Eq. (4.10):

$$N = aM^2 + bM + ci^2 + di + eV + f \tag{4.10}$$

where,

N	Heart rate growth rate (%);
M	Altitude (km);
V	Vehicle speed (km/h);
i	Longitudinal slope (%);
a, b, c, d, e, f	Undetermined coefficients.

Solve it with the nonlinear fitting tool in SPSS software, and get $a = -71.41$, $b = 555.248$, $c = 0.498$, $d = 6.026$, $e = 0.089$, $f = -1075.498$, substitute into Eq. (4.10) to get the final model formula (4.11):

$$N = -71.41M^2 + 555.248M - 0.498i^2 + 6.026i + 0.089V - 1075.498 \\ M \in (3613, 4020) \tag{4.11}$$

The goodness of fit (R square) of the model is 0.939, which is good.

Similarly, the model is established for the altitude of 4,036–4,727 m, get $a = 0.652$, $b = 4.906$, $c = -0.251$, $d = 3.255$, $e = -0.101$, $f = -3.503$, substitute into Eq. (4.10) to get the final model formula (4.12):

$$N = 0.652M^2 + 4.906M - 0.241i^2 + 3.255i - 0.101V - 3.503 \quad M \in (4036, 4727) \tag{4.12}$$

The goodness of fit (*R* square) of the model is 0.957, which is good.

On the whole, the characteristic indicators of driver reaction time and heart rate change can well reflect the rules of drivers' perception, operation ability and psychological and physiological state changes during driving in the low pressure and oxygen deficient environment at high altitude.

According to the analysis of the driver response time data collected from different altitude test sites, it can be found that: altitude and driving time will have an impact on the drivers' response ability, and the impact of driver age is not very obvious. Through the real vehicle test at different altitudes, the change of drivers' heart rate is collected, and the heart rate change models of drivers in the flat curve section, curve slope combination section, and longitudinal slope section are established respectively, and the model fitting is high.

4.2 Fatigue Characteristics of Drivers in Low Oxygen Environment

The oxygen content in high altitude areas is low, and drivers are more likely to enter fatigue state than those in plain areas. In order to obtain the fatigue characteristics of drivers in high altitude areas, driving simulation tests were carried out at three different altitudes to collect the psychological and physiological changes of drivers and analyze the fatigue characteristics of drivers at different altitudes.

4.2.1 *Test Plan Design*

(1) **Test Location**

The altitude of Golmud along the Qinghai-Tibet Highway is about 2,800 m, and then from Xidatan to Lhasa, the average altitude is more than 4,000 m. The altitude distribution along the line is shown in Fig. 4.19.

In the section Golmud Lhasa of the Qinghai-Tibet Highway, the mileage at an altitude of 3,000–4,000 m is less than 200 km, and the high altitude is mainly in the range of 4,000–5,000 m. In order to study the impact of altitude on driver fatigue, three locations at different altitudes are successively selected along G109 from Golmud to Tanggula Mountain for testing, respectively: Nachitai, with an average altitude of 3,500 m; Xidatan, averaged 4,200 m; Tuotuo River, averaged 4,600 m. The specific distribution locations of three different test points in are shown in Fig. 4.20.

Golmud is 2,800 m above sea level, with a population of 300,000 as of 2011. There altitude sickness is rare, and the medical and health conditions are relatively good. Even if an accident occurs, timely treatment can be carried out. Nachtai is located about 94 km southwest of Golmud City, with an altitude of about 3,500 m,

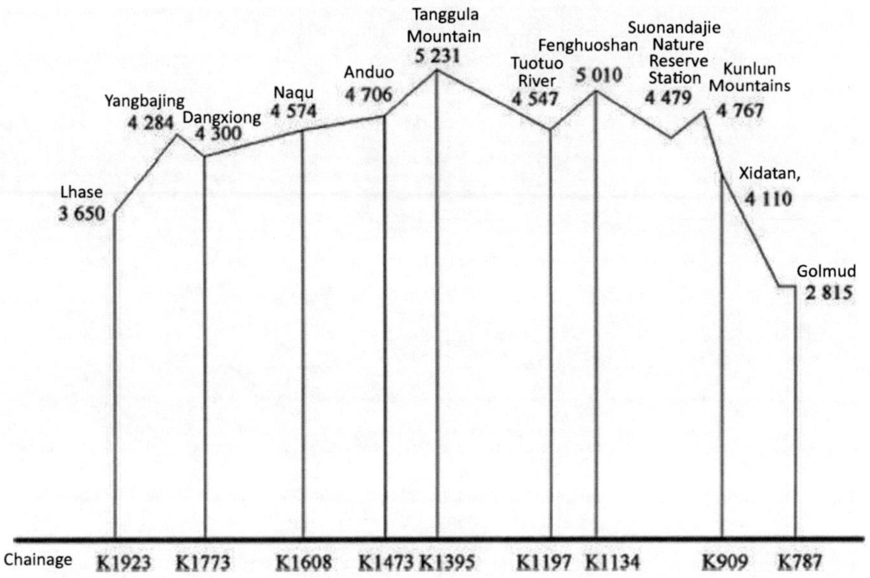

Fig. 4.19 Altitude change along G109 (Xining Lhasa)

and rise about 700 m than Golmud. It is located in the maintenance section of Nachtai Highway, power supply is convenient, so this point is selected for simulation test.

The average altitude of Xidatan section is 4,200 m, about 48 km away from Nachitai, and rise about 700 m than Nachitai. There is Yuzhu Peak Snow Mountain near Xidatan, with cold temperature and snow cover all the year round.

Tuotuohe Town, with an average altitude of 4,600 m, is the source of the Yangtze River, 178 km from Tanggula Mountain. With thin air, it is a place along the Qinghai-Tibet Highway with the worst climate and the most severe altitude sickness.

(2) **Test Equipment**

The wireless physiological recording and analysis system MP150, produced by Biopic Company of the United States, is used in this experiment, and the sampling frequency of heart rate is 2,000 Hz. It can collect ECG, EEG, EMG, EOG, blood oxygen saturation and other psychophysiological indicators. The pasting position of the data acquisition patch is shown in Fig. 4.21, and the data acquisition software interface is shown in Fig. 4.22.

The simulation equipment is composed of 3D display screen, steering wheel, accelerator and brake, gear, seat, etc. The simulator can generate different sounds according to the depth of the accelerator, and simulate the real driving environment as much as possible (Fig. 4.23).

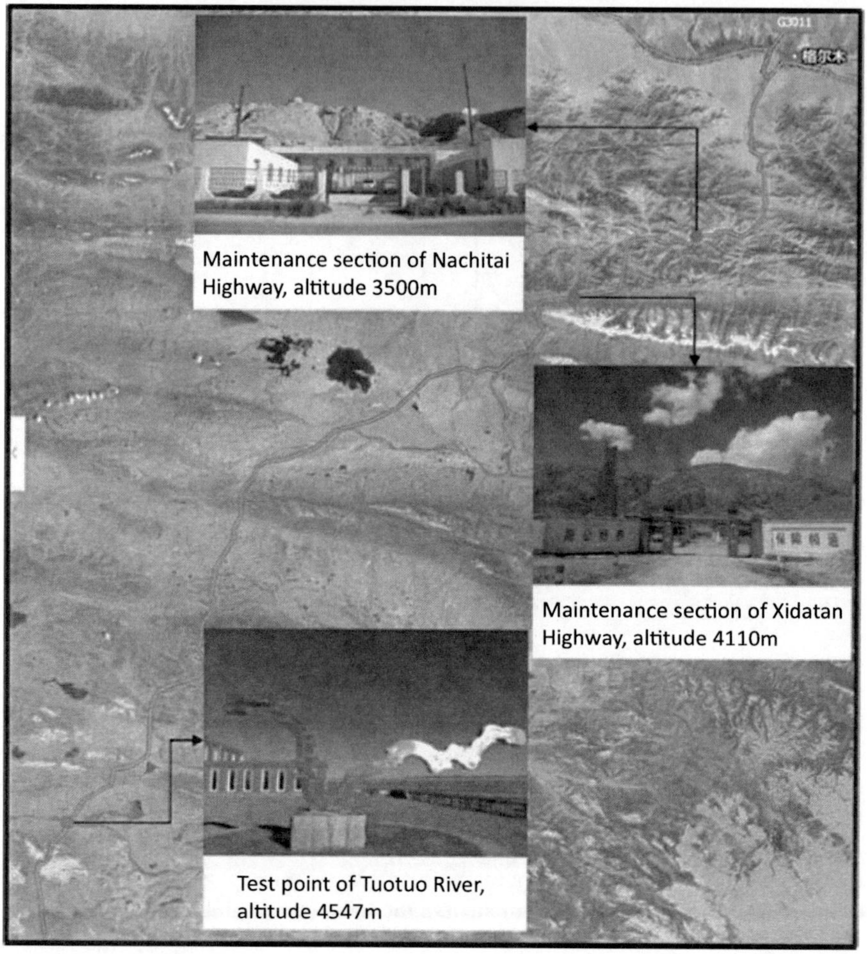

Fig. 4.20 Location of three different test points

(3) **Test Plan and Process**

In order to fully reflect the fatigue state of the driver, the test time shall be kept at about 2 h. If the driver is very sleepy, dizzy or even vomiting and cannot continue the test, the test can be terminated in advance.

During the test, the staff shall record the relevant details of the tested personnel in detail, especially the typical phenomena in the fatigue process, such as yawning, tears, napping, vehicles driving on the line and driving out of the road. In addition, it is necessary to record the driver's chat, phone calls, smoking, etc. After each test, a questionnaire shall be conducted on the driver's fatigue level. These records are one of the criteria for judging the level of driver fatigue. Figure 4.24 shows the test flow chart.

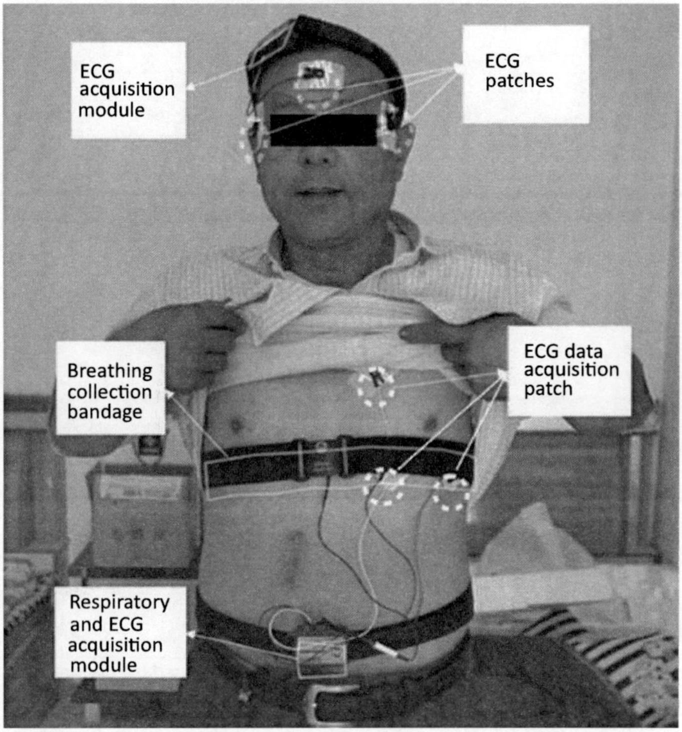

Fig. 4.21 Paste position of data acquisition patch

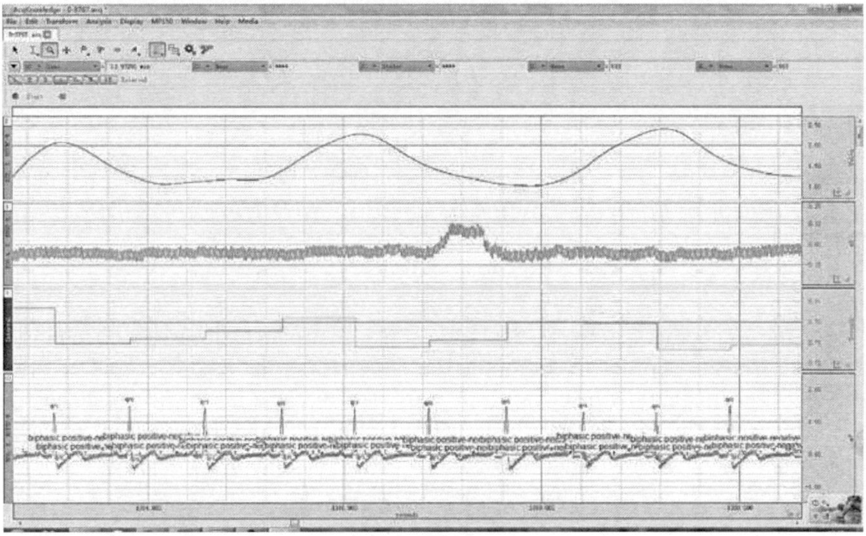

Fig. 4.22 Data acquisition software interface

Fig. 4.23 Driver simulation test scenario

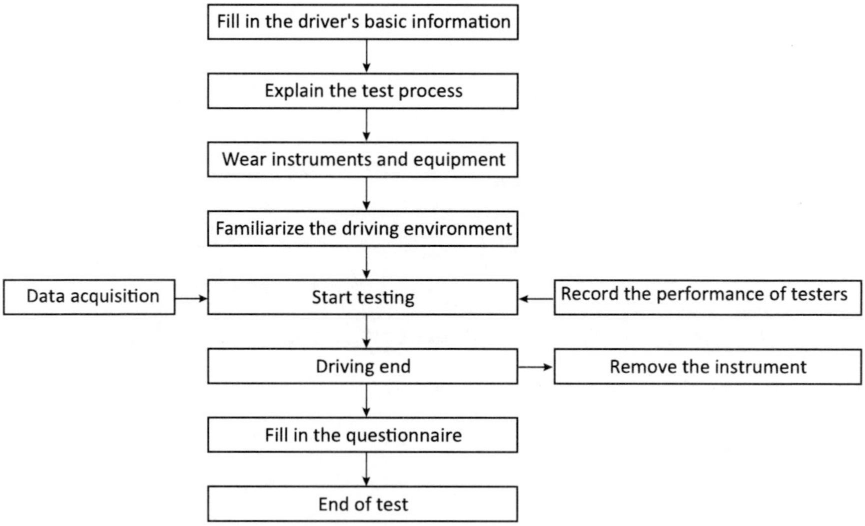

Fig. 4.24 Test flow chart

4.2.2 *Fatigue Characteristics of Driving at Different Altitudes*

(1) **Evaluation Index**

Due to the great otherness in physical fitness of each driver, their resistance to fatigue is also different. In order to overcome the otherness in the heartbeat rate of every driver, the heartbeat interval change rate is used to characterize the heart rate change of each driver. The average heartbeat interval in the previous 10 min is taken as the heartbeat interval in the awake state, and its calculation method is as follows:

$$R_i = \frac{T_i - T_{10}}{T_{10}} \tag{4.13}$$

Table 4.3 Comparison matrix (diagnostic index > cut-off point is abnormal)

Results of "gold standard"	Abnormal	Normal	Total number of cases
Diagnostic index > cut-off point	a	b	$a + b$
Diagnostic index < cut-off point	c	d	$c + d$
Total number of cases	$m = a + c$	$n = b + d$	$m + n$

Note a'', b, c, d are number of cases

where,

R_i Change rate of average heartbeat interval in i period;
T_{10} Mean heartbeat interval in the previous 10 min (s);
T_i Average heartbeat interval in i period (s).

From the above analysis, it can be concluded that 120 s is used as the interval for calculation.

(2) **ROC Curve Introduction**

ROC (Receiver Operator Characteristic) curve, originally used to evaluate radar performance. The ROC curve is used to describe and evaluate the accuracy of the target detection algorithm, which is applicable to the case of binary classification, that is, there are only "yes" or "no" results. Through literature search, it is found that ROC curve is widely used in the medical field and can well determine the discrimination threshold of different indicators (Table 4.3).

Sensitivity (se) = a/m; Specificity (sp) = d/n; Accuracy (AC) = $(a + d)/(m + n)$; Youden index (Yi) = $a/m - b/n$; Positive probability ratio (LR+) = a/b; Negative probability ratio (LR−) = c/d.

ROC curve is a curve drawn with sensitivity as the ordinate and "1-specificity" as the abscissa. Sensitivity and specificity change with the value of the cut-off point of diagnostic variables (indicators). The ROC curve analysis method is described as follows.

The position and shape of ROC curve determine the recognition ability of detection algorithm. Each action point in ROC space is a combination of sensitivity and specificity, as shown in Fig. 4.25. When the threshold value is taken on the straight line *AB* (such as point *C*), it means that the judgment result is that the sensitivity and specificity are identical, accounting for 50% respectively, which is of no practical significance. When the threshold value is taken on the straight line *AD* or *DB*, it indicates that the specificity is 1 or the sensitivity is 1, the discrimination accuracy is the highest, and there is no overlapping area between samples. When the threshold value is taken on the curve *GH*, it means that there is overlapping area between samples when this threshold value is used for judgment. The closer to the discrimination threshold at the upper left of the ROC curve, the smaller the overlapping area of the samples, which means the better the discrimination performance.

Fig. 4.25 Schematic diagram of ROC curve

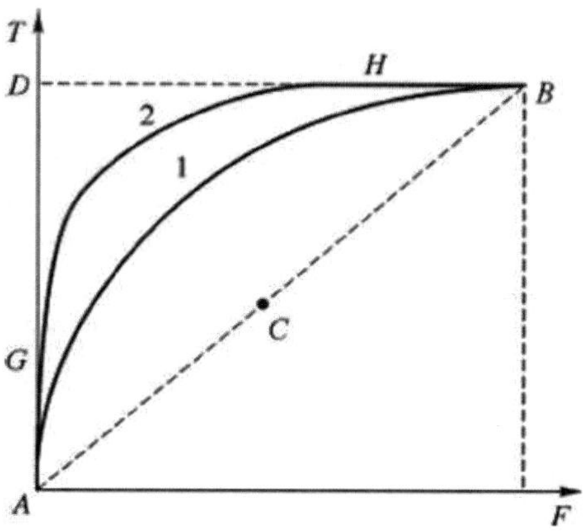

The area under the ROC curve (denoted as S) is a two-dimensional visual description of the comprehensive evaluation and discrimination algorithm. S reflects the accuracy of the threshold value. The value range of S is 0.5–1.0. When S is 0.5–0.7, it indicates that the accuracy is low. When S is 0.7–0.9, the accuracy is medium. When $S > 0.9$ it indicates high accuracy. Another potential function of ROC curve is to determine the optimal threshold value. For a certain discrimination method, the ROC curve contains all the combinations of sensitivity and specificity. The closer the point at the top left is, the better the discrimination effect will be. In application, you can select the Y index on the curve as close to the upper left as possible:

$$\text{Younden index} = \text{sensitivity} + \text{specificity} - 1 \tag{4.14}$$

In the medical field, when patient and non-patient indicators overlap a lot, there is also a certain rate of missed diagnosis and misdiagnosis, that is, miscalculation. Therefore, it is obviously arbitrary to diagnose patients and non-patients by only one point of tangency. It is necessary to discuss the range of suspicious values, which should be the overlapping area of patients and non-patients. At the same time, two diagnostic points are set at the overlap of patient and non-patient distributions, so that the positive and negative distributions are divided into three ranges, namely, positive, suspicious and negative ranges.

(3) **Selection of Test Duration**

As for the selection of the test duration, at least the driver needs to be in a fatigue state. For example, the test data of a tester A2, the test time is 3 h, but after 2 h, the anti-fatigue phenomenon appears, and the driver changes from fatigue state to sober state. Therefore, the determination of test time length must ensure two points: first, fatigue must occur; second, if the time is too long, the anti-fatigue phenomenon

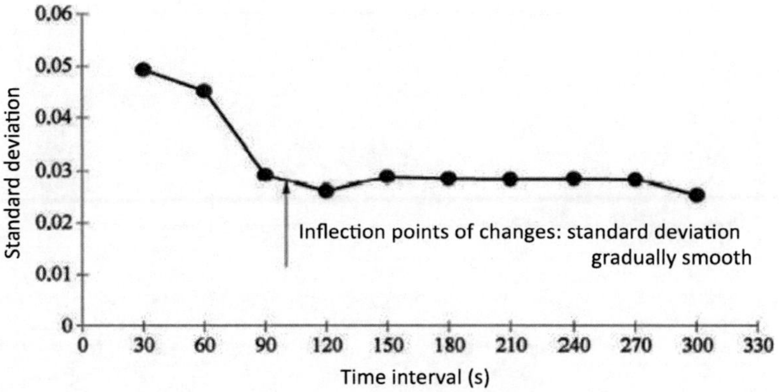

Fig. 4.26 Standard deviation under different analysis intervals

will appear, in order to avoid this phenomenon, the drivers' driving state must be observed in real time during the test.

(4) **Selection of Analysis Unit Duration**

The selection of analysis interval should follow three principles: (i) Reflect the details of data changes; (ii) Do not highlight the sudden change of data at a certain time; (iii) reflect the overall change trend of data.

From 30 to 300 s, every 30 s is an interval, which is divided into 10 groups of different data. The average values under different heartbeat intervals are calculated respectively, and then the standard deviation within the time period is calculated respectively, and the heartbeat interval is plotted as a curve. Figure 4.26 shows the standard deviation of real test data of a driver under different analysis intervals. It can be seen from Fig. 4.26 that the standard deviation change of heartbeat interval is not obvious after the analysis interval is 120 s, so the analysis interval is set as 120 s.

Average value calculation formula:

$$\overline{R} = \frac{R_1 + R_2 + \cdots + R_n}{n} \tag{4.15}$$

Standard deviation calculation formula:

$$\delta = \sqrt{\frac{1}{n}\sum_{i=1}^{n}\left(R_i - \overline{R}\right)^2} \tag{4.16}$$

where,

$\overline{R}$ Mean value of heartbeat interval (s);
R_i heartbeat interval at time i (s);
n Number of heartbeats;
δ Standard deviation.

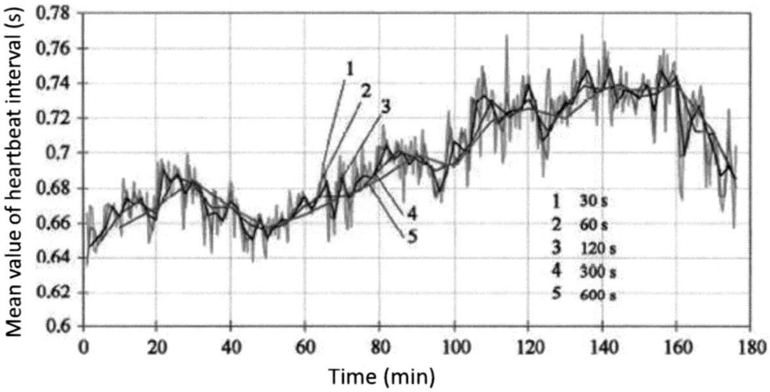

Fig. 4.27 Heartbeat interval under different time length

The heartbeat interval is calculated by using different time interval lengths, namely 30, 60, 120, 300 and 600 s. Taking the heartbeat data of driver A as an example, the curve is drawn as shown in Fig. 4.27.

It can be seen from Fig. 4.27 that the curve graph with an interval of 30 s fluctuates greatly, and the curve graph with an interval of 600 s is the gentlest. However, if the 600 s duration is taken as the analysis unit, on the one hand, the data volume is too small, even if the test duration is 1 h, only six test data will be generated; on the other hand, the range of fatigue judgment results is wide, the total test accuracy will be reduced, and it is impossible to determine whether the fatigue state occurs in the front or rear of the 10 min. With regard to the selection of analysis interval, it is necessary to reflect the details of RR interval changes without highlighting some random fluctuations, and to reflect the overall change law of data. Based on the above analysis, 120 s interval is finally adopted for relevant calculation.

(5) **Analysis of the Drivers' Heartbeat Data**

The analysis results of the heartbeat data at three different altitudes are as follows.

(1) Analysis of Heartbeat Data in Nachitai

In Fig. 4.28, A1 and A6 are local drivers who have lived on the plateau for generations to adapt to the cold and oxygen deficient environment. Their heart rate changes are also significantly different from those of non-local drivers. The specific information is shown in Tables 4.4 and 4.5.

It can be seen from Table 4.5 that the heartbeat interval of A4 varies the most. Because the tested person did not sleep the night before, he quickly showed a state of fatigue. The test time of A3 was in the early morning, so he also showed a significant

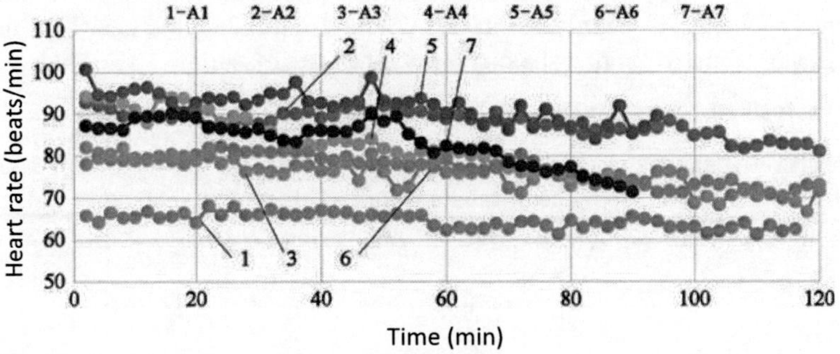

Fig. 4.28 Variation of drivers' heart rate with time in Nachitai

Table 4.4 Driver fatigue time and other related information (1)

Tester code	Type of driver's license	Fatigue time corresponding to critical point (min)	Optimal critical point (rate of change in heartbeat interval)	Running mileage (10,000 km)	High altitude driving experience	Non-local/ local	Professional driver
A3	C1	68	0.0895	5–10	No	Non-local	No
A4	A2	28	0.0681	Above 100	Yes	Local	Yes
A5	C1	58	0.0523	5–10	No	Non-local	No
A6	B2	97.6	0.087	5–10	Yes	Local	Yes
A7	C1	61	0.0729	5–10	Yes	Non-local	Yes
A2	B1	88	0.0826	60–100	Yes	Non-local	Yes
A1	B2	69	0.0202	0–5	Yes	Local	Yes
A8	A2	64	Fluctuation	60–100	Yes	Non-local	Yes
A9	A2	Anomaly	Anomaly	5–10	Yes	Local	Yes

Table 4.5 Heartbeat interval and heart rate changes (2)

Tester code	Heartbeat interval		% change (%)	Heart rate (beats/min)		Decreased value
	First 10 min	Last 10 min		First 10 min	Last 10 min	
A3	0.74	0.86	15.30	80.9	70.2	10.7
A4	0.64	0.80	24.20	93.2	75.1	18.2
A5	0.62	0.68	9.10	96.3	88.2	8.0
A7	0.69	0.81	17.70	87.3	74.1	13.1
A6	0.76	0.84	10.60	79.1	71.6	7.6
A2	0.66	0.73	10.40	91.3	82.7	8.6
A1	0.92	0.96	4.50	65.6	62.8	2.8

percentage change in heartbeat interval. At high altitudes, the heartbeat interval increases by about 10% after the driver enters a state of fatigue.

(2) Analysis of Heartbeat Data in Xidatan

The heart rate changes of the tested persons in Xidatan are shown in Fig. 4.29.

B1 is a local driver, so its heart rate is significantly slower than other testers. The relevant information about the driver is shown in Table 4.6.

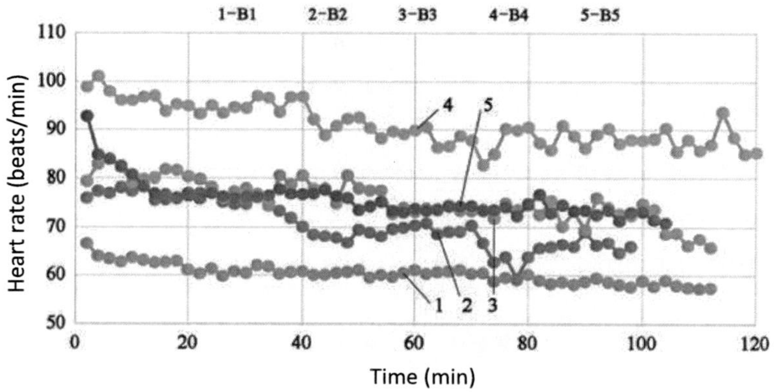

Fig. 4.29 Variation of drivers' heart rate with time in Xidatan

Table 4.6 Driver fatigue time and other related information (2)

No	Tester code	Type of driver's license	Time (min)	Optimal critical point (rate of change in heartbeat interval)	Running mileage (10,000 km)	High altitude driving experience	Ethnicity	Professional driver
1	B1	A2	69.0	0.077	More than 100	Yes	Tibetan	Yes
2	B2	A2	57.0	0.081	More than 100	Yes	Han	Yes
3	B3	A2	73.8	0.110	More than 100	Yes	Han	Yes
4	B4	C1	60.0	0.084	0–5	Yes	Han	No
5	B5	A2	70.0	0.152	10–30	Yes	Han	Yes
6	B7	C1	54.0	–	5–10	Yes	Han	Yes
7	B6	A2	66.0	–	60–100	Yes	Tibetan	Yes
8	B9	C1	40.0	–	Less than 1	No	Han	No
9	B8	C1	41.7	–	5–1	Yes	Han	No

Table 4.7 Heartbeat interval and heart rate changes (2)

Tester code	Heartbeat interval		% change (%)	Heart rate (beats/min)		Decreased value
	First 10 min	Last 10 min		First 10 min	Last 10 min	
B1	0.94	1.04	9.60	64.1	57.9	6.2
B2	0.78	0.90	12.00	77.2	66.6	10.6
B3	0.74	0.89	14.9	81.5	67.4	14.1
B4	0.61	0.68	6.80	98.0	87.9	10.1
B5	0.71	0.83	12.1	85.0	72.1	12.9

From Table 4.6, it can be seen that professional drivers show more robust anti-fatigue characteristics, and the fatigue time point is delayed.

In Table 4.7, B1 is a local driver who has lived on the plateau for generations and has adapted to the low pressure and oxygen deficient environment at high altitudes. The reduction in heart rate before and after fatigue was only 6.2 beats per min, and the number of heartbeats in the first 10 min and the last 10 min were also minimal. Tester B3 has the most significant change in heartbeat interval percentage at the fatigue point. B3 is a professional driver who drives on the Qinghai-Tibet Highway all year round and shows stronger fatigue resistance. B4 is a non-professional driver, often drives passenger cars and has lived in plain area for a long time, when he showed a state of fatigue, the change rate of the heartbeat interval is 6.8%, and the average heartbeat rate in the first 10 min is 98 beats/min, the highest among the five testers.

(3) Analysis of Heartbeat Data in Tuotuohe

The heart rate change in Tuotuohe is shown in Fig. 4.30, in which the heart rate of testers C6 and C9 fluctuates wildly and will not be analyzed for the time being.

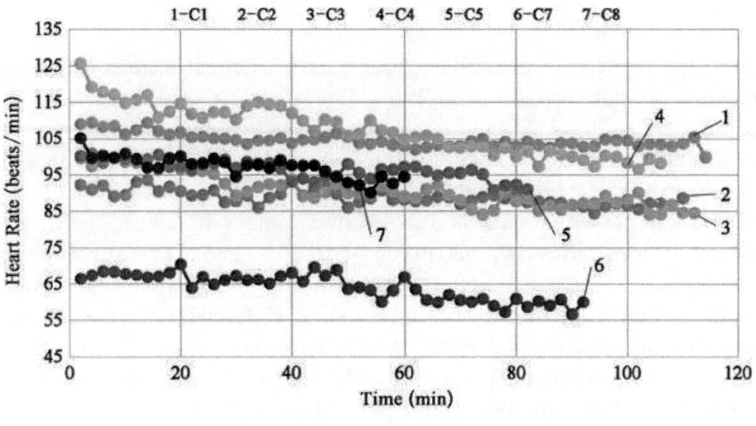

Fig. 4.30 Variation of drivers' heart rate with time in Tuotuohe

Table 4.8 Driver fatigue time and other related information (3)

No	Tester code	Type of driver's license	Time (min)	Optimal critical point (rate of change in heartbeat interval)	Running mileage (10,000 km)	High altitude driving experience	Ethnicity	Professional driver
1	C3	A2	70.0	0.115	More than 100	Yes	Han	Yes
2	C4	B2	56.8	0.126	0–5	Yes	Han	No
3	C8	B2	41.0	0.048	10–30	Yes	Han	Yes
4	C5	C1	34.0	0.039	0–5	No	Han	No
5	C7	A2	68.0	0.083	0–5	Yes	Han	Yes
6	C1	A2	49.0	0.038	30–60	Yes	Han	No
7	C2	C1	32.0	0.018	5–10	Yes	Han	No

Table 4.9 Heartbeat interval and heart rate changes (3)

Tester code	Heartbeat interval		% change (%)	Heart rate (beats/min)		Decreased value
	First 10 min	Last 10 min		First 10 min	Last 10 min	
C3	0.61	0.70	13.19	98.8	84.9	13.9
C4	0.50	0.61	16.88	118.9	98.5	20.5
C8	0.59	0.65	8.22	101.1	92.8	8.3
C5	0.60	0.65	7.41	99.7	92.3	7.4
C7	0.89	1.01	12.47	67.8	59.3	8.4
C1	0.55	0.58	4.81	108.3	103.1	5.3
C2	0.66	0.69	4.11	90.8	87.1	3.7

Due to the fluctuation of the heartbeat data of the tested persons C6 and C9, the statistics are not carried out in Table 4.8. It can be seen from Table 4.8 that professional drivers have stronger anti-fatigue traits than non-professional drivers.

Due to fluctuations in the heartbeat data of subjects C6 and C9, no statistics are made in Table 4.9. The test driver C4 is a non-professional driver from other places who often drive passenger cars, and its heartbeat interval between the first 10 min and the last 10 min varies by a maximum of 16.88%, and his heart rate per minute decreased by 20.5 beats per minute. In general, the percentage change in heart rate interval before and after 10 min showed a distinct difference, but most of the tested drivers' heart rate reduction values in the first 10 min and the last 10 min were within ten beats. The test environment here is in a plateau oxygen deficient state, so that the test results may differ from those in the plain area.

Table 4.10 Statistics of driver fatigue time

Location	Fatigue time (min)
Nachitai	68.8
Xidatan	65.2
Tuotuohe	50.1

(4) Statistical Results of Data

Since the heartbeat data of local drivers is significantly different from that of non-local drivers, only the fatigue time of non-local drivers in the driving simulator is analyzed here. The specific statistical results are shown in Table 4.10. The data for which ROC curve analysis has been performed, and the data without ROC curve analysis, fluctuate due to different reasons, such as phone calls, etc., and the heartbeat data fluctuates without regularity. Therefore, this type of data is not counted in the statistic of averaged fatigue time at different altitude.

It can be seen from Table 4.10 that with the increase in altitude, the fatigue time point appears earlier.

4.2.3 Fatigue Model When Driving at Different Altitudes

The results of driver fatigue are generally divided into fatigue or sobriety. The possible factors affecting driver fatigue are altitude, driving time, driver category, etc., where the driver category is a discrete variable. The results of driver fatigue are classified as ordinal variables, so the usual linear regression method is no longer applicable. Therefore, the binary Logit model is used to establish the fatigue model of the driver under different altitude conditions. The principle of the binary Logit model and the results of the model establishment are introduced as follows.

(1) Model Introduction

The principle of the binary Logit model and the related derivation process is introduced as follows.

Suppose the result of the dependent variable Y is only 0 or 1,

$$Y = \begin{cases} 1 \text{ not occur} \\ 0 \text{ occur} \end{cases}, \text{ independent variable } X_1, X_2, \ldots, X_n \tag{4.17}$$

The probability of occurrence and non-occurrence under the action of n independent variables is denoted as

$$\begin{aligned} &\text{Occur: } P_i = P_i(Y = 1|X_1, X_2, \ldots, X_n), 0 \le P_i \le 1 \\ &\text{Not Occur: } 1 - P_i,\ o \le P_i \le 1 \end{aligned} \tag{4.18}$$

Transform the probability of occurrence and non-occurrence, and it becomes into

$$\pi_i = \frac{P_i}{1 - P_i}, i = 1, 2, \ldots, n \tag{4.19}$$

Then the binary logit model is

$$\text{logit}(P_i) = \ln\left(\frac{P_i}{1 - P_i}\right) = \eta_i, i = 1, 2, \ldots, n \tag{4.20}$$

After the transformation, the value range of P_i is still 0–1, but the value range of Logit(P_i) is $(-\infty, +\infty)$, which overcomes the problem that the probability value exceeds the interval [0, 1].

η_i can be further expressed as a linear combination of a set of independent variables:

$$\eta_i = \beta_0 + \beta_1 x_{11} + \cdots + \beta_k x_{ik} = \beta_0 + \sum_{k=1}^{k} \beta_k x_{ik},\ i = 1, 2, \ldots, n \tag{4.21}$$

Inversely, the formula for calculating the probability P_i is:

$$\frac{P_i}{1 - P_i} = \exp\left(\beta_0 + \sum_{k=1}^{k} \beta_k x_{ik}\right), i = 1, 2, \ldots, n \tag{4.22}$$

$$P_i = \frac{\exp\left(\beta_0 + \sum_{k=1}^{k} \beta_k x_{ik}\right)}{1 + \exp\left(\beta_0 + \sum_{k=1}^{k} \beta_k x_{ik}\right)}, i = 1, 2, \ldots, n \tag{4.23}$$

The regression parameters of the Logit model were estimated using the maximum likelihood estimation method.

(2) Modeling

According to the above analysis, four variables of driving time (T), altitude (H), driver category (N) and heartbeat interval change rate (B) are initially selected to establish a binary Logit model:

$$\ln\left[\frac{P(Y = 1)}{1 - P(Y = 1)}\right] = \beta_0 + \beta_1 T + \beta_2 H + \beta_3 N + \beta_4 B \tag{4.24}$$

where,

$P(Y = 1)$	probability of driver fatigue;
β_0	constant;
β_1, β_2, β_3, β_4	independent variable coefficient;
T	the length of time (min);

H altitude, in this book, the altitude is divided into three sections according to the test points;

N driver category, in this book, is only divided into local driver $N = 1$ and non-local driver $N = 0$;

B the rate of change of the heartbeat interval. One data is collected every 2 min.

Regarding the interpretation of the regression coefficients, since the above driver fatigue model performs nonlinear Logit transformation on the dependent variables, the influence of the independent variables on the dependent variables in the model will be nonlinear and not applicable. When other independent variables remain unchanged if an independent variable X_j changes by one unit, the corresponding occurrence rate of the dependent variable changes by $\exp(\beta_j)$ units. Here, the (odds) are defined as the ratio of the probability of a specific outcome to the probability of another outcome (ratio). Using P to represent the probability of success and 1 − P to represent the probability of failure, the probability of success is the ratio $P/(1 - P)$. Then $\exp(\beta_j)$ represents the relative occurrence rate of one group relative to another group "successfully" experiencing an event, that is, the ratio of the occurrence rates of different groups.

Driver types are divided into two categories in this book, local and non-local drivers. Occupation types are divided into two categories, professional and non-professional drivers. Furthermore, altitudes are divided into three categories. These three parameters are dummy variables. Generally speaking, if a dummy variable contains m categories, $m - 1$ dummy variables need to be introduced. In model parameter calibration, the dummy variables need to be coded. Test point A is arranged as the reference category among the three categories, and the specific coding results are shown in Table 4.11.

The comprehensive test of the model coefficients is shown in Table 4.12.

The probability value in Table 4.12 is 0.000, less than the significance level of 0.05, which shows that the built model is reasonable.

(3) The Analysis of Model Results

It can be seen from Table 4.13 that all variables are less than the significant level of 0.05. Among the variable coefficients, only the coefficient of the driver type is negative, and the others are positive. That is, more easily fatigue occurs when time

Table 4.11 The coding of dummy variable

Dummy variable		Parameter encoding	
		(1)	(2)
Test point	Test point A	0	0
	Test point B	1	0
	Test point C	0	1
Driver category	Non-local driver	0	
	Local driver	1	

Table 4.12 The comprehensive test of the model coefficients

Item		Chi-Square	df	Sig.
Step 1	Step	562.270	5	0.000
	Block	562.270	5	0.000
	Model	562.270	5	0.000

Table 4.13 Variables in the equation

Variable		B	S.E	Wald	df	Sig.	Exp(B)
Step 1 a	Length of time	0.157	0.02	88.84	1.00	0.00	1.17
	Altitude			63.56	2.00	0.00	
	Altitude (1)	1.29	0.44	8.43	1.00	0.00	3.6
	Altitude (2)	4.669	0.60	59.77	1.00	0.00	106.59
	Heartbeat interval change rate	0.149	0.05	9.35	1.00	0.00	1.16
	Driver category (1)	−2.79	0.51	30.08	1.00	0.00	0.06
	Constant	−13.987	1.33	110.69	1.00	0.00	0.06

Notes
a—Variables entered in step 1: length of time, rate of change in heartbeat interval, altitude, driver type
B—The estimated value of the final model parameters
S.E—The corresponding standard error
Wald—The square of the ratio of the regression coefficient to the standard error
df—The degree of freedom
Sig.—The significance level, generally taking the confidence level of 0.05 as the standard for judging whether the independent variable is significant
Exp(B)—The result of the exponential operation on the regression coefficient B value

is longer, rate of heartbeat interval change is greater and altitude is higher, and non-local drivers feel fatigue more easily. The analysis results of the model coefficients are also consistent with the actual data analysis results.

Regarding the interpretation of the variable coefficients, the coefficient value for the length of time is 0.157, Exp(0.157) = 1.17. That is, every time the length of time increases by one unit, the occurrence rate of fatigue will be 1.17 times than that of the original. The heartbeat interval change rate coefficient is 0.149, Exp(0.149) = 1.16. That is, for every 1% increase in the heartbeat interval change rate, the occurrence rate of fatigue will be 1.209 times of the original.

This model uses test point A as a reference for the altitude. The fatigue occurrence rates of test points B and C relative to test point A are 3.63 and 106.59, respectively. At test point C, fatigue occurs more quickly than at test point A.

Overall, the two variables of time length and altitude have larger Wald test values. Moreover, it is Sig. value is less than 0.05, indicating that these two factors have an important influence on driver fatigue, and the Wald test value of time length is the largest, and its influence is the largest.

According to the analysis results of the above data, the Logit model of driver fatigue under different circumstances can be obtained. The fatigue model at test points A, B, and C is as follows.

Logit model of test point A:

$$\ln\left[\frac{P(Y=1)}{1-P(Y=1)}\right] = 0.157T + 0.149B - 2.790N - 13.987 \tag{4.25}$$

Logit model of test point B:

$$\ln\left[\frac{P(Y=1)}{1-P(Y=1)}\right] = 0.157T + 0.149B - 2.790N - 12.697 \tag{4.26}$$

Logit model of test point C:

$$\ln\left[\frac{P(Y=1)}{1-P(Y=1)}\right] = 0.157T + 0.149B - 2.790N - 9.318 \tag{4.27}$$

where,

$N = 0$, non-local drivers;
$N = 1$, local drivers.

The other variables have the same meaning as above. From the above formula, the fatigue probability of test point A can be deduced inversely:

$$P(Y=1) = \frac{\exp(0.157T + 0.149B - 2.790N - 13.987)}{1 + \exp(0.157T + 0.149B - 2.790N - 13.987)} \tag{4.28}$$

From the above formula, the fatigue probability of test point B can be deduced inversely:

$$P(Y=1) = \frac{\exp(0.157T + 0.149B - 2.790N - 12.679)}{1 + \exp(0.157T + 0.149B - 2.790N - 12.679)} \tag{4.29}$$

From the above formula, the fatigue probability of test point C can be deduced inversely:

$$P(Y=1) = \frac{\exp(0.157T + 0.149B - 2.790N - 9.318)}{1 + \exp(0.157T + 0.149B - 2.790N - 9.318)} \tag{4.30}$$

The variables in the above formula have the same meaning as above.

The fatigue change surface of the non-local driver ($N = 0$) is shown in Fig. 4.31.

The fatigue change surface of the local driver ($N = 1$) is shown in Fig. 4.32.

The heartbeat interval change rate at the critical point in high altitude areas is mainly concentrated in 6–9%, accounting for about 47.4% of the total, so the heartbeat

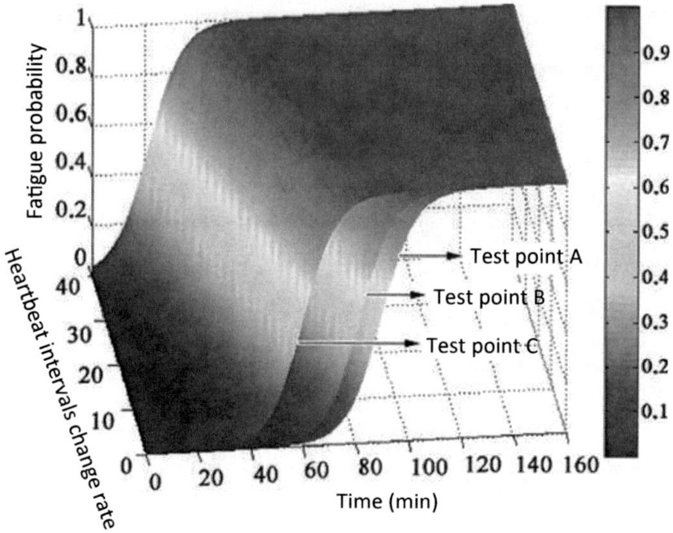

Fig. 4.31 Fatigue curve of non-local drivers

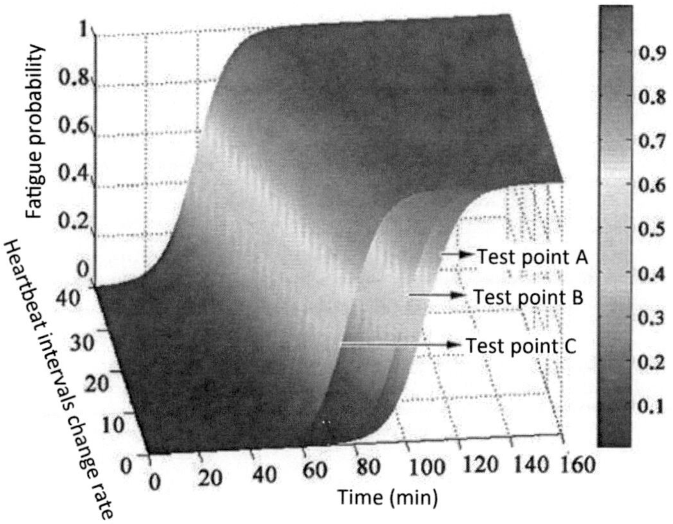

Fig. 4.32 Fatigue curve of local drivers

interval change rate is taken as 7.5%. For non-local drivers ($N = 0$) and Local drivers ($N = 1$), calculate their fatigue probability at two test points, respectively, and draw a curve as shown in Fig. 4.33.

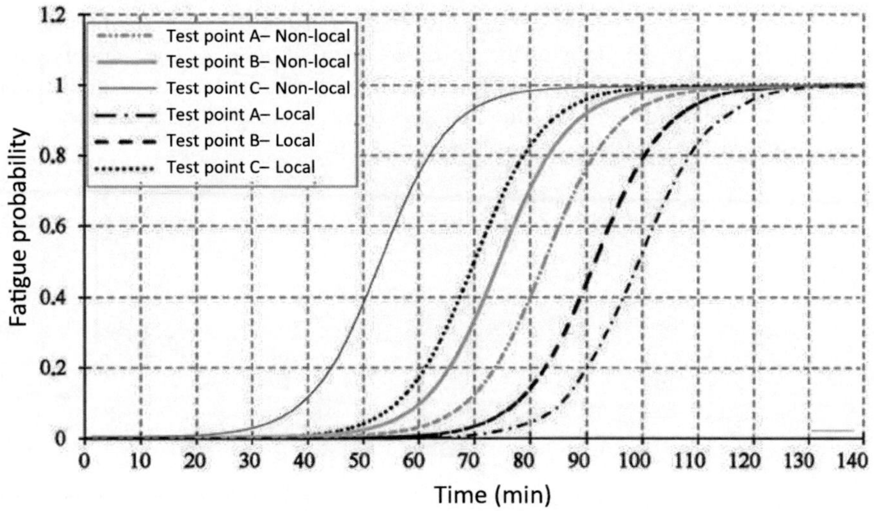

Fig. 4.33 The comparison of fatigue curves between non-local and local drivers

According to Fig. 4.33, the fatigue probability increases with time. The higher the altitude, the earlier the fatigue time point appears. The foreign and local drivers show apparent differences, and the local drivers have more robust anti-fatigue characteristics.

Take the fatigue probability of foreign drivers as 15% and 85%, respectively. The degree of driver fatigue is 15% and 85%, and the corresponding time is shown in Table 4.14.

Field simulation tests were carried out at three different altitude locations, the changes in the psychological and physiological data of the tested drivers and the actual performance status of the drivers were recorded, and the relevant data analysis was carried out within 2 min as the analysis unit. The fatigue threshold of each driver was determined one by one using the ROC curve, which is widely used in medicine.

The research results of driver fatigue characteristics show that with the increase in altitude, the driver fatigue time point gradually advances. Taking the Qinghai-Tibet Highway as an example, the fatigue time of the Tuotuohe is significantly earlier than that of Xidatan and Nachitai because the Tuotuohe is in the hinterland of the Qinghai-Tibet Plateau, with low oxygen content, and the climatic conditions are worse than

Table 4.14 Drivers' fatigue time

Test point	Fatigue time (min)	
	Degree of fatigue 15%	Degree of fatigue 85%
Test point A	71	92
Test point B	63	85
Test point C	41	63

the former. At the same time, through the analysis of the change law of heartbeat data, combined with the four factors of time length, altitude, driver category, and heartbeat interval change rate, and using the binary Logit model, the fatigue model of different types of drivers at different altitudes was successfully established.

Chapter 5
Operating Speed Model of Freeway in High Altitude Areas

According to the research of vehicle performance in plateau under low pressure and oxygen deficient environment, the dynamic performance of motor vehicles in high-altitude areas decreases by 7–8% when altitude increases 1,000 m, which indicates that the vehicle dynamic performance in high altitude areas is mainly affected by the highway vertical section alignment. In high altitude areas, although the horizontal alignment will affect the drivers' perception and operating ability to some extent, according to the research in Chap. 4, the horizontal alignment indicators have little impact on the drivers' perception and operating ability, and the highway horizontal alignment indicators are generally high. Therefore, when studying the highway operating speed model in high altitude areas, the impact of the vertical section alignment on the vehicle running speed should be considered. Therefore, according to the reduction law of vehicle dynamic performance, this chapter obtains the longitudinal slope with the same equilibrium velocity in the area of normal altitude (less affected by altitude factors), namely the equivalent slope, prepossessing a modified model of longitudinal slope in the operating speed model by analyzing the equilibrium speed of the highway at a certain actual longitudinal slope at different altitudes. Taking the existing high-grade highways in high-altitude areas (Lhasa-Nyingchi high-grade highway, Lhasa-Kongga airport highway) as the research object, the operation speed data of typical characteristic sections of straight section, curve section, longitudinal slope section, curved slope section and other sections were collected respectively, and the accuracy of the model was verified and analyzed in combination with the operation speed model of the current *Code for Safety Assessment of Highway Projects* and the high-altitude longitudinal slope correction model. Therefore, the prediction model of the operating speed distribution of two typical vehicles on the plateau freeway is proposed. The research content of this chapter provides the basis for the research on safety design methods such as design speed segmentation technology of freeway in high altitude areas.

J. Liu, *Technical Indicators and Safety Design of Freeway in High Altitude Area*,
https://doi.org/10.1007/978-981-99-0620-8_5

Table 5.1 Specific parameters of transmission

Type	MQ200-5F	Main drive reduction ratio		4.235
Gear	5th gear	Master transmission ratio	1st gear	3.769
Transmission efficiency	90%		2nd gear	2.095
Maximum gear ratio	3.545		3rd gear	1.433
Weight	33.5 kg		4th gear	1.079
Speedometer transmission ratio	13:22		5th gear	0.851
Type of external gear shifting mechanism	Pull lock type		Reverse gear	3.182
Maximum torque	200 N m			

Note Tire type: 205/55/R16, Tire radius r = 0.306 m

5.1 Representative Model

5.1.1 Representative Model of Passenger Cars

Takes Shanghai Volkswagen, Sagitar 2012, 1.6, manual fashion car as the research model. The main parameters are as follows.

Curb Weight is 1,275 kg, Full Load Mass is 1,650 kg, and Calculated Vehicle Weight is 1500 kg.

Engine Type: EA11, Output Volume: 1.6 L, Maximum Power: $P_{max} = 77$ kW (5600 r/min), Maximum Torque: $T_{tgmax} = 155$ N m (3500 r/min).

Transmission Type: MQ200; Final Drive Ratio: 4.235; Forward Gear Ratio: 3.769 for gear 1, 2.095 for gear 2, 1.433 for gear 3, 1.079 for gear 4, and 0.851 for gear 5; Transmission Efficiency: $\eta_T = 0.90$.

Table 5.1 shows specific parameters of transmission.

5.1.2 Representative Model of Trucks

Take Dongfeng Tianlong Tractor, DFL4251A9, combined with Tank-Type Semi-Trailer, Dongyue CQ9401GYY, as the research model for trucks. The main parameters are as follows.

Curb Weight is 8,500 kg, Total Quasi Traction Weight is 40,000 kg, and Standard Calculated Vehicle Weight is 28,000 kg.

Windward Area: $A = 7$ m^2.

Wheel Radius: $R = 0.53775$ m.

Maximum Engine Power (Revolution Speed): 250 kW (1900 r/min).

Maximum Engine Torque (Revolution Speed): 1500 N m (1200–1300 r/min).

Table 5.2 Transmission ratio and rotating mass conversion coefficient

Gear	1	2	3	4	5	6	7
Transmission ratio	12.1	9.41	7.31	5.71	4.46	3.48	2.71
Conversion coefficient	1.5947	1.3811	1.2411	1.1586	1.1079	1.0769	1.0573
Gear	8	9	10	11	12	R1	R2
Transmission ratio	2.11	1.64	1.28	1.00	0.78	11.56	2.59
Conversion coefficient	1.0445	1.0371	1.0322	1.0280	1.0248	–	–

Axle Speed Ratio (Final Drive Ratio): $i_0 = 4.1$.

Transmission Ratio of every gear, and Rotating Mass Conversion Coefficient δ shows in Table 5.2.

Relevant parameters of brake are as follows:

Number of equipment: 12;

Brake Drum Mass: $m_g = 80.05$ kg;

Specific Heat Capacity of Brake Drum: $c_g = 482$ J/(kg ^{0}C);

External Area of Brake Drum: $A_g = 0.39$ m^2.

5.2 Influence of Power and Resistance Reduction on Vehicle Operating Speed

Equilibrium Velocity refers to the speed at which the vehicle loses acceleration ability under driving conditions such as corresponding gear, accelerator force, longitudinal slope, and altitude. Its characteristic is that the vehicle acceleration resistance is zero, that is, the driving force in the vehicle driving equation is equal to the driving resistance. This section focuses on the influence of altitude on vehicle equilibrium velocity and establishes corresponding calculation model. According to the driving force balance equation, when the vehicle loses its acceleration ability, the acceleration resistance $F_j = 0$. So, the balance equation of automobile force can be simplified as:

$$F_t = F_i + F_w + F_f \tag{5.1}$$

Substitute the expressions of driving force and resistance to get:

$$\frac{\eta_T k_P k_T T_{tq} i_0 i_g}{r} = mgi + k_a m_0 \left(5.3 \times 10^{-6} \upsilon^2\right) + m\left(3.07 \times 10^{-4} \upsilon + 0.0536\right) \tag{5.2}$$

where,

Naturally Aspirated Engine

$$k_{\text{p}} = \frac{1 - 1.2360 \times 10^{-4} H + 5.2960 \times 10^{-9} H^2}{0.9363}$$

Intercooled Supercharged Engine

$$k_{\text{P}} = \frac{1 - 7.6238 \times 10^{-5} H^2}{0.95977}$$

$$k_{\text{a}} = \frac{1 - 1.0028 \times 10^{-4} H + 3.3795 \times 10^{-9} H^2}{0.948}$$

Let $k_{\text{m}} = \text{m}_0/\text{m}$, the formula is transformed into

$$i = \frac{1}{g}\left(\frac{\eta_{\text{T}} k_{\text{p}} k_{\text{T}} T_{\text{tq}} i_0 i_{\text{g}}}{mr} - 5.3 \times 10^{-6} k_{\text{a}} k_{\text{m}} v^2 - 3.07 \times 10^{-4} v - 0.0536\right) \quad (5.3)$$

The engine reference torque:

$$T_{\text{tq}} = -9.325 \times 10^{-4} n^2 + 2.460n + 27.120$$

Engine torque load factor:

$$k_{\text{T}} = 1$$

Mechanical efficiency:

$$\eta_{\text{T}} = 0.838$$

The gradeability evaluated here is the gradeability of the engine in non-intense state, so the engine revolution speed is set as 2000 r/min.

5.3 Relationship Between Equilibrium Velocity and Equivalent Slope

According to the relationship model between equilibrium velocity and altitude obtained in the previous section, and in combination with the balance equation of driving force, the longitudinal slope of vehicles at different altitudes with the same equilibrium velocity can be obtained. This paper puts forward the equivalent slope theory based on altitude. By determining the equivalent slope of different altitudes in high altitude areas relative to the normal altitude, the application method of this theory in the operating speed prediction model is proposed.

5.3.1 *Equivalent Slope and Slope Offset Less Than 80 km/h*

(1) Equivalent Slope and Slope Offset

If the slope i_1 at the general altitude (527.65 m) is the same as the equilibrium velocity corresponding to the actual slope i_2 at the corresponding altitude, the slope i_1 is called the equivalent slope of the actual slope i_2 at the corresponding altitude. At the same time $\Delta i = i_1 - i_2$ is called slope offset or slope correction.

The slope offset at the corresponding altitude can be used as the main technical basis for grade reduction (technical standard).

(2) Equivalent Slope Equations (Equilibrium Velocity ν)

The relationship between engine revolution speed n and speed is as follows

$$\upsilon = \frac{2\pi nr \times 60}{1\,000 i_0 i_g} = \frac{3\pi nr}{25 i_0 i_g}$$

Then,

$$\frac{i_0 i_g}{r} = \frac{3\pi n}{25\upsilon}$$

Therefore, under the altitude of 527.65 m and altitude H, the balance equations of vehicle forces corresponding to the equilibrium velocity of the uphill section are:

$$\frac{3\pi n\eta_T k_T T_{tq}}{25m\upsilon} = gi_1 + 5.3 \times 10^{-6}\frac{m_0}{m}\upsilon^2 + \left(3.07 \times 10^{-4}\upsilon + 0.0536\right) \quad (5.4)$$

$$\frac{3\pi n\eta_T k_p k_T T_{tq}}{25m\upsilon} = gi_2 + 5.3 \times 10^{-6}k_a\frac{m_0}{m}\upsilon^2 + \left(3.07 \times 10^{-4}\upsilon + 0.0536\right) \quad (5.5)$$

The equations formed by two simultaneous equations are called equivalent slope equations.

The relevant variables and parameters are calculated as follows:

$$m_0 = 56000 \text{ kg};$$
$$g = 9.8 \text{ m/s}^2;$$
$$n = 2000 \text{ r/min};$$
$$\eta_T = 0.838;$$
$$k_T = 1;$$
$$T_{tq} = -9.325 \times 10^{-4}n^2 + 2.460n + 27.120 \text{ N m};$$

Naturally Aspirated Engine:

$$k_p = \frac{1 - 1.2360 \times 10^{-4}H + 5.2960 \times 10^{-9}H^2}{0.9363}$$

Intercooled Supercharged Engine:

$$k_P = \frac{1 - 7.6238 \times 10^{-5} H^2}{0.95977}$$

$$k_a = \frac{1 - 1.0028 \times 10^{-4} H + 3.3795 \times 10^{-9} H^2}{0.948}$$

where,

- H Altitude (m);
- m Actual Vehicle Weight (kg);
- v Equilibrium Velocity (m/s);
- i_1 Equivalent Slope;
- i_2 Actual Slope.

At the corresponding altitude, only one of the three variables i_1, i_2, v, in the equivalent slope equation group is independent.

(3) Slope offset based on equilibrium velocity v

The equivalent slope equation group is subtracted, and the transformation is as follows:

$$\frac{3\pi n\eta_T k_T T_{tq}}{25mv}\left(1 - k_p\right) = g(i_1 - i_2) + 5.3 \times 10^{-6}(1 - k_a)\frac{m_0}{m}v^2 \quad (5.6)$$

Let $\Delta i = i_1 - i_2$, the calculation formula of slope offset based on equilibrium velocity is:

$$\Delta i = \frac{1}{mg}\left[\frac{3\pi n\eta_T k_T T_{tq}}{25v}(1 - k_P) - 5.3 \times 10^{-6}(1 - k_a)m_0 v^2\right] \quad (5.7)$$

Table 5.3 shows the slope offset corresponding to the equilibrium velocity of 80 km/h calculated by Formula (5.7).

Table 5.4 shows the slope offset corresponding to the equilibrium velocity of 80 km/h.

Table 5.5 shows the slope offset corresponding to the equilibrium velocity of 75 km/h calculated by Formula (5.7).

The slope offset (equilibrium velocity 75 km/h) corresponding to weight 30t vehicle in Table 5.5 is shown in Fig. 5.1.

The formula is as follows:

$$\Delta i = -2.061 \times 10^{-3} + 4.004 \times 10^{-6} H - 1.770 \times 10^{-10} H^2 \quad (5.8)$$

Table 5.6 shows the slope offset corresponding to the equilibrium velocity of 75 km/h.

Table 5.3 Slope offset corresponding to equilibrium velocity of 80 km/h

Weight/altitude (kg/m)	Base slope	527.65	1000	1500	2000	2500	3000
49,000	0.81%	0.00%	0.10%	0.20%	0.29%	0.38%	0.46%
28,000	2.01%	0.00%	0.17%	0.34%	0.51%	0.66%	0.81%
10,000	7.07%	0.00%	0.48%	0.96%	1.42%	1.86%	2.27%
20,000	3.14%	0.00%	0.24%	0.48%	0.71%	0.93%	1.14%
30,000	1.83%	0.00%	0.16%	0.32%	0.47%	0.62%	0.76%
40,000	1.17%	0.00%	0.12%	0.24%	0.36%	0.47%	0.57%
50,000	0.78%	0.00%	0.10%	0.19%	0.28%	0.37%	0.45%
60,000	0.51%	0.00%	0.08%	0.16%	0.24%	0.31%	0.38%
Weight/altitude (kg/m)	Base slope	3500	4000	4500	5000	5500	6000
49,000	0.81%	0.54%	0.62%	0.69%	0.75%	0.81%	0.86%
28,000	2.01%	0.95%	1.08%	1.20%	1.31%	1.42%	1.51%
10,000	7.07%	2.66%	3.02%	3.36%	3.67%	3.96%	4.23%
20,000	3.14%	1.33%	1.51%	1.68%	1.84%	1.98%	2.11%
30,000	1.83%	0.89%	1.01%	1.12%	1.22%	1.32%	1.41%
40,000	1.17%	0.66%	0.76%	0.84%	0.92%	0.99%	1.06%
50,000	0.78%	0.53%	0.60%	0.67%	0.73%	0.79%	0.85%
60,000	0.51%	0.44%	0.50%	0.56%	0.61%	0.66%	0.70%

Note The reference slope (Altitude: 527.65 m) minus the slope offset is the slope corresponding to the equilibrium velocity at the corresponding altitude

The slope offset corresponding to the equilibrium speed of 70 km/h calculated by formula (5.7) is shown in Table 5.7.

Table 5.8 shows the slope offset corresponding to the equilibrium velocity of 70 km/h.

(4) **Application of Slope Offset Values**

The slope offset value can be used as an approximation for calculating the operating speed based on altitude. For example, to calculate the operating speed at an altitude of 5,000 m, only 1.31% of the slope offset corresponding to the typical calculated vehicle weight of 28,000 kg needs to be added to the actual slope, so that it can be approximately calculated according to the normal operating speed model; (0 at 5,000 m altitude is equivalent to 1.31% at 527.65 m altitude; 0.69% is equivalent to 2% at 527.65 m altitude). The trucks' longitudinal slope on a flat section (−2.0%, 2.0%) will change to (−3.31%, 0.69%) at 5,000 m, the longitudinal slope of 1.31% plays the role of dynamic compensation in fact.

(5) **Realistic Slope Offset Values**

The slope offset values related to mass m and velocity v were derived in the previous section, but in practical applications the known variables are not the equilibrium

Table 5.4 Slope offset corresponding to equilibrium velocity of 80 km/h

Weight/altitude (kg/m)	527.65	1000	1500	2000	2500	3000
49,000	0.81%	0.71%	0.61%	0.52%	0.43%	0.35%
28,000	2.01%	1.84%	1.67%	1.50%	1.35%	1.20%
10,000	7.07%	6.59%	6.11%	5.65%	5.21%	4.80%
20,000	3.14%	2.90%	2.66%	2.43%	2.21%	2.00%
30,000	1.83%	1.67%	1.50%	1.35%	1.21%	1.07%
40,000	1.17%	1.05%	0.93%	0.81%	0.71%	0.60%
50,000	0.78%	0.68%	0.58%	0.49%	0.40%	0.32%
60,000	0.51%	0.43%	0.35%	0.28%	0.20%	0.14%
Weight/altitude (kg/m)	3500	4000	4500	5000	5500	6000
49,000	0.27%	0.19%	0.12%	0.06%	0.00%	−0.05%
28,000	1.06%	0.93%	0.81%	0.70%	0.60%	0.50%
10,000	4.42%	4.05%	3.71%	3.40%	3.11%	2.84%
20,000	1.81%	1.63%	1.46%	1.30%	1.16%	1.02%
30,000	0.94%	0.82%	0.71%	0.60%	0.50%	0.42%
40,000	0.51%	0.41%	0.33%	0.25%	0.18%	0.11%
50,000	0.24%	0.17%	0.10%	0.04%	−0.02%	−0.07%
60,000	0.07%	0.01%	−0.05%	−0.10%	−0.15%	−0.19%

Table 5.5 Slope offset corresponding to equilibrium velocity of 75 km/h

Weight/altitude (kg/m)	Base slope	527.65	1000	1500	2000	2500	3000
49,000	1.01%	0.00%	0.11%	0.22%	0.32%	0.42%	0.51%
28,000	2.35%	0.00%	0.19%	0.38%	0.56%	0.73%	0.90%
10,000	7.98%	0.00%	0.53%	1.06%	1.57%	2.05%	2.51%
20,000	3.60%	0.00%	0.26%	0.53%	0.79%	1.03%	1.25%
30,000	2.14%	0.00%	0.18%	0.35%	0.52%	0.68%	0.84%
40,000	1.41%	0.00%	0.13%	0.27%	0.39%	0.51%	0.63%
50,000	0.97%	0.00%	0.11%	0.21%	0.31%	0.41%	0.50%
60,000	0.68%	0.00%	0.09%	0.18%	0.26%	0.34%	0.42%
Weight/altitude (kg/m)	Base slope	3500	4000	4500	5000	5500	6000
49,000	1.01%	0.60%	0.68%	0.76%	0.83%	0.89%	0.95%
28,000	2.35%	1.05%	1.19%	1.33%	1.45%	1.56%	1.67%
10,000	7.98%	2.93%	3.34%	3.71%	4.06%	4.38%	4.68%
20,000	3.60%	1.47%	1.67%	1.86%	2.03%	2.19%	2.34%
30,000	2.14%	0.98%	1.11%	1.24%	1.35%	1.46%	1.56%
40,000	1.41%	0.73%	0.83%	0.93%	1.01%	1.10%	1.17%
50,000	0.97%	0.59%	0.67%	0.74%	0.81%	0.88%	0.94%
60,000	0.68%	0.49%	0.56%	0.62%	0.68%	0.73%	0.78%

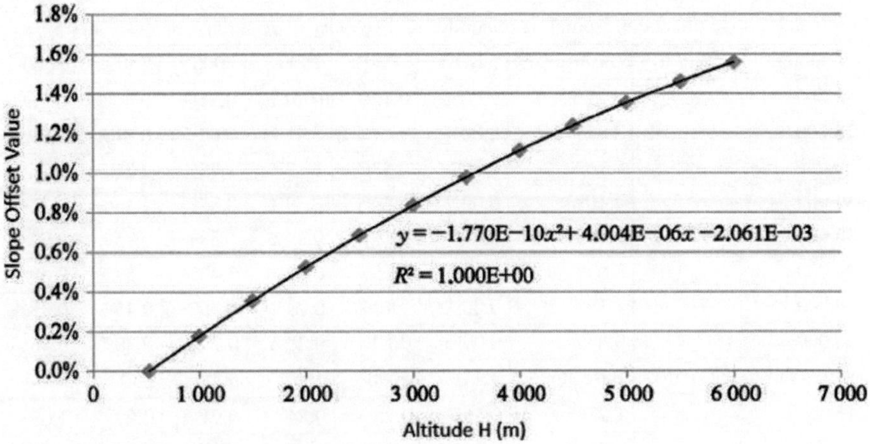

Fig. 5.1 Slope offset (30t, 75 km/h)

Table 5.6 Slope offset corresponding to equilibrium velocity of 75 km/h

Weight/altitude (kg/m)	527.65	1000	1500	2000	2500	3000
49,000	1.01%	0.90%	0.79%	0.69%	0.59%	0.49%
28,000	2.35%	2.16%	1.97%	1.79%	1.61%	1.45%
10,000	7.98%	7.45%	6.91%	6.41%	5.93%	5.47%
20,000	3.60%	3.33%	3.07%	2.81%	2.57%	2.34%
30,000	2.14%	1.96%	1.78%	1.61%	1.45%	1.30%
40,000	1.41%	1.28%	1.14%	1.02%	0.90%	0.78%
50,000	0.97%	0.86%	0.76%	0.66%	0.56%	0.47%
60,000	0.68%	0.59%	0.50%	0.42%	0.34%	0.26%
Weight/altitude (kg/m)	3500	4000	4500	5000	5500	6000
49,000	0.41%	0.33%	0.25%	0.18%	0.11%	0.05%
28,000	1.30%	1.16%	1.02%	0.90%	0.78%	0.68%
10,000	5.04%	4.64%	4.27%	3.92%	3.60%	3.30%
20,000	2.13%	1.93%	1.74%	1.57%	1.41%	1.26%
30,000	1.16%	1.03%	0.90%	0.79%	0.68%	0.58%
40,000	0.67%	0.57%	0.48%	0.39%	0.31%	0.24%
50,000	0.38%	0.30%	0.23%	0.16%	0.09%	0.03%
60,000	0.19%	0.12%	0.06%	0.00%	−0.05%	−0.10%

Table 5.7 Slope offset corresponding to equilibrium velocity of 70 km/h

Weight/altitude (kg/m)	Base slope	527.65	1000	1500	2000	2500	3000
49,000	1.22%	0.00%	0.12%	0.24%	0.35%	0.46%	0.56%
28,000	2.71%	0.00%	0.21%	0.42%	0.62%	0.81%	0.99%
10,000	8.96%	0.00%	0.58%	1.17%	1.73%	2.26%	2.76%
20,000	4.10%	0.00%	0.29%	0.59%	0.87%	1.13%	1.38%
30,000	2.48%	0.00%	0.19%	0.39%	0.58%	0.75%	0.92%
40,000	1.67%	0.00%	0.15%	0.29%	0.43%	0.57%	0.69%
50,000	1.18%	0.00%	0.12%	0.23%	0.35%	0.45%	0.55%
60,000	0.85%	0.00%	0.10%	0.20%	0.29%	0.38%	0.46%
Weight/altitude (kg/m)	Base slope	3500	4000	4500	5000	5500	6000
49,000	1.22%	0.66%	0.75	0.84	0.91	0.99	1.05
28,000	2.71%	1.16%	1.31	1.46	1.60	1.73	1.84
10,000	8.96%	3.24%	3.68	4.09	4.48	4.84	5.16
20,000	4.10%	1.62%	1.84	2.05	2.24	2.42	2.58
30,000	2.48%	1.08%	1.23	1.36	1.49	1.61	1.72
40,000	1.67%	0.81%	0.92	1.02	1.12	1.21	1.29
50,000	1.18%	0.65%	0.74	0.82	0.90	0.97	1.03
60,000	0.85%	0.54%	0.61	0.68	0.75	0.81	0.86

Table 5.8 Slope offset corresponding to equilibrium velocity of 70 km/h

Weight/altitude (kg/m)	527.65	1000	1500	2000	2500	3000
49,000	1.22%	1.10%	0.98%	0.87%	0.76%	0.65%
28,000	2.71%	2.50%	2.29%	2.09%	1.90%	1.72%
10,000	8.96%	8.38%	7.79%	7.23%	6.70%	6.20%
20,000	4.10%	3.81%	3.51%	3.23%	2.97%	2.71%
30,000	2.48%	2.28%	2.09%	1.90%	1.72%	1.55%
40,000	1.67%	1.52%	1.37%	1.23%	1.10%	0.97%
50,000	1.18%	1.06%	0.94%	0.83%	0.73%	0.63%
60,000	0.85%	0.76%	0.66%	0.57%	0.48%	0.39%
Weight/altitude (kg/m)	3500	4000	4500	5000	5500	6000
49,000	0.56%	0.47%	0.38%	0.30%	0.23%	0.16%
28,000	1.55%	1.39%	1.25%	1.11%	0.98%	0.86%
10,000	5.72%	5.28%	4.87%	4.48%	4.12%	3.80%
20,000	2.48%	2.26%	2.05%	1.86%	1.68%	1.52%
30,000	1.40%	1.25%	1.11%	0.98%	0.86%	0.75%
40,000	0.86%	0.75%	0.64%	0.55%	0.46%	0.37%
50,000	0.53%	0.44%	0.36%	0.28%	0.21%	0.15%
60,000	0.32%	0.24%	0.17%	0.11%	0.05%	−0.01%

velocity v, but the actual slope i_2, so it is necessary to derive the equivalent slope i_1 and the slope offset value $\Delta i = i_1 - i_2$ based on the actual slope i_2. For this reason, the analysis continues by returning to the set of equivalent slope equations.

(5–5) − (5–4) × k_a, is

$$\frac{3\pi n\eta_T(k_P - k_a)k_T T_{tq}}{25m} = gvi_2 - k_a gvi_1 + (1 - k_a)\left(3.07 \times 10^{-4}v + 0.0536\right)v$$

So there is

$$Av^2 + Bv + C = 0 \tag{5.9}$$

where,

$$A = 3.07 \times 10^{-4}(1 - k_a)$$
$$B = gi_2 - k_a gi_1 + 0.053\,6(1 - k_a)$$
$$C = -\frac{3\pi n\eta_T(k_P - k_a)k_T T_{tq}}{25m}$$

Solving for the variable v and rounding off the increasing roots according to the analysis yields

$$v = \frac{-B - \sqrt{B^2 - 4AC}}{2A} \tag{5.10}$$

Substituting (5.10) into (5.4), (5.5) or (5.6) will form the equation for the equivalent slope i_1. The following equation is substituted into (5.6) equation form.

$$\frac{3\pi n\eta_T k_T T_{tq}}{25mv}(1 - k_P) = g(i_1 - i_2) + 5.3 \times 10^{-6}(1 - k_a)\frac{m_0}{m}v^2 \tag{5.11}$$

where,

$$v = \frac{-B - \sqrt{B^2 - 4AC}}{2A}$$

$$A = 3.07 \times 10^{-4}(1 - k_a)$$
$$B = gi_2 - k_a gi_1 + 0.0536(1 - k_a)$$
$$C = -\frac{3\pi n\eta_T(k_P - k_a)k_T T_{tg}}{25m}$$

The v in this equation can also be downgraded, substituting $v^2 = -\frac{Bv+C}{A}$ into (5.11):

$$\frac{3\pi n\eta_T k_T T_{tq}}{25\upsilon}(1-k_P) = mg(i_1-i_2) - \frac{5.3\times10^{-6}(1-k_a)m_0(B\upsilon+C)}{A}$$

Continue to transform

$$\frac{3\pi n\eta_T k_T T_{tq}}{25}(1-k_P)A = Amg(i_1-i_2)\upsilon - 5.3\times10^{-6}(1-k_a)m_0\left(B\upsilon^2+C\upsilon\right)$$

Then substituting $\upsilon^2 = -\frac{B\upsilon+C}{A}$ into this equation,

$$\frac{3\pi n\eta_T k_T T_{tq}}{25}(1-k_P)A^2 = A^2mg(i_1-i_2)\upsilon - 5.3\times10^{-6}(1-k_a)m_0\left(AC\upsilon - B^2\upsilon - BC\right) \quad (5.12)$$

Thus, given the vehicle weight m, altitude H, and slope i_2, the equivalent slope i_1 can be calculated by using the root of the equation according to Eq. (5.11) or Eq. (5.12).

It should be noted that programming calculation can be less tedious, just substitute the actual slope i_2 into (5.5) to solve for the equilibrium velocity v and then into (5.4) to calculate i_1; or to substitute the equilibrium velocity v into (5.7) to calculate the slope offset value Δi.

(6) **Relevant Parameter Range**

In the set of equivalent slope equations for the three variables i_1, i_2, and v, the relevant parameters range as follows.

Equilibrium speed v range: [v_{min}, 80 km/h], where v_{min} is determined by the equivalent slope $i_1 = 10\%$.

Equivalent slope i_1 range: [i_{1min}, 10%], where i_{1min} is determined by the equilibrium speed $v = 80$ km/h.

Actual slope i_2 range: [i_{2min}, i_{2max}], where i_{2min} is determined by the equilibrium speed $v = 80$ km/h and i_{2max} is determined by the equilibrium speed $v = v_{min}$ (equivalent to the equivalent slope $i_1 = 10\%$).

5.3.2 *Equivalent Slope and Slope Offset Values Greater Than 80 km/h*

(1) **Power Surplus and Engine Torque Load Factor**

The methods of "calculating the slope offset by given the equilibrium speed" and "calculating the equivalent slope by given the actual slope" are derived above, and the equilibrium speed range for these two methods is [v_{min}, 80 km/h], where v_{min} is determined by the equivalent slope $i_1 = 10\%$.

According to the previous article "corresponding slope at equilibrium speed of 80 km/h", at an altitude of 6,000 m the equilibrium speed of a 28t vehicle at 0.5% of slope has reached 80 km/h; the equilibrium speed of a 50t vehicle will reach 80 km/h only at the downhill (−0.07%) at the same altitude.

When the slope is less than "the slope corresponding to the balance speed of 80 km/h", the balance speed will be greater than 80 km/h, and the actual running speed will not exceed this speed, so there will be a power surplus. And the surplus power is not output, so the engine torque load factor will be reduced accordingly.

(2) **Equivalent Slope Equation System (Desired Speed v_e)**

When the equilibrium speed is greater than 80 km/h, the equilibrium equations of the car force corresponding to the torque load factor of the car engine at the usual altitude (527.65 m) and the altitude H are as follows:

$$\frac{3\pi n\eta_T k_T T_{tq}}{25mv_e} = gi_1 + 5.3\times10^{-6}\frac{m_0}{m}v_e^2 + \left(3.07\times10^{-4}v_e + 0.0536\right) \quad (5.13)$$

$$\frac{3\pi n\eta_T k_P k_T T_{tq}}{25mv_e} = gi_2 + 5.3\times10^{-6}k_a\frac{m_0}{m}v_e^2 + \left(3.07\times10^{-4}v_e + 0.0536\right) \quad (5.14)$$

The system of equations formed by combining the two equations is also called the equivalent system of slope equations.

The relevant variables and parameters are calculated as follows.

$$m_0 = 56000 \text{ kg};$$

$$g = 9.8 \text{ m/s}^2;$$

$$n = 2000 \text{ r/min};$$

$$\eta_T = 0.838;$$

$$v_e = 80 \text{ km/h};$$

$$T_{tq} = -9.325\times10^{-4}n^2 + 2.460n + 27.120(\text{N m});$$

$$k_P = \frac{1 - 1.2360\times10^{-4}H + 5.2960\times10^{-9}H^2}{0.936\,3};$$

$$k_a = \frac{1 - 1.0028\times10^{-4}H + 3.3795\times10^{-9}H^2}{0.948};$$

where,

H Altitude (m).

m Actual vehicle weight (kg).
k_T Engine torque load factor (corresponding to the desired speed v_e).
i_1 Equivalent slope.
i_2 Actual slope.

At the corresponding elevation, only one of the three variables i_1, i_2, k_T in the set of equivalent slope equations is independent.

(3) **Slope Offset Value Based on Load Factor**

The two equations of the equivalent slope equation set are subtracted and transformed as follows.

$$\frac{3\pi n\eta_T k_T T_{tq}}{25mv_e}(1-k_P) = g(i_1 - i_2) + 5.3\times 10^{-6}(1-k_a)\frac{m_0}{m}v_e^2 \quad (5.15)$$

Make $\Delta i = i_1 - i_2$, so that the slope offset value based on the engine torque load factor is calculated as:

$$\Delta i = \frac{1}{mg}\left[\frac{3\pi n\eta_T k_T T_{tq}}{25v_e}\left(1-k_p\right) - 5.3\times 10^{-6}(1-k_a)m_0 v_e^2\right] \quad (5.16)$$

The corresponding slope offset values for the load factor 1.0 using Eq. (5.16) are shown in Table 5.9.

The data in this table are consistent with the slope offset values corresponding to the equilibrium speed of 80 km/h.

The slope offset values corresponding to the load factor 0.5 are calculated using Eq. (5.16) in Table 5.10.

The slope offset values corresponding to the load factor 0.0 are calculated using Eq. (5.16) in Table 5.11.

The slope offset values corresponding to the load factor −0.5 are calculated using Eq. (5.16) in Table 5.12.

(4) **Realistic Slope Offset Values**

The slope offset values related to the mass m and the engine torque load factor are derived in the previous section, but in practice the known variables are not the engine torque load factor but the actual slope i_2. Therefore, it is necessary to derive the equivalent slope i_1 and the slope offset $\Delta i = i_1 - i_2$ based on the actual slope i_2.

According to Eq. (5.14), the transformation gives

$$k_T = \frac{gi_2 + 5.3\times 10^{-6}k_a\frac{m_0}{m}v_e^2 + \left(3.07\times 10^{-4}v_e + 0.053\,6\right)}{\frac{3\pi n\eta_T k_p T_{tg}}{25mv_e}}$$

Table 5.9 Slope offset values (load factor 1.0)

Weight of vehicle/ altitude (kg/m)	Benchmark slope	527.65	1000	1500	2000	2500	3000
49,000	0.81%	0.00%	0.10%	0.20%	0.29%	0.38%	0.46%
28,000	2.01%	0.00%	0.17%	0.34%	0.51%	0.66%	0.81%
10,000	7.07%	0.00%	0.48%	0.96%	1.42%	1.86%	2.27%
20,000	3.14%	0.00%	0.24%	0.48%	0.71%	0.93%	1.14%
30,000	1.83%	0.00%	0.16%	0.32%	0.47%	0.62%	0.76%
40,000	1.17%	0.00%	0.12%	0.24%	0.36%	0.47%	0.57%
50,000	0.78%	0.00%	0.10%	0.19%	0.28%	0.37%	0.45%
60,000	0.51%	0.00%	0.08%	0.16%	0.24%	0.31%	0.38%
Weight of vehicle/ altitude (kg/m)	Benchmark slope	3500	4000	4500	5000	5500	6000
49,000	0.81%	0.54%	0.62%	0.69%	0.75%	0.81%	0.86%
28,000	2.01%	0.95%	1.08%	1.20%	1.31%	1.42%	1.51%
10,000	7.07%	2.66%	3.02%	3.36%	3.67%	3.96%	4.23%
20,000	3.14%	1.33%	1.51%	1.68%	1.84%	1.98%	2.11%
30,000	1.83%	0.89%	1.01%	1.12%	1.22%	1.32%	1.41%
40,000	1.17%	0.66%	0.76%	0.84%	0.92%	0.99%	1.06%
50,000	0.78%	0.53%	0.60%	0.67%	0.73%	0.79%	0.85%
60,000	0.51%	0.44%	0.50%	0.56%	0.61%	0.66%	0.70%

The equivalent slope i_1 can be obtained by substituting into Eq. (5.13) and calculating the offset value $\Delta i = i_1 - i_2$.

(5) **Relevant Parameter Range**

In the set of equivalent slope equations for the three variables i_1, i_2, and k_T, the relevant parameters range as follows.

Load factor k_T range: $[-k_{Tmin}, 1.0]$, where k_{Tmin} is determined by engine braking performance.

Equivalent slope i_1 range: $[i'_{1\,min}, i'_{1max}]$, where $i'_{1max} = i_{1min}$ is determined by the equilibrium speed $v = 80$ km/h, and $i'_{1\,min}$ is determined by k_{Tmin}.

Actual slope i_2 range: $[\, i'_{2\,min}, i'_{2max}]$, where $i'_{2max} = i_{2min}$ is determined by the equilibrium speed $v = 80$ km/h, and $i'_{2\,min}$ is determined by k_{Tmin}.

See i_{1min}, i_{2min} from Sect. 5.3.1 of the book.

Table 5.10 Slope offset values (load factor 0.5)

Weight of vehicle/ altitude (kg/m)	Benchmark slope	527.65	1000	1500	2000	2500	3000
49,000	−0.19%	0.00%	0.04%	0.08%	0.12%	0.15%	0.19%
28,000	0.26%	0.00%	0.07%	0.14%	0.21%	0.27%	0.33%
10,000	2.17%	0.00%	0.19%	0.39%	0.57%	0.75%	0.91%
20,000	0.69%	0.00%	0.10%	0.19%	0.29%	0.37%	0.46%
30,000	0.19%	0.00%	0.06%	0.13%	0.19%	0.25%	0.30%
40,000	−0.06%	0.00%	0.05%	0.10%	0.14%	0.19%	0.23%
50,000	−0.20%	0.00%	0.04%	0.08%	0.11%	0.15%	0.18%
60,000	−0.30%	0.00%	0.03%	0.06%	0.10%	0.12%	0.15%
Weight of vehicle/ altitude (kg/m)	Benchmark slope	3500	4000	4500	5000	5500	6000
49,000	−0.19%	0.22%	0.25%	0.27%	0.30%	0.32%	0.34%
28,000	0.26%	0.38%	0.43%	0.48%	0.52%	0.56%	0.60%
10,000	2.17%	1.07%	1.21%	1.34%	1.46%	1.58%	1.68%
20,000	0.69%	0.53%	0.60%	0.67%	0.73%	0.79%	0.84%
30,000	0.19%	0.36%	0.40%	0.45%	0.49%	0.53%	0.56%
40,000	−0.06%	0.27%	0.30%	0.34%	0.37%	0.39%	0.42%
50,000	−0.20%	0.21%	0.24%	0.27%	0.29%	0.32%	0.34%
60,000	−0.30%	0.18%	0.20%	0.22%	0.24%	0.26%	0.28%

5.4 Operating Speed Prediction Model Considering Equivalent Slope

According to the equivalent slope theory, the operating speed prediction model for high altitude areas is used as follows.

i. Calculate the equilibrium velocity corresponding to the actual slope. According to the equilibrium velocity equation:

$$\frac{\eta_T k_P k_T T_{tq^2} i_0 i_g}{r} = mgi + k_a m_0 \left(5.3 \times 10^{-6} \upsilon^2\right) + m\left(3.07 \times 10^{-4} \upsilon + 0.053\,6\right) \tag{5.17}$$

Solve the equation to obtain the equilibrium velocity *v*. The relevant parameters are as follows.

$$m_0 = 56000 \text{ kg};$$

$$g = 9.8 \text{ m/s}^2;$$

Table 5.11 Slope offset values (load factor 0.0)

Weight of vehicle/altitude (kg/m)	Benchmark slope	527.65	1000	1500	2000	2500	3000
49,000	−1.19%	0.00%	−0.02%	−0.04%	−0.06%	−0.07%	−0.09%
28,000	−1.49%	0.00%	−0.03%	−0.07%	−0.10%	−0.13%	−0.16%
10,000	−2.74%	0.00%	−0.09%	−0.19%	−0.28%	−0.36%	−0.45%
20,000	−1.77%	0.00%	−0.05%	−0.09%	−0.14%	−0.18%	−0.22%
30,000	−1.44%	0.00%	−0.03%	−0.06%	−0.09%	−0.12%	−0.15%
40,000	−1.28%	0.00%	−0.02%	−0.05%	−0.07%	−0.09%	−0.11%
50,000	−1.19%	0.00%	−0.02%	−0.04%	−0.06%	−0.07%	−0.09%
60,000	−1.12%	0.00%	−0.02%	−0.03%	−0.05%	−0.06%	−0.07%
Weight of vehicle/Altitude (kg/m)	Benchmark slope	3500	4000	4500	5000	5500	6000
49,000	−1.19%	−0.11%	−0.12%	−0.14%	−0.15%	−0.17%	−0.18%
28,000	−1.49%	−0.19%	−0.22%	−0.24%	−0.27%	−0.29%	−0.31%
10,000	−2.74%	−0.53%	−0.60%	−0.68%	−0.75%	−0.81%	−0.88%
20,000	−1.77%	−0.26%	−0.30%	−0.34%	−0.37%	−0.41%	−0.44%
30,000	−1.44%	−0.18%	−0.20%	−0.23%	−0.25%	−0.27%	−0.29%
40,000	−1.28%	−0.13%	−0.15%	−0.17%	−0.19%	−0.20%	−0.22%
50,000	−1.19%	−0.11%	−0.12%	−0.14%	−0.15%	−0.16%	−0.18%
60,000	−1.12%	−0.09%	−0.10%	−0.11%	−0.12%	−0.14%	−0.15%

$$n = 2000\ \text{r/min};$$

$$\eta_{\text{T}} = 0.838;$$

$$k_{\text{T}} = 1;$$

$$T_{\text{tq}} = -9.325 \times 10^{-4} n^2 + 2.460n + 27.120(\text{N} \cdot \text{m});$$

$$k_{\text{P}} = \frac{1 - 1.2360 \times 10^{-4} H + 5.2960 \times 10^{-9} H^2}{0.9363}$$

$$k_{\text{a}} = \frac{1 - 1.0028 \times 10^{-4} H + 3.3795 \times 10^{-9} H^2}{0.948}$$

where,

H Altitude (m).

Table 5.12 Slope offset values (load factor −0.5)

Weight of vehicle/altitude (kg/m)	Benchmark slope	527.65	1000	1500	2000	2500	3000
49,000	−2.19%	0.00%	−0.08%	−0.16%	−0.23%	−0.30%	−0.37%
28,000	−3.24%	0.00%	−0.13%	−0.27%	−0.40%	−0.53%	−0.64%
10,000	−7.64%	0.00%	−0.38%	−0.76%	−1.13%	−1.47%	−1.81%
20,000	−4.22%	0.00%	−0.19%	−0.38%	−0.56%	−0.74%	−0.90%
30,000	−3.08%	0.00%	−0.13%	−0.25%	−0.38%	−0.49%	−0.60%
40,000	−2.51%	0.00%	−0.09%	−0.19%	−0.28%	−0.37%	−0.45%
50,000	−2.17%	0.00%	−0.08%	−0.15%	−0.23%	−0.29%	−0.36%
60,000	−1.94%	0.00%	−0.06%	−0.13%	−0.19%	−0.25%	−0.30%
Weight of vehicle/altitude (kg/m)	Benchmark slope	3500	4000	4500	5000	5500	6000
49,000	−2.19%	−0.43%	−0.49%	−0.55%	−0.60%	−0.65%	−0.70%
28,000	−3.24%	−0.76%	−0.86%	−0.96%	−1.06%	−1.14%	−1.22%
10,000	−7.64%	−2.12%	−2.42%	−2.69%	−2.96%	−3.20%	−3.43%
20,000	−4.22%	−1.06%	−1.21%	−1.35%	−1.48%	−1.60%	−1.71%
30,000	−3.08%	−0.71%	−0.81%	−0.90%	−0.99%	−1.07%	−1.14%
40,000	−2.51%	−0.53%	−0.60%	−0.67%	−0.74%	−0.80%	−0.86%
50,000	−2.17%	−0.42%	−0.48%	−0.54%	−0.59%	−0.64%	−0.69%
60,000	−1.94%	−0.35%	−0.40%	−0.45%	−0.49%	−0.53%	−0.57%

Note The negative engine torque load factor refers to the engine providing resistance, i.e., engine braking (including exhaust braking)

m Actual vehicle weight (kg).
v Equilibrium speed (km/h).

ii. Calculate the equivalent slope corresponding to the actual slope based on the equilibrium speed. When the equilibrium speed is less than the desired speed (120 km/h for passenger cars and 80 km/h for trucks), the difference between the equivalent slope i_1 and the actual slope i_2, that is $\Delta i = i_1 - i_2$ (slope offset value), is

$$\Delta i = \frac{1}{mg}\left[\frac{3\pi n\eta_T k_T T_{tq}}{25v}(1 - k_P) - 5.3 \times 10^{-6}(1 - k_a)m_0 v^2\right] \tag{5.18}$$

The equivalent slope i_1 is obtained by substituting the slope offset value formula.

iii. When the equilibrium speed is greater than the desired speed (truck $v_e = 80$ km/h),

$$\Delta i = \frac{1}{mg}\left[\frac{3\pi n \eta_{\mathrm{T}} k_{\mathrm{T}} T_{\mathrm{tq}}}{25 \upsilon_{\mathrm{e}}}(1 - k_{\mathrm{P}}) - 5.3 \times 10^{-6}(1 - k_{\mathrm{a}}) m_0 \upsilon_{\mathrm{e}}^2\right] \tag{5.19}$$

where,

$$k_{\mathrm{T}} = \frac{g i_2 + 5.3 \times 10^{-6} k_{\mathrm{a}} \frac{m_0}{m} \upsilon_{\mathrm{e}}^2 + \left(3.07 \times 10^{-4} \upsilon_{\mathrm{e}} + 0.053\,6\right)}{\frac{3\pi n \eta_{\mathrm{T}} k_{\mathrm{P}} T_{\mathrm{tg}}}{25 m \upsilon_{\mathrm{e}}}}$$

where $v_{\mathrm{e}} = 80$ km/h. The equivalent slope i_1 is obtained by substituting the slope offset value formula.

iv. Based on the equivalent slope it is possible to analyze and calculate the corresponding operating speed at high altitude using the prediction model for operating speed at usual altitude.
v. It must be emphasized that the equivalent slope conversion should be carried out before the route segmentation, which means that in terms of the actual slope, the flat section at high altitude is no longer fixed at −3.0% to 3.0%, but changes with the variation of altitude.
vi. Passenger cars also have the problem of equivalent slope conversion. Since the engine torque load factor of passenger cars is flexible, modes of "naturally aspirated equivalent" and "intercooled supercharged equivalent" are proposed, which can be used with the "naturally aspirated" and "intercooled supercharged" modes of trucks respectively.
vii. Complete operating speed prediction analysis.

5.5 Validation Analysis of Operating Speed Prediction Model

In order to verify the accuracy of the highway operating speed model in high altitude areas, the actual operating speed of highway in high altitude areas was collected on site. In Tibetan region, there are two high-grade highways: one is the Lhasa Kongga International Airport Highway, which was opened to traffic on July 17, 2011, with a total length of 37.8 km, it has been in operation for about 5 years by the time of speed measurement. Its management experience is maturing, and in order to prevent vehicles from speeding, the Lagoon airport highway uses both interval and spot speed measurement for speed control. The other one is the Linzhi-Lhasa highway (hereinafter referred to as Lin-La Highway), which is currently under construction, of which the construction of Lhasa-Mozhukongka section has been completed, and will be put into trial operation in the second half of 2015. The combination of fixed-point speed measurement and mobile speed measurement is used for traffic supervision, and only passenger cars are allowed to pass during the trial operation phase, and trucks are prohibited. Figure 5.2 shows the speed measuring equipment laid on the special highway of Lhasa Kongga International Airport.

Fig. 5.2 Laying speed measuring equipment

The velocities of three typical sections were collected, which are: straight section, curved section and combined section with curved slope. Tables 5.13 and 5.14 show the locations of stakes and related information of the speed collection equipment deployed on the Lhasa Gongga International Airport Highway and the Lin-La Highway.

(1) **Results and Comparison of Speed Collection**

The actual collected the 85th percentile speed were compared with the predicted results of the operating speed model, and the results are shown in Tables 5.15, 5.16 and 5.17.

From Fig. 5.3, it can be seen that the actual speed of the Lagoon airport highway is close to the predicted results, and the actual speed is less than the predicted speed, with an average error percentage of 3.45% for the straight section and 6.76% for the curved section.

From Fig. 5.4, it can be seen that the actual speed of the Lagoon airport highway is close to the predicted results, and the actual speed is greater than the predicted speed, with an average error percentage of 3.41% for the straight section and 3.75% for the curved section.

Table 5.13 MC layout of Lhasa Gongga International Airport Highway

No. of MC	Chain age	Sites	Slope (%)	Radius of flat curve	No. of samples
MC1	K23 + 400	Starting point of straight line	0.446	–	6886
MC2	K26 + 300	Ending point of straight line	−0.300	7000	6494
MC3	K27 + 400	Connecting point of straight and curve line	0.217	–	5963
MC4	K28 + 000	Middle point of curve line	−0.360	1100	3202
MC5	K28 + 700	Connecting point of curve and straight line	0.210	–	5488

Table 5.14 MC layout of Lin-La highway

No. of MC	Stakes	Sites	Slope (%)	Radius of flat curve	No. of samples
MC1	4595 + 900	Top of slope	3.4	1800	1994
MC2	4596 + 410	Bottom of slope			1929
MC3	4607 + 430	Connecting point of curve and straight line	2.5	1129	1718
MC4	4607 + 529	Middle point of curve line			1647
MC5	4608 + 530	Connecting point of straight and curve line			1568

Table 5.15 Actual speed of passenger cars on Lhasa Gongga International Airport Highway (km/h)

No. of MC	Test sites for passenger cars	Max. value	Min. value	Average	Standard deviation	85th percentile speed
MC1	Starting point of straight line	157	17	84	16	108
MC2	Ending point of straight line	152	15	91	16	105
MC3	Connecting point of straight and curve line	142	10	88	16	100
MC4	Middle point of curve line	152	12	90	13	103
MC5	Connecting point of curve and straight line	146	25	85	14	97

Table 5.16 Actual speed of trucks on Lhasa Gongga International Airport Highway (km/h)

No. of MC	Test sites for trucks	Max. value	Min. value	Average	Standard deviation	85th percentile speed
MC1	Starting point of straight line	132	17	77	14	91
MC2	Ending point of straight line	116	32	73	13	86
MC3	Connecting point of straight and curve line	118	34	72	12	83
MC4	Middle point of curve line	120	35	73	12	85
MC5	Connecting point of curve and straight line	120	33	71	11	81

Table 5.17 Actual speed of passenger cars on Lin-La highway (km/h)

No. of MC	Test sites for trucks	Max. value	Min. value	Average	Standard deviation	85th percentile speed
MC6	Connecting point of straight and curve line	159	21	81	17	97
MC5	Middle point of curve line	154	22	82	15	95
MC3	Connecting point of curve and straight line	155	15	81	7	95
MC2	Top of slope	159	11	86	18	102
MC1	Bottom of slope	123	19	70	12	81

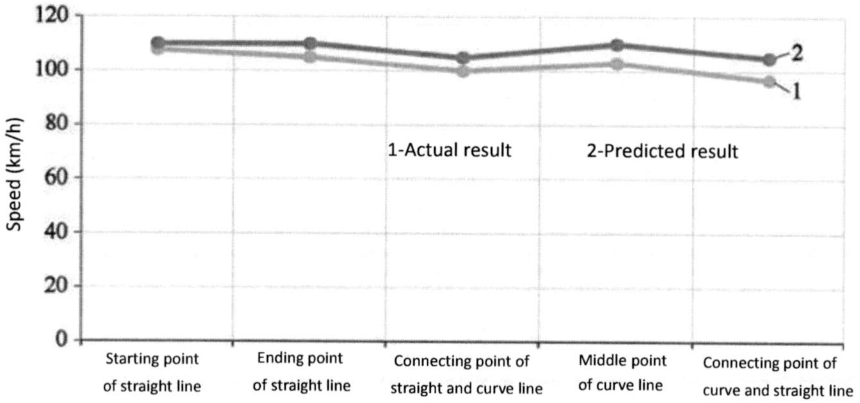

Fig. 5.3 Comparison of actual speed and predicted speed of Lhasa Kongga International Airport Highway (passenger cars)

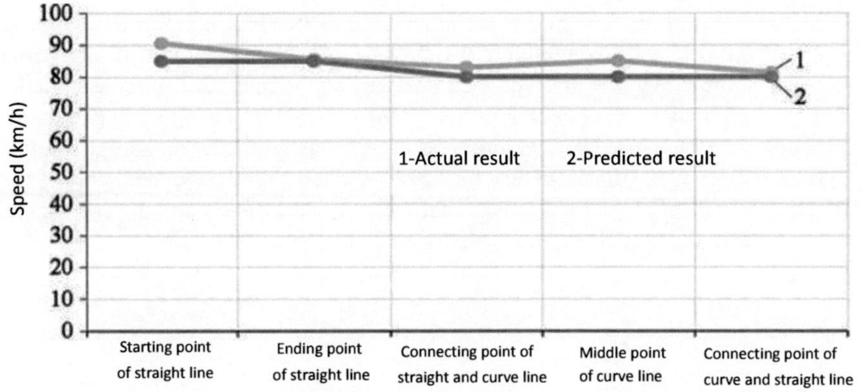

Fig. 5.4 Comparison of actual speed and predicted speed of Lhasa Kongga International Airport Highway (trucks)

From Fig. 5.5, it can be seen that the actual speed of Lin-La Highway is close to the predicted results, and the actual speed is less than the predicted speed, with an average error percentage of 6.2% for the curved section and 4.7% for the longitudinal section.

(2) **Analysis of the Results**

From the comparison analysis of actual collected speed and predicted speed, it can be concluded that: the average error of prediction for both trucks and passenger cars is less than 10%, and the error is basically 3–7%. In order to avoid over-speeding, both highways adopt the speed limit management method, so the vehicles cannot drive at free state, and the actual running speed will have some difference with the predicted speed.

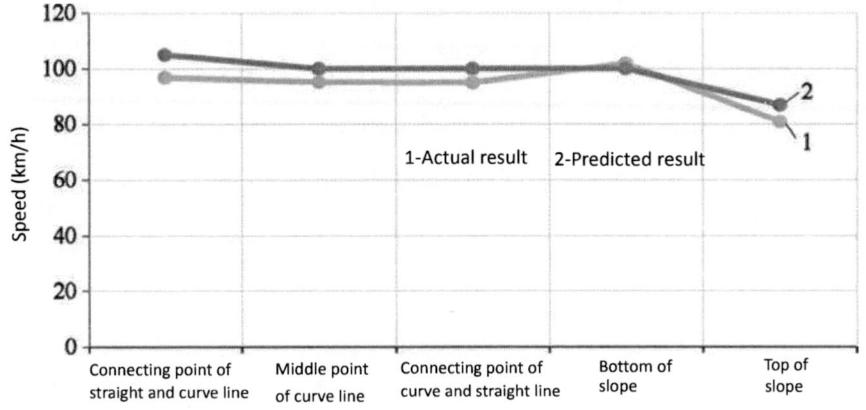

Fig. 5.5 Comparison of actual speed and predicted speed of Lin-La highway (passenger cars)

Due to the constraints of the current situation, only two high-grade highways have been built in Tibetan region, among which the Lin-La Highway is still in the trial operation stage at the time of speed measurement, so the speed collection and model accuracy verification are only carried out for the typical sections of these two roads. In the future, with the accelerated progress of highway construction in high altitude areas of China, it is still necessary to further expand the sample size, collect the operating speeds of typical sections of highways under different altitude conditions, further modify the model, and improve its prediction accuracy.

Chapter 6
Safety Geometric Design Techniques for Freeway in High Altitude Areas

In this chapter, according to the research of the vehicle performance change, the driver's psychological and physiological change characteristics, and the highway operating speed prediction model under the low pressure and oxygen deficient environment in the high-altitude area, the horizontal, vertical and horizontal geometric design indicators and parameters suitable for the highway in the high-altitude area are proposed. According to the service function and driving fatigue characteristics of highways in high-altitude areas, the types of service facilities and corresponding reasonable spacing are proposed. Based on factors such as highway function, topography and operating speed distribution, the design speed segmentation technology and speed transition design method of highways in the high-altitude area are proposed.

According to the different topographic and geological conditions and design speed selection criteria in the Qinghai-Tibet Plateau, this paper proposes a design process for optimizing the safety of highway routes in high-altitude areas and finally establishes the principles for selecting technical standards for highways in high-altitude areas. This chapter is the core research of this book. The research results of the design speed segmentation technology of highways in high-altitude areas, horizontal, vertical and horizontal geometric design indicators and parameters, and the spacing of service facilities are also directly applied to the preparation of "Safety Design Guidelines for Highways on Qinghai-Tibet Plateau".

Finally, the speed control plan of highways in a unique environment in high altitude areas is studied. Through the actual investigation of the vehicle speed of the Qinghai-Tibet Highway in high-altitude areas and the use of a large number of MetroCount traffic detector data, the influence of air oxygen content on the speed difference between trucks and passenger cars on straight sections and grade sections was explored. Under the road section, the speed difference gradually increases with the decrease of oxygen content; when the altitude is higher than 4350 m, the speed difference between trucks and passenger cars exceeds 20 km/h. The rule of influence of oxygen content on the variation is obtained through the research on the speed

J. Liu, *Technical Indicators and Safety Design of Freeway in High Altitude Area*,
https://doi.org/10.1007/978-981-99-0620-8_6

difference and operating speed characteristics of trucks and passenger cars on the Qinghai-Tibet Highway under different oxygen content.

6.1 Main Geometric Indicators and Parameters of Freeways in High Altitude Areas

The leading geometric indicators of the highway route include the radius of the circular curve, the grade and the length of the slope, the width of each component of the cross-section, the sight distance, and the super-elevation.

Based on the research results in Chaps. 3 and 4, this section studies the impact of factors such as vehicle performance, drivers' psychological and physiological changes, road surface conditions under ice and snow conditions, and road service functions on road geometric design indicator parameters in high-altitude, low-pressure, oxygen deficient environments in high-altitude areas. To establish the primary technical indicator system of highway geometric design that meets the requirements of safe operation in the plateau areas, low-pressure and oxygen deficient environment.

6.1.1 Research on Horizontal Alignment Indicator

The unique climatic conditions in the high-altitude area significantly impact the driver's driving. At the same time, the typical roads (such as G109) in the high-altitude area were constructed early and renovated several times in the later period. The alignment indicator of many road sections is challenging to meet the requirements of the current specification. In addition, when driving on the plateau, the driver's speed is generally lower than that in the plain area due to the poor road conditions and the high-altitude oxygen deficient environment. Therefore, this study analyzed the alignment indicator of high-altitude roads. They were combined with the driver's psychological and physiological characteristics, plateau environmental factors and running speed, hoping to obtain reasonable alignment indicator parameters that meet the safety of driving in unique environments in high-altitude areas.

(1) **Cut-off point of driver's psycho-physiological response**

According to the existing research, the corresponding heart rate growth rates under the three psychological states of driver comfort, tension and fear are 18%, 27% and 39%, respectively. Environmental factors change the driver's physiological threshold in a high-altitude and oxygen deficient environment. The cut-off value of each alignment is discussed through the four-variable model formula of heart rate growth rate, altitude, alignment indicator and vehicle speed.

(2) **Horizontal curve section**

(1) **Altitude 3567 ~ 3957 m**

By calculating the change law of the heart rate growth rate at the highest altitude, the limit value of the radius of the horizontal curve that meets the driving requirements in this altitude interval is obtained. Substitute V = 80 km/h, M = 3957 m into formula (4.4), and obtain the N-R two-factor relational formula:

$$N = 0.208R^{-0.002} \quad (V = 80\,\text{km/h},\ M = 3957\,\text{m}) \tag{6.1}$$

Substitute radius R into the above formula with an interval of 50 m and a range of [50, 1000], and draw the result as shown in Fig. 6.1.

Figure 6.1 shows that the function is convex and monotonically decreasing, there is no extreme value, and the curve tends to steep left and slow to right. Therefore, the maximum curvature of the curve is calculated by the following equation to obtain the cut-off point between the steep curves:

$$K = \frac{|N''|}{\left[1 + (N')^2\right]^{3/2}} \tag{6.2}$$

It was calculated that the curvature maximum point appeared near R = 300 m, corresponding to the heart rate growth rate of 20.62%. When the radius is less than 300 m, the heart rate growth rate is fast; when the radius is more significant than

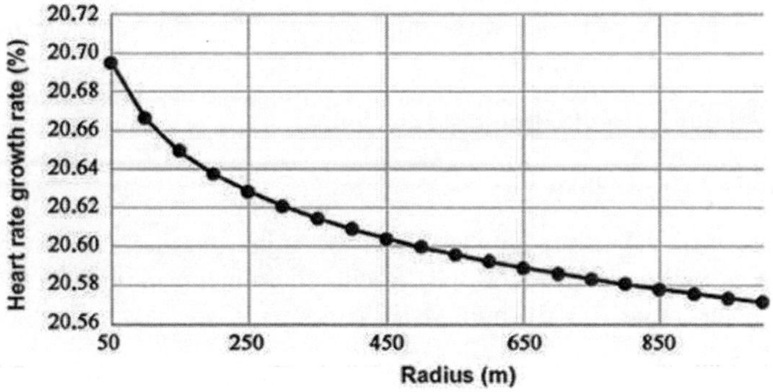

Fig. 6.1 Heart rate growth rate versus radius (1)

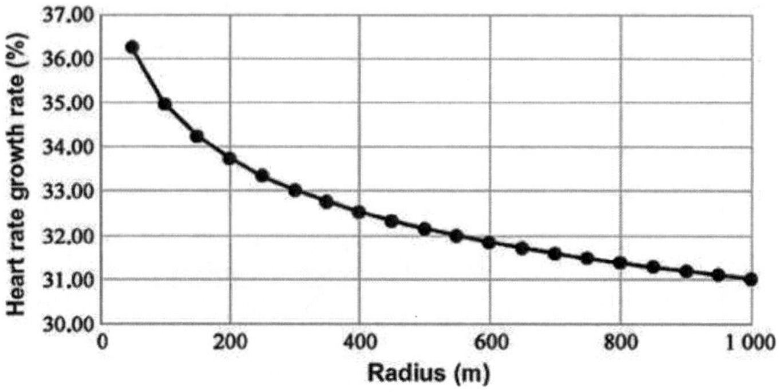

Fig. 6.2 Heart rate growth rate versus radius (2)

300 m, the heart rate growth rate is slow. It is dangerous to the driver when their heart rate growth rate changes abruptly.

(2) **Altitude 4100–4702 m**

By calculating the heart rate growth rate at the highest altitude, the limit value of the radius of the horizontal curve that meets the driving requirements in this altitude interval is obtained. Substitute V = 80 km/h, M = 4702 m into formula (4.5), and get N-R Two-factor relationship:

$$N = 0.444R^{0.052} \quad (V = 80\,\text{km/h},\ M = 4702\,\text{m}) \tag{6.3}$$

Substitute radius R into the above formula with an interval of 50 m and a range of [50, 1000], and draw the result as shown in Fig. 6.2. It was calculated that the maximum point of curvature change appeared at R = 350 m, corresponding to a heart rate growth rate of 32.77%. When the radius is less than 350 m, the heart rate growth rate is fast; when the radius is more significant than 350 m, the heart rate growth rate is slow. It is dangerous to the driver when the heart rate growth rate changes abruptly.

(3) **Curved slope combination road section**

(1) **Altitude 3616–3998 m**

By calculating the heart rate growth rate at the highest altitude, the limit value of the alignment indicator that meets the driving requirements in this altitude interval is obtained. Substitute V = 80 km/h, M = 3.616 km into formula (4.7), and get N-W two-factor relationship:

$$N = -1.224 \times 10^{-5} W^2 + 0.008W + 22.51 \quad (V = 80\,\text{km/h},\ M = 3.998\,\text{km}) \tag{6.4}$$

Substituting i = 20‰ into the above formula, we get

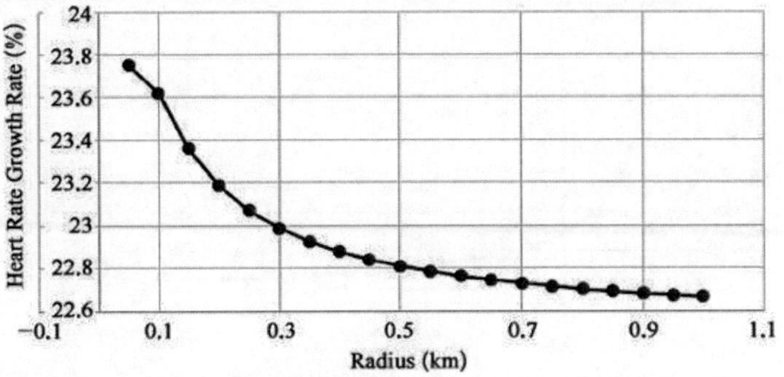

Fig. 6.3 Heart rate growth rate versus radius (3)

$$N = -4.896 \times 10^{-3} R^{-2} + 0.16R^{-1} + 22.51 \tag{6.5}$$

Substitute radius R into the formula (6.5) with the interval of 0.05 km and the range [0.05,1], and draw the result as shown in Fig. 6.3.

It is calculated that the maximum curvature change point appears at R = 300 m, corresponding to the alignment combination value W = 66.67‰/km, and the heart rate growth rate is 22.99%, which is consistent with the research results of the horizontal curve radius in the same altitude range.

(2) **Altitude 4127–4630 m**

By calculating the heart rate growth rate at the highest altitude, the limit value of the alignment indicator that meets the driving requirements in this altitude interval is obtained. Substitute V = 80 km/h, M = 4.630 km into formula (4.9), and obtain N-W two-factor relationship:

$$N = 0.008W + 32.4128 \quad (V = 80\,\text{km/h},\, M = 4.630\,\text{km}) \tag{6.6}$$

Substituting i = 20‰ into (6.6), we get

$$N = 0.16R^{-1} + 32.4128 \tag{6.7}$$

Substitute radius R into the formula (6.7) with the interval of 0.05 km and the range [0.05,1], and draw the result as shown in Fig. 6.4.

(4) **Recommended values of highway alignment indicator parameters on the plateau**

Summarize the calculated alignment indicator parameters and heart rate growth rate corresponding to the cut-off point of the driver's psychological and physiological response, and obtain the recommended values of the highway alignment indicator parameters on the plateau, as shown in Table 6.1.

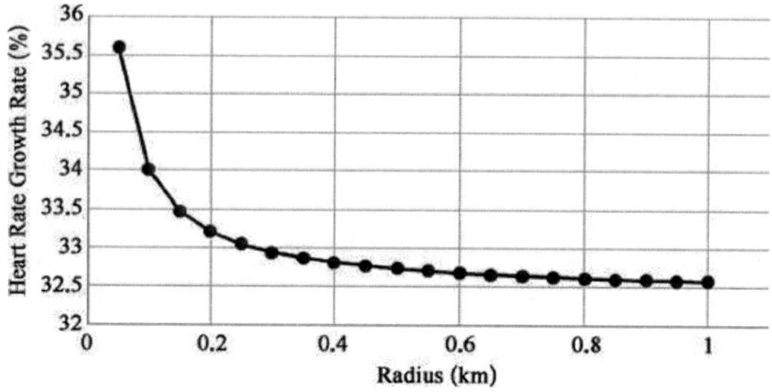

Fig. 6.4 Heart rate growth rate versus radius (4)

Table 6.1 Recommended values of highway alignment indicators on the plateau

Section	Horizontal curve radius R'(m)	Heart rate growth rate N'(%)	Alignment combination value W (‰km)
Altitude 3 567 ~ 3957 m horizontal curve	300	20.62	\
Altitude 4100 ~ 4 702 m horizontal curve	350	32.77	\
Altitude 3616 ~ 3 998 m curved slope combination	300	22.99	66.67
Altitude 4 127 ~ 4 630 m curved slope combination	250	33.05	80

According to previous research results, the heart rate growth rates for the three psychological states of driver comfort, tension and fear are 18%, 27% and 39%, respectively. The N' corresponding to R' can be considered as the physiological threshold of the driver under different road sections in the altitude interval.

The degree of change of the driver's heart rate growth rate before and after the corresponding R' differs for each road section. When $R < R'$, the heart rate growth rate increases significantly as the radius of the horizontal curve decreases; when $R > R'$, it increases with the radius of the horizontal curve, and the heart rate growth rate decreases slowly. Therefore, it is advisable to control the radius of the horizontal curve to be greater than R' to ensure driving safety.

6.1.2 Design Indicators of Gradient of Highway

In high-altitude areas, the oxygen content is low, and the vehicle's driving dynamics in this area are weakened. The design of the longitudinal gradient indicator must be re-determined in combination with reducing the vehicle's dynamic performance.

(1) **The calculation method for balance speed, acceleration and deceleration slope length**

The vehicle is driving on a road section with a particular slope. If the initial speed is too high, the vehicle's speed gradually decreases with the driving distance increase, and the driver adopts a lower gear. Then the vehicle drives at a constant speed on the longitudinal gradient. If the initial speed is too low, the vehicle accelerates, and after reaching a certain speed, the vehicle starts to drive at a constant speed.

(1) Calculation method of balance speed

In the uphill section, the vehicle driving force equilibrium equation is

$$F_t = F_i + F_w + F_f + F_j \tag{6.8}$$

When the car loses its ability to accelerate, the acceleration resistance Fj = 0. So the equilibrium equation can be simplified as

$$F_t = F_i + F_w + F_f \tag{6.9}$$

where

Vehicle driving force: $F_t = \frac{3\pi n \eta_T k_P k_T T_{tg}}{25\upsilon}$

Slope resistance: $F_i = mgi$

Air resistance: F_w adopts the corresponding fitting formula of each model;

Rolling resistance: F_f using the corresponding fitting formula of each model;

Rotating mass conversion factor: δ adopts the corresponding parameters of the model.

The above F_t F_w F_f can be expressed as a function of the vehicle speed 0, so according to the vehicle equation, *i*the corresponding equilibrium speed v_p can be calculated when the uphill gradient is given.

(2) Calculation method of acceleration and deceleration slope length

From the vehicle running equation $F_j = F_t - F_i - F_f - F_w$ and the acceleration resistance formula, we can obtain the following:

$$\delta m \frac{du}{dt} = F_t - F_i - F_f - F_w \tag{6.10}$$

After variable substitutions and unit conversions

$$dS = \frac{\delta m u_a}{12960(F_t - F_i - F_f - F_w)} du_a \tag{6.11}$$

Conduct integration

$$S=\int_{u_{a0}}^{u_a}\frac{\delta m u_a}{12960(F_t-F_i-F_f-F_w)}du_a\quad(\text{unit}:\text{km})\tag{6.12}$$

The speed is represented by υ, and the distance unit is converted to m, then there are

$$S=\int_{v_0}^{v}\frac{\delta m\upsilon}{12.96(F_t-F_i-F_f-F_w)}dv\quad(\text{unit}:\text{m})\tag{6.13}$$

where

Vehicle driving force: $F_t=\frac{3\pi n\eta_T k_P k_T T_{tg}}{25\upsilon}$

Slope resistance: $F_i=mgi$

Air resistance: F_w adopts the corresponding fitting formula of each model;

Rolling resistance: F_f using the corresponding fitting formula of each model;

Rotating mass conversion factor: δ adopts the corresponding parameters of the model.

(2) **Balance speed calculation at different altitudes**

The basic parameters of the four-axle trucks are as follows: the weight of the test vehicle is 30.99t, the calculated speed is 1400 r/min, the calculated torque is 1500 N·m, and the transmission efficiency is 0.89. The balancing speed, acceleration slope length, and deceleration slope length of any given vehicle weight can be simulated and calculated using the practical processing method. When the vehicle weight is 30t, the balance speed is shown in Table 6.2.

The balance speed curve is shown in Fig. 6.5.

(3) **Calculation of acceleration slope length at different altitudes**

Taking the four-axle trucks as the research object, the acceleration slope length value is determined. The basic parameters of the four-axle trucks are as follows: the weight of the test vehicle is 30.99 t, the calculated speed is 1400 r/min, the calculated torque is 1500 N·m, and the transmission efficiency is 0.89. The calculation results of the acceleration slope length curve at an altitude of 3000 m are shown in Table 6.3.

The acceleration slope length represents the travel distance required to accelerate from a lower speed to a higher speed under certain altitude and gradient conditions, where the balance speed is the maximum speed that can be achieved.

The acceleration slope length curve is shown in Fig. 6.6.

The calculation results of the acceleration slope length curve at an altitude of 4000 m are shown in Table 6.4, and the acceleration slope length curve is shown in Fig. 6.7.

The calculation results of the acceleration slope length curve at an altitude of 5000 m are shown in Table 6.5, and the acceleration slope length curve is shown in Fig. 6.8.

Table 6.2 Balance speed under different altitudes

Gradient (%)	527.65	1000 m	1500 m	2000 m	2500 m	3000 m	3500 m	4000 m	4500 m	5000 m	5500 m	6000 m
0.00	105.98	105	103.84	102.56	101.14	99.57	97.83	95.92	93.81	91.5	88.96	86.19
0.50	95.02	93.72	92.22	90.59	88.82	86.9	84.82	82.57	80.14	77.53	74.72	71.7
1.00	84.99	83.46	81.73	79.87	77.89	75.78	73.52	71.13	68.59	65.9	63.05	60.05
1.50	76.01	74.36	72.5	70.54	68.48	66.3	64.02	61.62	59.11	56.5	53.77	50.93
2.00	68.14	66.44	64.55	62.58	60.52	58.38	56.16	53.85	51.46	49	46.45	43.84
2.50	61.33	59.64	57.78	55.85	53.86	51.81	49.7	47.52	45.29	43.01	40.67	38.28
3.00	55.48	53.84	52.04	50.2	48.31	46.37	44.39	42.36	40.29	38.19	36.04	33.87
3.50	50.47	48.89	47.18	45.44	43.66	41.84	39.99	38.11	36.2	34.26	32.3	30.31
4.00	46.17	44.67	43.05	41.41	39.74	38.04	36.32	34.57	32.81	31.02	29.21	27.39
4.50	42.47	41.05	39.52	37.98	36.41	34.82	33.22	31.6	29.96	28.31	26.64	24.96
5.00	39.26	37.92	36.48	35.03	33.56	32.08	30.58	29.07	27.s5	26.02	24.47	22.92
5.50	36.47	35.2	33.85	32.48	31.10	29.71	28.31	26.9	25.48	24.06	22.62	21.17
6.00	34.02	32.83	31.55	30.26	28.97	27.66	26.35	25.03	23.7	22.36	21.02	19.67
6.50	31.87	30.74	29.53	28.31	27.09	25.86	24.63	23.39	22.14	20.88	19.63	18.36
7.00	29.96	28.88	27.74	26.59	25.44	24.28	23.11	21.94	20.77	19.59	18.4	17.22
7.50	28.25	27.23	26.15	25.06	23.97	22.87	21.77	20.66	19.55	18.44	17.32	16.2
8.00	26.73	25.76	24.73	23.69	22.66	21.62	20.57	19.52	18.47	17.42	16.36	15.3
8.50	25.35	24.43	23.45	22.46	21.48	20.49	19.49	18.5	17.5	16.5	15.49	14.49
9.00	24.11	23.23	22.29	21.35	20.41	19.47	18.52	17.57	16.62	15.67	14.72	13.76
9.50	22.97	22.13	21.24	20.34	19.45	18.54	17.64	16.74	15.83	14.92	14.01	13.1
10.00	21.94	21.14	20.28	19.42	18.57	17.7	16.84	15.98	15.11	14.24	13.37	12.5

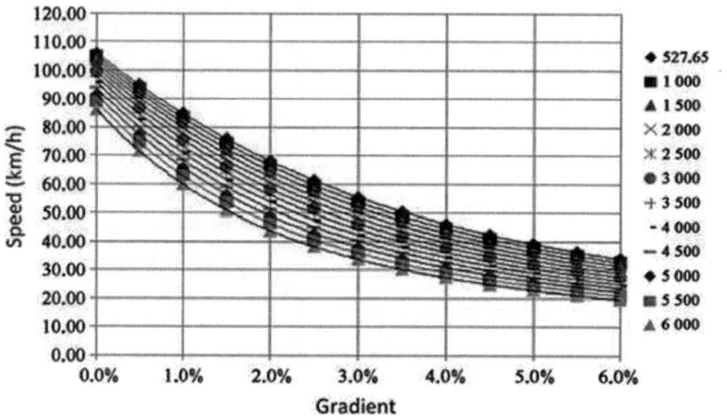

Fig. 6.5 Balance speed curve

(4) Calculation of Deceleration Slope Length under Different Altitudes

Taking the four-axle trucks as the research object, the deceleration slope length, the basic parameters of the four-axle trucks and the vehicle type used for acceleration slope length are determined. Deceleration slope length means that under certain altitude and slope conditions, the vehicle travels upward at a higher speed on the grade section of a certain slope. Even if the accelerator is fully opened, the running speed will still decrease since the driving resistance is greater than the vehicle power, until the speed is reduced to a balanced speed and the vehicle starts to run at a uniform speed on the uphill section.

Table 6.6 shows the calculation results of deceleration slope length curve at 3000 m above sea level. The deceleration slope length curve is shown in Fig. 6.9.

The calculation results of deceleration slope length curve at 4000 m above sea level are shown in Table 6.7, and the deceleration slope length curve is shown in Fig. 6.10.

The calculation results of deceleration slope length curve at 5000 m above sea level are shown in Table 6.8, and the deceleration slope length curve is shown in Fig. 6.11.

(5) Determination of Grade Length and Grade Reduction in High Altitude Area

Based on the above research results, it can be concluded that when the equilibrium speed is equal to the minimum allowable speed, the corresponding gradient is the maximum gradient. The corresponding slope length needs to be limited under the maximum slope. Tables 6.9 and 6.10 shows the relevant research results.

(1) Maximum Grade and Gradient Reduction

The vehicle power decreases with the increase of altitude, so the smaller maximum grade should be adopted in the grade design. For highways with design speed less

Table 6.3 Acceleration slope lengths of different gradients (altitude 3000 m)

Speed(km/h)	0.0%	0.5%	1.0%	1.5%	2.0%	2.5%	3.0%	3.5%	4.0%	4.5%	5.0%	5.5%	6.0%
0	0	0	0	0	0	0	0	0	0	0	0	0	0
5	0.2	0.2	0.2	0.2	0.3	0.3	0.3	0.3	0.3	0.3	0.3	0.3	0.3
10	2	2	2	2	2	2	2	2	2	2	2	2	2
15	5	6	6	6	7	7	7	7	8	8	8	9	9
20	14	14	15	15	16	17	18	19	21	22	24	26	29
25	26	28	30	32	34	36	40	43	48	54	63	75	97
30	46	49	53	58	63	70	79	91	108	136	195	1060	926
35	75	81	89	99	112	129	154	194	276	1547	1383		
40	114	126	142	162	190	233	309	518	1996				
45	168	189	217	257	320	437	830	2489					
50	238	274	325	405	553	1036	4058						
55	330	389	481	645	1092	4437							
60	448	546	712	1091	7928								
65	602	763	1080	2442									
70	802	1071	1755	9555									
75	1066	1534	4005										
80	1421	2319	9999										
BalanceSpeed(km/h)	99.6	86.9	75.8	66.3	58.4	51.8	46.4	41.8	38.0	34.8	32.1	29.7	27.7

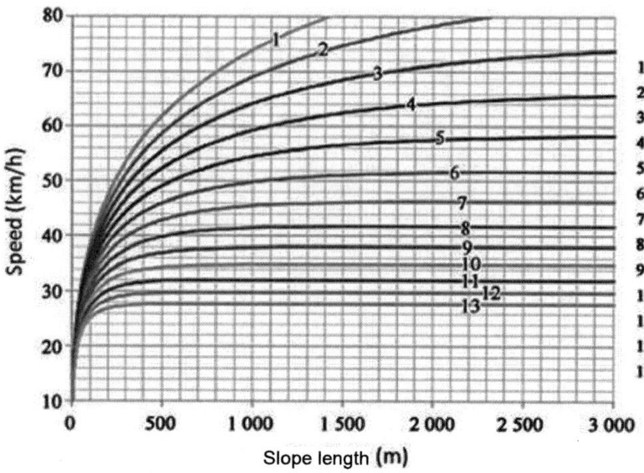

Fig. 6.6 Acceleration slope length curve (Altitude 3000 m)

than or equal to 80 km/h and located in plateau areas with altitude above 3500 m, the maximum grade shall be reduced according to Table 6.11.

(2) Maximum Slope Length

The limitation of maximum slope length refers to the distance traveled when the vehicle speed drops to the minimum allowable speed and the vehicle is controlled to drive on the ramp. Table 6.12 shows the limitation of slope length for different slopes required under different altitudes.

6.1.3 Definition and Standard of Continuous Long and Steep Downhills

(1) Definition of Continuous Long and Steep Downhills of Freeways

Continuous long and steep downhill refers to the downhill direction of vehicles on freeways and Class-I-highways. The average grade (average grade and slope length) of a single grade or a combination of multiple grade sections exceeds the above indicators. According to relevant tests and investigations, after the grade and slope length exceed the above range, corresponding to the comprehensive performance, safety equipment conditions and driver behavior characteristics of actual trucks in China, the brake drum temperature may be too high after the continuous use of the main brake, which may lead to partial or complete loss of braking efficiency.

Table 6.4 Acceleration slope lengths of different slopes (altitude 4000 m)

Speed(km/h)	0.0%	0.5%	1.0%	1.5%	2.0%	2.5%	3.0%	3.5%	4.0%	4.5%	5.0%	5.5%	6.0%
0	0	0	0	0	0	0	0	0	0	0	0	0	0
5	0.3	0.3	0.3	0.3	0.3	0.3	0.3	0.3	0.3	0.3	0.3	0.3	0.3
10	2	2	2	2	2	2	2	2	2	3	3	3	3
15	6	7	7	7	8	8	8	9	9	9	10	11	11
20	15	16	17	18	19	20	21	23	25	27	30	34	39
25	30	31	34	36	39	43	47	53	61	72	89	123	400
30	52	56	61	67	75	85	99	119	155	240	9999	833	833
35	84	93	103	117	135	163	207	303	1783	1249			
40	129	145	166	195	239	316	520	4463					
45	190	219	258	319	428	733	8588						
50	272	320	394	524	857	6653							
55	378	461	599	908	5495								
60	518	657	928	2017									
65	702	937	1531	9999									
70	945	1360	3392										
75	1272	2070	9999										
80	1731	3687											
Balance Speed(km/h)	95.9	82.6	71.1	61.6	53.8	47.5	42.4	38.1	34.6	31.6	29.1	26.9	25

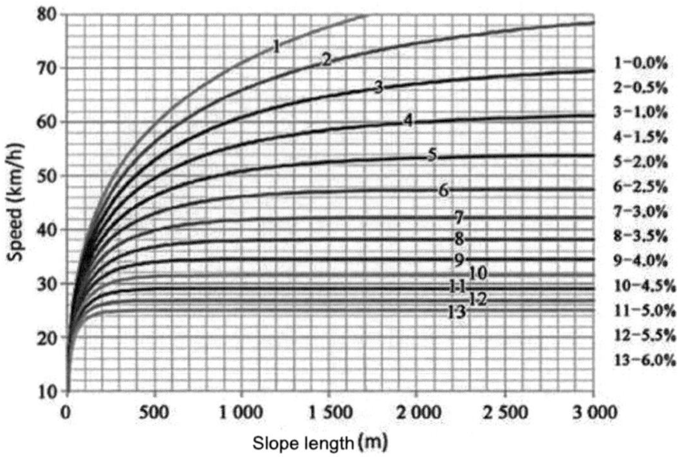

Fig. 6.7 Acceleration slope length curve (Altitude 3000 m)

Relevant working conditions proposed by this indicator include the following aspects:

I. Test vehicle type (six-axle articulated vehicle) under standard load (full load 49t);
II. The vehicle continuously goes downhill at a speed of 60 km/h;
III. Both main brake braking and engine braking are used (it will be safer if exhaust braking and other auxiliary braking systems are used);
IV. The brake drum temperature will not exceed 200 °C, and there is no obvious loss of braking performance.

The continuous long and steep downhill indicator is the critical value of the maximum slope length corresponding to different average grades of the continuous downhill section of the highway under the above working conditions, when the continuous downhill of the vehicle can ensure that the brake drum temperature is within the safe and effective range and there is no obvious loss of brake drum braking efficiency.

(2) **Main Factors Affecting Ambient Brake Temperature in Plateau Area**

According to the principle of energy conservation, during the downhill process of the car, the potential energy of the car is converted through the work done by air resistance, rolling resistance, engine braking resistance, main brake resistance and other resistances; In order to maintain the normal running speed, the residual sliding force that cannot be overcome by other resistances is borne by the main brake of the vehicle and is converted into heat energy through friction. On the other hand, due to the temperature difference between the main brake and the surrounding air, part of the heat energy is transferred to the surrounding air flow through convection.

Table 6.5 Acceleration slope length of different slopes (altitude 5000 m)

Speed(km/h)	0.0%	0.5%	1.0%	1.5%	2.0%	2.5%	3.0%	3.5%	4.0%	4.5%	5.0%	5.5%	6.0%
0	0	0	0	0	0	0	0	0	0	0	0	0	0
5	0.3	0.3	0.3	0.3	0.3	0.3	0.3	0.3	0.3	0.3	0.3	0.3	0.4
10	2	2	2	2	3	3	3	3	3	3	3	3	3
15	7	8	8	8	9	9	10	10	11	11	12	13	14
20	17	18	19	20	22	24	26	28	31	35	40	48	61
25	34	36	39	43	47	52	59	69	84	109	176	746	576
30	59	65	72	80	92	108	133	179	317	1338	880		
35	97	108	123	143	173	223	335	2347	1308				
40	149	171	201	246	325	527	2451						
45	221	260	319	422	679	4446							
50	317	386	502	762	6104								
55	445	565	804	1799									
60	615	827	1374	9999									
65	843	1227	3313										
70	1153	1909	9999										
75	1588	3499											
80	2239	9999											
Balance Speed(km/h)	91.5	77.5	65.9	56.5	49	43	38.2	34.3	31	28.3	26	24.1	22.4

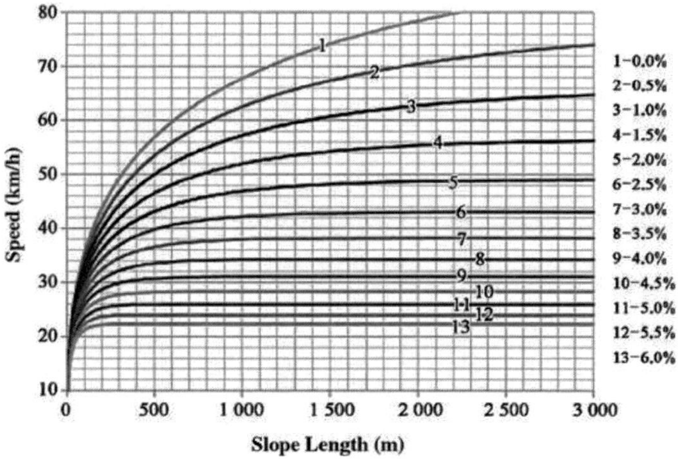

Fig. 6.8 Acceleration slope length curve (altitude: 5000 m)

According to Newton's cooling formula, the corresponding differential equation can be obtained:

$$m_{\mathrm{B}}C\frac{\mathrm{dT}}{\mathrm{dt}} = P_{\mathrm{B}} - hA_{\mathrm{C}}(T - T_{\mathrm{a}}) \tag{6.14}$$

where,

m_B effective thermal mass of braking system;
C specific heat capacity of brake drum;
P_B input power of braking system;
h effective convection conductivity coefficient;
A_C effective heat dissipation area of braking system;
T_a environment temperature.

The brake system input power P_B is closely related to the vehicle gravity power, vehicle rolling resistance power and vehicle air resistance power. The typical characteristics of high-altitude and low-pressure and oxygen deficient areas are that with the increase of altitude, the oxygen content decreases, the air density decreases, and the corresponding air resistance also decreases, while the input power of the braking system increases. Compared with the low-altitude plain area, the load borne by the brake during the downhill process of the vehicle further increases. In addition, air convection is positively related to air density (for example, energy cannot be transferred by thermal convection under vacuum). The coefficient of air relative density (relative to the altitude of 527.65 m) is:

$$k_{\mathrm{a}} = \frac{1 - 1.0028 \times 10^{-4}H + 3.3444 \times 10^{-9}H^2}{0.9480} \tag{6.15}$$

Table 6.6 Deceleration slope length of different grades (altitude: 3000 m)

Speed (km/h)	0.0%	0.5%	1.0%	1.5%	2.0%	2.5%	3.0%	3.5%	4.0%	4.5%	5.0%	5.5%	6.0%
80	0	0	0	0	0	0	0	0	0	0	0	0	0
75	0	0	9999	583	308	209	159	128	107	92	81	72	65
70				1579	683	442	328	260	216	185	161	143	129
65				9999	1208	715	512	400	328	279	242	214	192
60					2370	1068	724	551	446	375	324	285	254
55					7712	1672	990	723	573	476	407	356	316
50						6017	1417	939	718	585	495	430	380
45							4850	1293	905	713	592	508	445
40								8128	1255	886	709	596	516
35									3824	1533	895	712	601
30										2721	3682	1083	741
25												1888	3403
Balanced Speed (km/h)	99.6	86.9	75.8	66.3	58.4	51.8	46.4	41.8	38.0	34.8	32.1	29.7	27.7

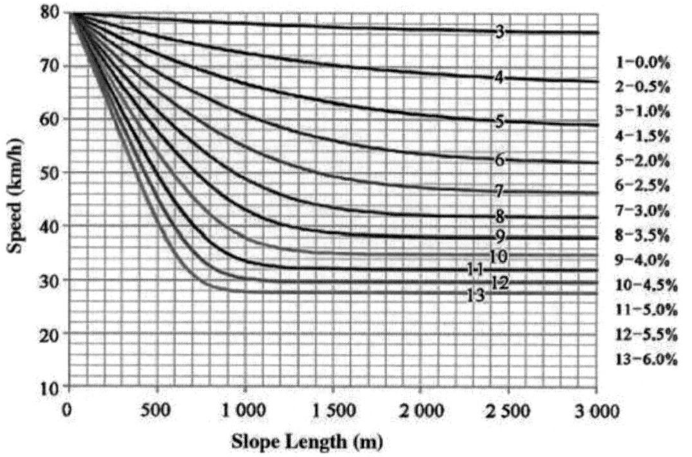

Fig. 6.9 Deceleration slope length curve (altitude: 3000 m)

However, the heat conduction of the brake is surface to surface contact, so the ratio of the relative air density on the surface is $k_a^{2/3}$, then the effective convection conductivity coefficient is

$$h_R = k_a^{2/3} h_{R0} \tag{6.16}$$

In which, the effective convection conductivity coefficient at normal altitude

$$h_{R0} = 5.3345 + 1.53258\upsilon - 0.003653\upsilon^2 \tag{6.17}$$

(3) Temperature control model of automobile brake considering altitude

In the Ministry of Transport's transportation science and technology project "Research on Design Methods and Key Indicators of Freeway Grade", the calculation model of automobile brakes at ordinary altitudes has been studied in depth. The corresponding models of this project have considered automobile weight, windward area, brake parameters, initial temperature, ambient temperature, grade parameters, automobile speed, automobile friction power (divided into air resistance and mechanical resistance) The engine braking power and exhaust braking power are widely applicable and practical, and can be used as the basic calculation model for further research of this project.

Combining Eq. (3.47) and Eq. (6.9), the brake temperature calculation model considering altitude factor is as follows:

$$T(t) = T_0 + (T_a - T_0 + k_2 P_B)\left(1 - e^{-k_1 t}\right) \tag{6.18}$$

Table 6.7 Deceleration slope length of different grades (altitude: 4000 m)

Speed (km/h)	0.0%	0.5%	1.0%	1.5%	2.0%	2.5%	3.0%	3.5%	4.0%	4.5%	5.0%	5.5%	6.0%
80	0	0	0	0	0	0	0	0	0	0	0	0	0
75	0	0	1242	424	257	185	144	118	100	87	77	69	62
70			9999	986	550	383	294	239	201	174	153	136	123
65				1962	906	603	453	363	303	260	228	203	183
60				9999	1407	859	626	494	408	348	303	269	241
55					2636	1195	824	635	517	437	379	334	299
50					7057	1808	1078	796	636	531	456	400	357
45						5842	1520	1004	773	633	538	468	415
40							4557	1399	957	755	629	541	476
35								3200	1507	939	745	626	543
30									2517	2363	1004	753	629
25											1952	1439	1306
Balanced Speed (km/h)	95.9	82.6	71.1	61.6	53.8	47.5	42.4	38.1	34.6	31.6	29.1	26.9	25.0

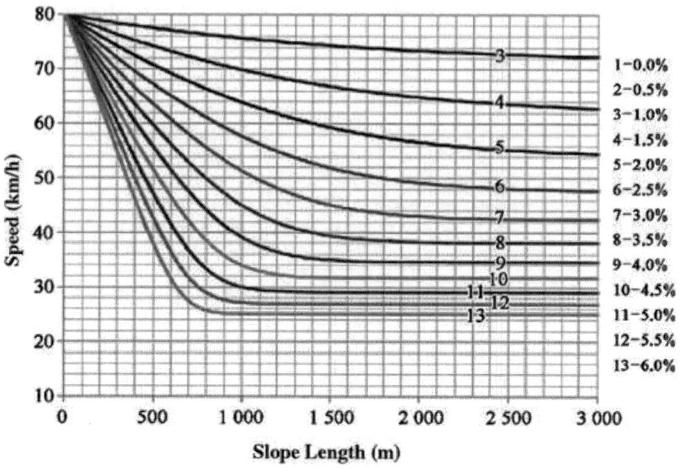

Fig. 6.10 Deceleration slope length curve (altitude: 4000 m)

where,

k_1 reciprocal of thermal time parameter, $k_1 = \frac{h_R A_g}{m_g C_g}(1/s)$
K_2 reciprocal of total heat transfer parameters, $k_2 = \frac{k_z}{n_z h_R A_g}$ (°C/W)
h_R effective convection conductivity coefficient, $h_R = k_a^{2/3}(5.3345 + 1.53258\upsilon - 0.003653\upsilon^2)$
P_B power borne by brake;
t – vehicle running time (s);
T_0 initial temperature (°C);
T_a ambient temperature (°C);
T calculation temperature (°C).

(4) Qualitative Indicator of Continuous Long and Steep Downhill of Freeway in High Altitude Area

When the test vehicle (six-axle articulated vehicle) is used and the vehicle continues to descend the slope at a speed of 60 km/h, and the main brake and engine brake are used at the same time, the slope length corresponding to the corresponding slope is determined according to the brake drum temperature of 200 °C.

In addition to considering the altitude factor, the above parameters are adopted. See Table 6.13 for qualitative indicators of slope length boundary when the brake reaches 200 °C at different altitudes and slopes.

The slope length curve corresponding to the predetermined temperature of 200 °C at each altitude with the slope as the independent variable is shown in Fig. 6.12.

It can be seen from Fig. 6.12 that with the increase of the grade, the slope lengths reaching the predetermined temperature at different altitudes tend to be consistent, which indicates that the steeper the grade is, the smaller the influence of altitude factors on the brake temperature is.

Table 6.8 Deceleration slope length of different grades (altitude: 5000 m)

Speed (km/h)	0.0%	0.5%	1.0%	1.5%	2.0%	2.5%	3.0%	3.5%	4.0%	4.5%	5.0%	5.5%	6.0%
80	0	0	0	0	0	0	0	0	0	0	0	0	0
75	0	9999	679	333	221	165	132	110	94	82	73	66	60
70			1800	726	460	338	267	220	188	164	145	130	118
65			9999	1238	730	521	406	332	281	244	216	193	175
60				2102	1055	722	552	447	376	324	285	255	230
55				9121	1508	956	710	566	472	405	354	315	284
50					2644	1263	891	695	572	486	423	375	337
45					6350	1852	1125	843	680	572	494	435	389
33,740						6885	1578	1039	807	666	568	497	442
35							4459	1533	989	781	653	564	497
30								3144	2373	993	770	646	560
25										2156	1661	818	655
Balanced Speed (km/h)	91.5	77.5	65.9	56.5	49.0	43.0	38.2	34.3	31.0	28.3	26.0	24.1	22.4

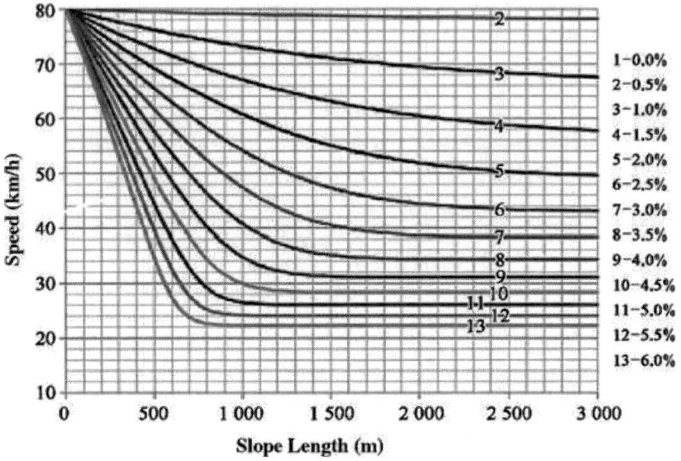

Fig. 6.11 Deceleration slope length curve (altitude: 5000 m)

Table 6.9 Grade with unlimited slope length under different altitudes

Design Speed(km/h)	3000 m (%)	3500 m (%)	4000 m (%)	4500 m (%)	5000 m (%)	5500 m (%)
120(80)	0.8	0.7	0.6	0.5	0.4	0.3
110(65)	1.6	1.5	1.4	1.3	1.2	1.1
80(50)	2.7	2.5	2.3	2.1	1.9	1.7

Table 6.10 Maximum grade corresponding to the maximum allowable speed under different altitudes (slope length shall be limited)

Design Speed(km/h)	3000 m (%)	3500 m (%)	4000 m (%)	4500 m (%)	5000 m (%)	5500 m (%)
120(60)	1.9	1.7	1.6	1.5	1.3	1.2
110(50)	2.6	2.5	2.3	2.1	2.0	1.8
80(40)	3.6	3.5	3.3	3.0	2.8	2.5

Table 6.11 Grade reduction under different altitudes

Object	3500 ~ 4500 m	4500 ~ 5500 m	Above 5500 m
Grade reduction	1.5	2.0	2.5

Table 6.12 Maximum slope length of different elevations and grades

Altitude		3000 ~ 4000 m				4000 ~ 5000 m				Above 5000 m			
Design speed(km/h)		120	100	80	60	120	100	80	60	120	100	80	60
Grades (%)	2	2370	7710	\	\	1405	2625	7055	\	1055	1510	2645	\
	3	725	990	1415	\	625	825	1080	4555	550	710	890	1580
	4	445	570	720	1255	410	515	635	960	375	470	570	805
	5	\	405	495	710	\	380	455	630	\	355	425	570
	6	\	\	380	515	\	\	355	475	\	\	335	440

Note The minimum allowable value of design speed during climbing corresponds to the length requirement

Table 6.13 Indicators for continuous long and steep downhill with consideration of altitudes (km)

Gradient (%)	527.65 m	1000 m	2000 m	3000 m	4000 m	5000 m	5500 m
−2.5	43.4	31.1	21.3	17.1	14.8	13.4	12.9
−3.0	14.8	13.4	11.4	10.2	9.3	8.8	8.5
−3.5	9.3	8.8	7.9	7.3	6.8	6.5	6.4
−4.0	6.8	6.5	6.0	5.7	5.4	5.2	5.1
−4.5	5.4	5.2	4.9	4.6	4.5	4.3	4.3
−5.0	4.4	4.3	4.1	3.9	3.8	3.7	3.7
−5.5	3.8	3.7	3.5	3.4	3.3	3.2	3.2
−6.0	3.3	3.2	3.1	3.0	2.9	2.9	2.9

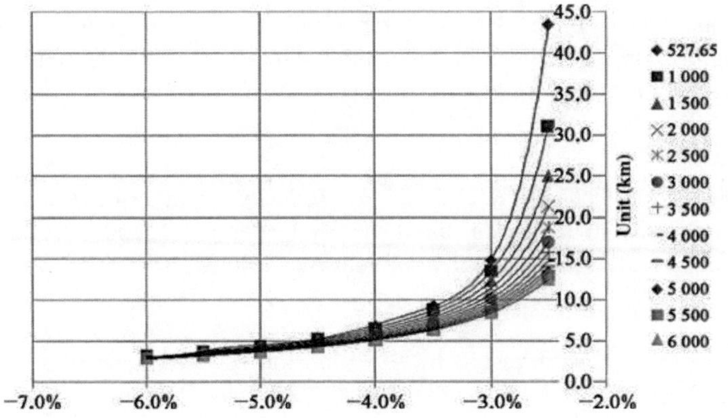

Fig. 6.12 Limit slope length with various altitudes (200 °C)

6.1.4 Cross-section Composition and Size

The standard cross-section composition of the subgrade of the freeway shall include the carriageway, the median, the left and right marginal strips, the left and right hard shoulders, and the left and right earth shoulders.

(1) **Number of Lanes and Lane Width**

According to the traffic volume prediction value, traffic capacity and service level checking calculation of each section of Qinghai-Tibet Highway, the project adopts the standard of four lane freeway.

The relevant provisions in China's Technical Standards for Highway Engineering (JTGB01-2014) also point out that: "after verification the 3.5 m lane width can be used for highways with a design speed of 80 km/h and above, mainly for medium and small passenger vehicles"; "For roads with more than four lanes, the lane width shall meet the width required for parallel driving of vehicles".

The traffic volume of highways in Qinghai-Tibet region is relatively low. Taking G109 as an example, the annual average daily traffic volume in 2014 is less than 5000 vehicles/d, and the minimum width of the inner lane can be taken as 3.5 m according to the main passenger cars in the inner lane.

(2) **Width of Median**

In the "National Planning for Scientific and Technological Programs during China's 12th Five-Year Plan", the composition of cross-section of multi-lane freeway was studied. According to the form and characteristics of combined corrugated beam guardrail, rigid guardrail, cable guardrail and other guardrails, the width of the median meeting the safety design requirements was analyzed without designing the left hard shoulder (Table 6.14). It should not only meet the minimum width required for arranging various facilities in the median, but also consider the requirements of curb and safety *C* value.

(3) **Left Curbs**

According to the *Technical Standards for Highway Engineering*, the left curbs is required to be 0.75 m in general, and 0.5 m can be demonstrated for sections limited by terrain and features. The width argument is as follows:

(1) Considering traffic safety

According to the research results of "Relationship between Lane Width and Operating Speed" from "Method and Standard for Freeway Operationg Speed Design" in the Highway Construction Standard and Specification Plan Program of Ministry of Communications in 2000, the pavement composition that has an impact on the actual running speed is: the width of the lane, curb and hard shoulder. The influence on operating speed by lane width change is greater than that from shoulder and curb

Table 6.14 Proposed median widths with various design speeds

Items		120 km/h	100 km/h	80 km/h
Rigid guardrails (m)	General	1.60	1.50	1.40
	Minimum	0.80	0.70	0.60
Composed corrugated beam guardrails (m)	General	1.90	1.80	1.70
	Minimum	1.20	1.10	1.00
Cable guardrails (m)	General	5.40	5.30	5.20
	Minimum	4.70	4.60	4.50

Note
i. When the design speed is 120 km/h, the width of the inner curb is 0.75 m
ii. When the design speed is 100 km/h, the width of the curb is 0.75 m in general and 0.5 m in minimum

width change. In the ideal state, when the width of the downlink lane is 3.75 m, the width of the left curb is 0.5 m, and the width of the right shoulder is 2.5 m. If the width of the actual road cross-section component is greater than this width, it is considered that its cross-section factors do not affect the free flow speed. See Table 6.15 for the operating speed values of passenger cars under various cross-section conditions.

It can be seen from Table 6.15 that, considering the operating speed, when the width of the cross section carriageway is 3.75 m and the width of the left curb is 0.5 m, the operating speed of passenger cars will not be affected; However, when the lane width remains unchanged and the curb width and shoulder width decrease, the operating speed of passenger cars will be affected.

At present, passenger cars on the Qinghai-Tibet Freeway account for nearly 50% of the total traffic volume. With the opening of the freeway, the proportion of passenger cars entering Tibet for tourism and local transportation will increase day by day. Their small cars will drive on the inner lane. When the median is 2.0 m, the minimum width of the left curb is 0.5 m.

(2) Considering the deformation control and stability of permafrost subgrade

Most of the road sections in the Qinghai-Tibet Plateau are located in permafrost sections. The design and construction technologies obtained from the research on permafrost subgrade over the past 40 years are aimed at solving the problem of maintaining the thermal stability of the subgrade when the width of the highway subgrade is only 10 m and the width of the railway subgrade is 7 m. However, compared with Class II highway, the width of freeway subgrade is relatively large, the total heat

Table 6.15 Operating speeds of passenger cars

Items	Lane widths (m)	Curb width (m)	Operating speed (km/h)
Inner lane	3.75	0.50	120
	3.75	0.20	116
	3.50	0.50	114

absorption area has increased by 1.5–2.0 times, and the total volume heat capacity of embankment filling materials has increased by 2.0–2.5 times. Preliminary calculation shows that the thaw disk under freeway permafrost subgrade will widen and deepen, which will have a serious impact on the water and heat balance of permafrost. Therefore, considering the deformation control and stability of the permafrost subgrade, the width of the left curb should be 0.5 m as low as possible on the premise of meeting the safe operation of vehicles.

(4) Right Hard Shoulder

The right hard shoulder shall have sufficient width to ensure full play of its functions. The minimum width required for hard shoulder to realize different functions is shown in Table 6.16. The width of hard shoulder shall be determined comprehensively according to various functions.

If law enforcement management, maintenance and temporary parking are not considered, the minimum value of hard shoulder on the right side can also be 1.5 m, but emergency parking strips should be added. The spacing of emergency parking strips should not be greater than 500 m and the effective length should not be less than 40 m.

(5) Cross Section Form

For permafrost regions, separate subgrade cross-section is recommended; In non-permafrost regions, it is recommended to adopt integral subgrade cross-section, and bridge tunnel transition sections can also adopt separate cross-section.

Table 6.16 Minimum width of right hard shoulder for various functions

Functions	General minimum width (m)	Applicable situations
For temporary parking of large trucks, due to emergency	3.0	Sufficient for parking trucks with five axles and six axles
Management and maintenance	2.5	Sufficient for parking law enforcement vehicles
Laying facilities and underground pipelines	1.0	For general
Bypassing accident place	2.5	Sufficient for passenger cars to bypass accident place
Returning to carriageway if off the lane	2.5	For general
Improving sight distance	0.5	Determined by specific environment and alignment
Drainage	1.0	Determined by rainfall and drainage capabilities
Improving traffic capacity	1.8	For general
Upgrading in the future	3.75	Rebuilding the hard shoulder to a lane
Emergency rescue	3.5	Sufficient for rescue vehicles

Table 6.17 Minimum widths of cross section with various speeds of highways in high altitude area

Item	120 km/h	100 km/h	80 km/h
Median width (m)	0.80	0.70	0.60
Carriageway width (m)	3.75 (3.50)	3.75 (3.50)	3.75 (3.50)
Left curb width (m)	0.5	0.5	0.5
Hard shoulder width (m)	1.5	1.5	1.5
Earth shoulder width (m)	0.75	0.75	0.75

Note The width of inside lane is indicated in brackets

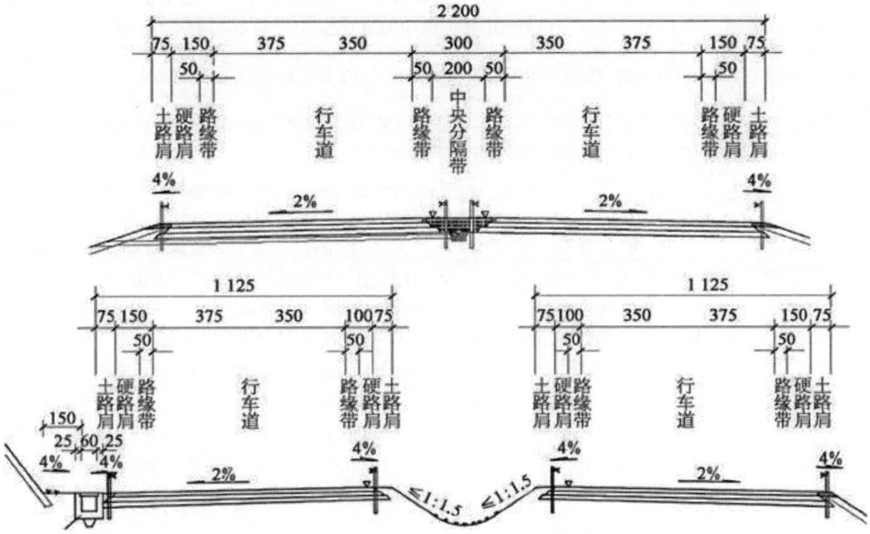

Fig. 6.13 The minimum width of cross-section composition

Through the above comprehensive analysis, the minimum width of cross-section composition under different design speeds is obtained, as shown in Table 6.17 and Fig. 6.13.

6.1.5 Crown Cross Slope and Pavement Superelevation

The high-altitude area has low temperature and is in the state of snow freezing all the year round, especially in winter, which will have a great impact on driving safety. According to the historical records of natural disasters in 26 administrative units along the Qinghai-Tibet Railway, Liu Fenggui and others made a statistical analysis of the data for 50 years from 1951 to 2000. They found that the frequency of snow disasters in the section from Kunlun Mountain to Tanggula Mountain was significantly higher than that in other sections. In the 50 years of statistical data, the frequency of snow disasters was more than 40 times (/50 years), which means that snow disasters occur almost every year. Under ice and snow conditions, the lateral friction coefficient decreases, and the driving stability of the vehicle in the superelevation section of the curve becomes worse. If the running speed is too fast, it is possible to drive out of the road; if the running speed is too slow, it is possible to slide to the inside of the curve.

Since the route trend of Qinghai-Tibet Freeway is basically the same as that of Qinghai-Tibet Railway, the section from Kunlun Mountain to Tanggula Mountain should be specially treated as ice and snow area, and the superelevation and radius of the route should take into account the influence of ice and snow conditions.

(1) **Value of Friction Coefficient under Ice and Snow Conditions**

The lateral friction coefficient is directly related to the driving stability when the vehicle is driving in a curved section. Therefore, the value of the lateral friction coefficient of the pavement under ice and snow conditions is discussed first.

According to the regulations in Green Book of the United States, under normal road conditions, the lateral friction coefficient has a certain relationship with the speed: when the speed is 80 km/h, the lateral friction coefficient is about 0.14; when the speed is 112 km/h, the lateral friction coefficient is about 0.10. The relationship between lateral friction coefficient and speed in Green Book is shown in Fig. 6.14.

See Table 6.18 for the lateral friction coefficient values adopted by different countries when the speed is 120 km/h.

Cheng Guozhu and others applied non-contact fifth wheel to conduct braking tests on ice and snow roads. When the friction factor observation value range of snow-covered roads is 0.04–0.31, the snow compaction has a great impact on the friction factor of snow-covered roads.

In line with the provisions of the textbook *Automobile Theory* (Reference [7] in this book), the values of sliding friction coefficient under ice and snow conditions are shown in Table 6.19.

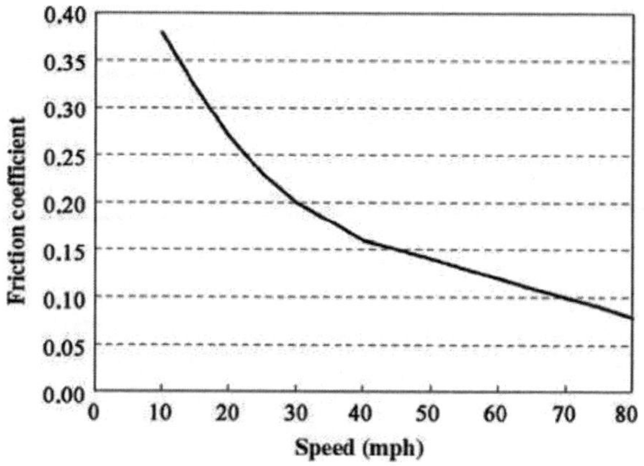

Fig. 6.14 Relationship between lateral friction coefficient and speed (1mph = 1.61 km/h)

Table 6.18 Lateral friction coefficient values of different countries at 120 km/h

Countries	Maximum lateral friction coefficient	Countries	Maximum lateral friction coefficient
Australia	0.11	Japan	0.06
Austria	0.10	Luxembourg	0.10
Belgium	0.10	Netherlands	0.08
Canada	0.09	Portugal	0.10
The French	0.10	Spain	0.10
Germany	0.07	The Swiss	0.10
The Irish	0.12	Britain	0.09
Italy	0.10	The United States	0.09

Table 6.19 Friction coefficient under different road conditions

Pavement status	Max. coefficient of adhesion	Max. coefficient of sliding
Snow (compacted)	0.20	0.15
ice	0.10	0.07

Through the analysis of the measured data, Liu Jing et al. see that the lateral friction coefficient ϕ provided by the road under the ice and snow conditions is 0.084 ~ 0.098 for trucks and 0.096 ~ 0.112 for cars. After comparative analysis, it is found that the friction coefficient between trucks and the road is relatively small under the ice and snow conditions.

Li Song ling et al. divided the pavement into three different statuses: snow (soft), snow (compacted) and ice. See Table 6.20 for the maximum adhesion coefficient and maximum sliding coefficient.

Wang Zhengyuan et al. used the braking distance method and the test vehicle method to test the conditions of snow clearing, soft snow, ice and snow crust, and complete icing respectively, and obtained the anti-sliding performance indicators of the corresponding conditions (Table 6.21).

Based on passed practice verification, the classical formula describing the relationship between the road transverse and longitudinal adhesion coefficient is

$$\mu_H = (0.6 \sim 0.7)\mu$$

Table 6.20 Friction coefficient under different pavement conditions

Pavement status	Max. coefficient of adhesion	Max. coefficient of sliding
Snow (soft)	0.30	0.20
Snow (compacted)	0.20	0.15
Ice	0.10	0.07

Table 6.21 Friction coefficient under different pavement conditions

Pavement status	Passenger cars	Trucks
Soft snow pavement	0.434	0.300
Snow crust	0.281	0.219
Complete icing	0.215	0.172
After clearing	0.756	0.371

Then the lateral and longitudinal adhesion coefficients are as follows:

$$\begin{cases} \mu_H = 0.6\mu \\ \mu_H^2 + \mu_V^2 = \mu^2 \end{cases}$$

where,

μ_H Lateral adhesion coefficient;
μ_V Longitudinal adhesion coefficient;
μ Road friction coefficient.

According to the above formula, the lateral friction coefficient value is calculated, as shown in Table 6.22.

See Table 6.23 for the calculated value of the adhesion coefficient used by Li Yunhui in calculating the speed limit value of the horizontal curve section.

Through the above analysis, it can be concluded that the value of the lateral friction coefficient of the pavement is various under different conditions of ice and snow pavement. The status of ice and snow pavement can be roughly divided into three

Table 6.22 Transverse friction coefficient under different pavement conditions

Pavement status	Passenger cars	Trucks
Soft snow pavement	0.26	0.18
Ice and snow crust	0.17	0.13
Complete icing	0.13	0.10
After clearing	0.45	0.22

Table 6.23 Adhesion coefficient of snow and ice pavement

Pavement status	Range of μ	Mean value of μ	Lateral friction coefficient
Ice film	0.06 ~ 0.14	0.10	0.060
Snow pulp	0.16 ~ 0.16	0.16	0.096
Ice crust	0.15 ~ 0.21	0.18	0.108
Snow crust	0.20 ~ 0.24	0.22	0.132
Ice and snow crust	0.22 ~ 0.28	0.25	0.150
Soft snow	0.30 ~ 0.43	0.37	0.222

categories: snow (soft), snow (compacted) and pavement icing. Through literature retrieval, the values of lateral friction coefficient under these three different conditions are summarized, as shown in Table 6.24.

The value range of lateral friction coefficient of pavement ice, snow crust and soft snow is 0.06 ~ 0.13, 0.13 ~ 0.17 and 0.18 ~ 0.26 respectively. Generally, soft snow will appear in the early snow, with the rolling of vehicles and other factors, it will gradually become snow crust, and then become ice after melting and freezing. In the icing status, the lateral friction coefficient is the minimum, and the vehicle is in the most unfavorable situation. The values of the lateral friction coefficient under the above three conditions are summarized in Table 6.25.

(2) **Determination of Curve Superelevation under Ice and Snow Conditions**

(1) Value of running speed

Vehicles on the road decrease its running speed due to the ice and snow conditions, so a lower speed can be used for calculation. On the one hand, in China the minimum speed limit of freeways is 60 km/h; on the other hand, considering the large proportion of trucks on the Qinghai-Tibet Freeway, in case of sudden snowfall, the vehicle

Table 6.24 Summary of values of transverse friction coefficient under different pavement conditions

Author or source of literature	Pavement status	Passenger cars	Trucks
Li Songling	Soft snow pavement	0.26	0.18
	Ice and snow crust	0.17	0.13
	Complete icing	0.13	0.10
Automobile Theory Fourth Edition	Snow (compacted)	0.15	
	Ice	0.07	
Wang Zhengjun	Soft snow pavement	0.26	0.18
	Ice and snow crust	0.17	0.13
	Complete icing	0.13	0.10
Li Yunhui	Soft snow	0.222	
	Snow crust	0.132	
	Ice film	0.060	

Table 6.25 Lateral friction coefficient under ice and snow road condition

Pavement status	Lateral friction coefficient
Soft snow pavement	0.18 ~ 0.26
Ice and snow crust	0.13 ~ 0.17
Icing	0.06 ~ 0.13

running speed is likely to be lower than the minimum speed limit of freeways of 60 km/h, so the minimum calculation speed used in the calculation of superelevation value in this area is set as 40 km/h.

(2) Control Conditions

On the one hand, it is necessary to consider the smooth driving of the vehicle without slip; on the other hand, the comfort of drivers and passengers shall be guaranteed. If the vehicle run too fast it may slip out of the road:

$$i_0 < \frac{v^2}{gr} - \varphi_{y\max} \tag{6.19}$$

That is, when the speed is too fast, if the superelevation value at the curve section is less than this value, vehicles may slip out.

When driving too slowly or even stopping at the curve section, if the superelevation value is too large, the vehicle may slip to the inside of the curve:

$$i_0 > \frac{v^2}{gr} + \varphi_{y\max} \tag{6.20}$$

That is, when the superelevation at the curve section is greater than the superelevation value, the vehicle may slip to the inside of the curve.

In order to make drivers and passengers have certain comfort when driving on curves, the following formula must also be met:

$$\mu = \frac{v^2}{gr} - i_0 < \text{Critical value} \tag{6.21}$$

The critical value is the minimum value to ensure the comfort of passengers and drivers during turning.

(3) Value of Curve Radius

In the *Design Specification for Highway Alignment* (JTGD20-2006), the general and limit values of curve radius under different design speeds are specified in Table 6.26.

The plateau area is sparsely populated and widely covered, so in the design of horizontal curves, the extreme minimum radius is not recommended. If the terrain conditions are limited, the general value of the minimum radius can be used.

(3) **Superelevation Safety Net Worth**

When the vehicle travels slowly at the curve radius, it may slip to the inside of the curve due to the low lateral friction coefficient; when the vehicle runs at a high speed at the radius of the curve, it may also slip to the outside of the curve because the lateral

Table 6.26 Minimum radius of circular curve at different design speeds

Item		120 km/h	100 km/h	80 km/h	60 km/h	40 km/h	30 km/h	20 km/h
Min. radius of circular curve (m)	General value	1000	700	400	200	100	65	30
	Limit value	650	400	250	125	60	30	15

Note "General value" refers to the value adopted under normal conditions; "Limit value" refers to the adopted value when conditions are limited

friction coefficient is too small. Next, the concept of net safety value is adopted to demonstrate the safety degree of vehicles at different speeds at the curve radius.

(1) Maximum Superelevation Net Worth

Firstly, the situation of vehicles sliding towards the inner side of the curve is analyzed. Subtract actual design superelevation value from the critical superelevation value of vehicles which slide towards the inner side of the curve at different speeds, we get the net safety value of superelevation. The net safety value corresponding to the maximum superelevation is called the maximum superelevation net safety value, the one corresponding to the minimum superelevation is called the minimum superelevation net safety value. The calculation formula of the maximum superelevation net safety value is as follows:

$$S_{内} = i_{0max} - i_0 \tag{6.22}$$

where,

S_{in} Maximum superelevation safety net value (slip inward);
i_{0max} Critical superelevation value for sliding towards the inside of the curve. If it is greater than the superelevation value, the vehicle will slide towards the inside of the curve;
i_0 Design value of actual road superelevation.

Taking 6% as the actual superelevation value, the superelevation safety net value under different speed conditions is plotted as shown in Fig. 6.15.

It can be seen from Fig. 6.15 that at the same speed, the smaller the curve radius is, the larger the superelevation net safety value is; under the condition of the same curve radius, the smaller the speed is, the smaller the superelevation net safety value is, because the smaller the speed is, the smaller the corresponding outward centrifugal force is, and the easier it is to slip toward the inner side of the curve. The 2014 version of China's *Technical Standards for Highway Engineering* stipulates that the highway with a high proportion of large freight vehicles should adopt a smaller maximum superelevation (value), and the maximum superelevation in ice and snow areas is 6%. As the running speed of trucks at curve section is low, in order to ensure

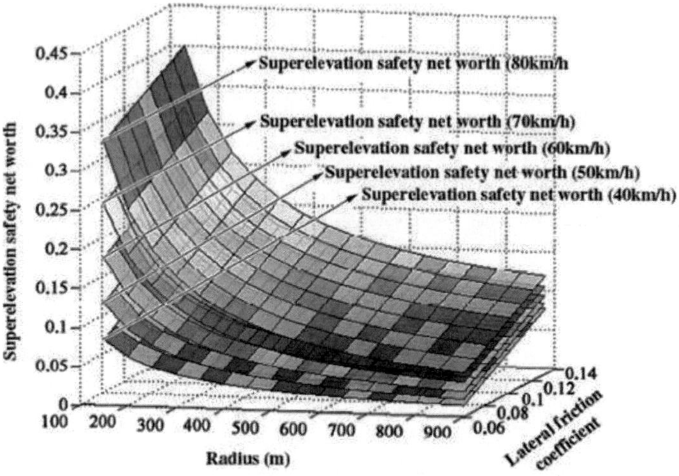

Fig. 6.15 Superelevation safety net worth. Superelevation safety net worth, Superelevation safety net worth (80 km/h), Superelevation safety net worth (70 km/h), Superelevation safety net worth (60 km/h), Superelevation safety net worth (50 km/h), Superelevation safety net worth (40 km/h), Lateral friction coefficient, Radius (m)

sufficient safety net value, small superelevation value should be adopted, which is consistent with the idea in the current specification.

When the superelevation value is large and the running speed is small, the net safety value gradually decreases, and there are certain potential safety hazards. Although in our country it stipulates that the minimum running speed of freeways is 60 km/h, according to the actual situation, trucks often have a running speed less than 60 km/h. When the lateral friction coefficient is 0.06, that is, when the road surface is in the most unfavorable ice status, the maximum superelevation values are calculated at speeds of 40 km/h, 50 km/h, 60 km/h, 70 km/h, and 80 km/h respectively, the actual superelevation value is 6%, and the net safety value of superelevation is calculated.

It can be seen from Fig. 6.16 that under the most unfavorable ice surface condition, with speed of 40 km/h, superelevation of 6%, curve radius of 700 m, its net safety value is less than 0.02, which should be avoided as far as possible. It can also be seen that the maximum superelevation value in snow and ice areas stipulated of 6% in the current specification can basically meet the safety requirements. In snow and ice areas, considering the slow running speed of trucks under full load, when the curve radius is large, it is recommended to adopt a smaller superelevation value. Figure 6.17 shows the phenomenon of vehicles sliding toward the inner side of the curve when driving on ice and snow roads with excessive superelevation.

(2) Minimum Superelevation Net Value

If the vehicle runs too fast and the friction coefficient is small under ice and snow conditions, the vehicle is likely to slide out of the road. If the actual superelevation

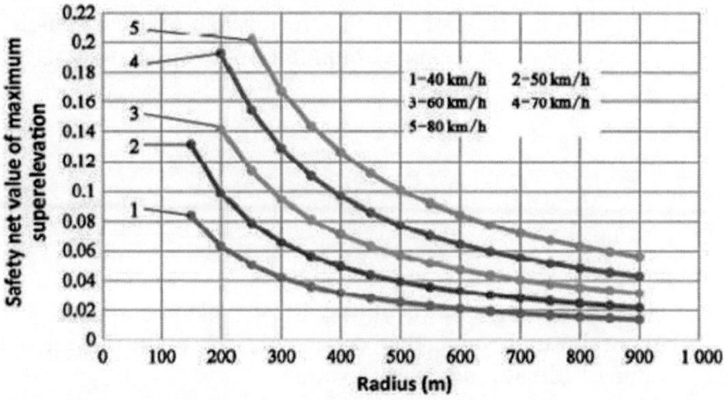

Fig. 6.16 Safety net value of maximum superelevation. Safety net value of maximum superelevation, Radius (m)

Fig. 6.17 Vehicle sliding to the inside of curve. Icing on pavement

of the road is 6%, when the critical value is greater than 6%, the vehicle will slide out of the road. The calculation formula of minimum superelevation net safety value is as follows:

$$S_{外} = i_0 - i_{0min} \tag{6.23}$$

where,

- S_{out} Minimum superelevation net safety value (sliding outward);
- i_{0min} Critical superelevation value for sliding to the outside of the curve. If it is less than the superelevation value, the vehicle will slide to the outside of the curve;
- i_0 Design value of actual road superelevation.

The critical value of superelevation and its net safety value under different speed conditions are plotted as shown in Fig. 6.18.

It can be seen from Table 6.18 that when the curve radius is fixed, the greater the speed, the smaller the lateral friction coefficient, and the smaller the net safety value, the easier it is for the vehicle to drive out of the road. Taking the lateral friction coefficient of the road surface of 0.06 and superelevation value of 6%, that is, the road surface is in the most unfavorable condition, the specific results are shown in Fig. 6.19.

At a speed of 80 km/h, superelevation of 6%, and the lateral friction coefficient of 0.06, that is, the road is in icing status, when the curve radius is less than 425 m, the

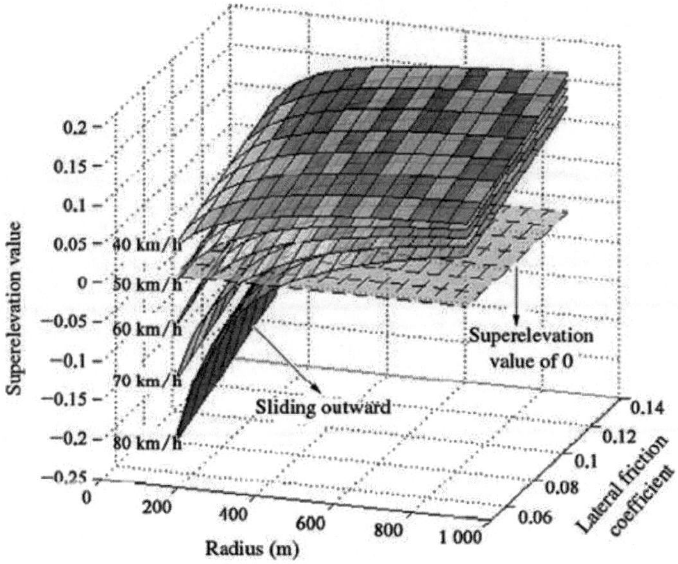

Fig. 6.18 Superelevation critical value and its net safety value under different speed conditions. Superelevation value, Superelevation value of 0, Sliding outward, Lateral friction coefficient, Radius (m)

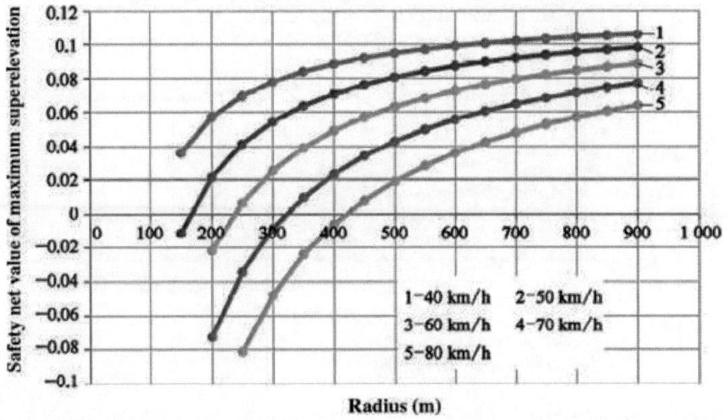

Fig. 6.19 Superelevation net safety value at different speeds. Safety net value of maximum superelevation, Radius (m)

vehicle drives out of the road; when the curve radius is 800 m, the net safety value of superelevation is about 6%. Under the same superelevation and running speed, the greater the curve radius, the greater the net safety value of superelevation.

(4) **Determination of Superelevation Value**

The safety degree of minimum and maximum superelevation under different speed conditions is demonstrated respectively above, and their net safety values are calculated. Next, based on calculation results of maximum superelevation and minimum superelevation, the safe value of superelevation under ice and snow conditions is demonstrated. On the one hand, the superelevation value should not be too small to prevent the vehicle from sliding out of the road; on the other hand, it should avoid the vehicle sliding towards the inside of the curve due to too slow speed and too large superelevation value. It is necessary to balance the maximum and minimum superelevation.

In order to prevent the vehicle from sliding out of the road, the superelevation should be designed as large as possible, while in order to prevent the vehicle from sliding toward the inside of the curve, the superelevation should be designed as small as possible. However, there is no contradiction between the two, because under the ice and snow conditions, the running speed of the vehicle will be significantly lower than the speed value under normal conditions.

At a speed of 40 km/h, calculate the maximum and minimum critical value of superelevation under the most unfavorable condition - road icing, the lateral friction coefficient is 0.06, and the calculation results are shown in Fig. 6.20.

The shaded area in Fig. 6.20 represents the safety range of superelevation. Outside the shaded area, vehicles may slide out of the road or toward the inside of the curve. Under the most unfavorable condition of road icing, the value of transverse friction coefficient is the minimum. Under the condition of superelevation of 6%, it is recommended that the minimum radius of circular curve should be 120 m, and

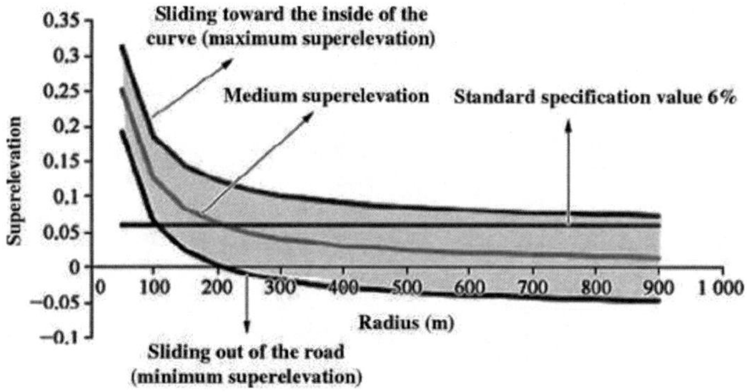

Fig. 6.20 Maximum and minimum superelevation critical values under the most unfavorable conditions (40 km/h). Superelevation, Sliding toward the inside of the curve (maximum superelevation), medium superelevation, standard specification value 6%, radius, sliding out of the road (minimum superelevation)

under that of 4%, the minimum radius should be 160 m. Since the intermediate value between the maximum and the minimum superelevation value is the most ideal one, the superelevation and radius values under different curve radii are recommended in Table 6.27.

When the curve radius is less than 200 m, its ideal superelevation value is greater than 6%. Therefore, it is recommended that when the speed is 40 km/h, in order to avoid vehicles driving out of the road under ice and snow conditions, the radius can be greater than or equal to 200 m if conditions permit. See Table 6.28 for the limit radius under different superelevation conditions at 40 km/h.

Similarly, the superelevation values at 60 km/h and 80 km/h can be obtained.

The shaded area in Fig. 6.21 represents the safety range of superelevation. It is suggested that at a design speed of 60 km/h, the limit radius of the curve is 240 m when superelevation is 6%, and the limit radius is 290 m when the superelevation is 4%.

Table 6.27 Values of ideal superelevation and radius at 40 km/h

Radius (m)	50	100	150	200	250	300	350
Recommended superelevation	\	\	\	6%	5%	4%	4%
Radius (m)	400	450	500	550	600	650	700
Recommended superelevation(%)	3	3	3	2	2	2	2

Table 6.28 Limit radius under different superelevation conditions at 840 km/h

Limit value (superelevation 6%)	Standard	60 m
	Proposal	120 m
Limit value (superelevation 4%)	Standard	65 m
	Proposal	160 m

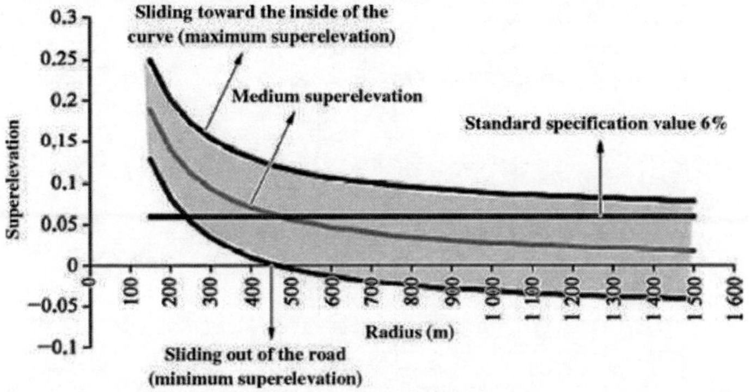

Fig. 6.21 Maximum and minimum superelevation under the most unfavorable conditions (60 km/h). Superelevation, sliding toward the inside of the curve (maximum superelevation), Medium superelevation, standard specification value 6%, Radius, Sliding out of the road (minimum superelevation)

Since the intermediate value between the maximum and the minimum superelevation value is the most ideal one, the superelevation value recommendations under different curve radii are shown in Table 6.29.

When the curve radius is less than 450 m, its ideal superelevation value is greater than 6%. Therefore, it is recommended that when the speed is 60 km/h, in order to prevent vehicles from driving out of the road under ice and snow conditions, the radius can be greater than or equal to 450 m if conditions permit. See Table 6.30 for the limit radius under different superelevation conditions at 60 km/h.

The ideal superelevation value at 80 km/h is shown in Fig. 6.22.

The shaded area in Fig. 6.22 represents the safety range of superelevation. Outside the shaded area, vehicles may slide out of the road or toward the inside of the curve. Under the most unfavorable condition of road icing, the value of transverse friction coefficient is the minimum. Under the condition of superelevation of 6% and 4%, it is recommended that the minimum radius of circular curve limit should be 430 m and 510 m respectively. Since the median value between the maximum and the minimum superelevation value is the most ideal one, the superelevation values under different curve radius are recommended in Table 6.31.

When the curve radius is less than 750 m, its ideal superelevation value is greater than 6%, so it is recommended that when the speed is 80 km/h, the radius can be greater than or equal to 750 m if conditions permit. See Table 6.32 for the limit radius under different superelevation conditions at 80 km/h.

The ideal superelevation value at the speed of 100 km/h is shown in Fig. 6.23.

The shaded area in Fig. 6.23 represents the safety range of superelevation. Outside the shaded area, vehicles may slide out of the road or toward the inside of the curve. Under the most unfavorable condition of road icing, the value of transverse friction coefficient is the minimum. Under the condition of superelevation of 6 and 4%, it is suggested that the minimum radius of circular curve should be 660 m and 800 m

Table 6.29 Ideal superelevation values under different curve radius at 60 km/h

Radius (m)	250	300	350	400	450	500	550	600	650	700	750	800	850
Recommended superelevation	–	–	–	–	6%	6%	5%	5%	4%	4%	4%	4%	3%
Radius (m)	900	950	1000	1050	1100	1150	1200	1250	1300	1350	1400	1450	1500
Recommended superelevation(%)	3	3	3	3	3	2	2	2	2	2	2	2	2

Table 6.30 Limit radius under different superelevation conditions at 60 km/h

Limit value (superelevation 6%)	Standard	135 m
	Proposal	240 m
Limit value (superelevation 4%)	Standard	150 m
	Proposal	290 m

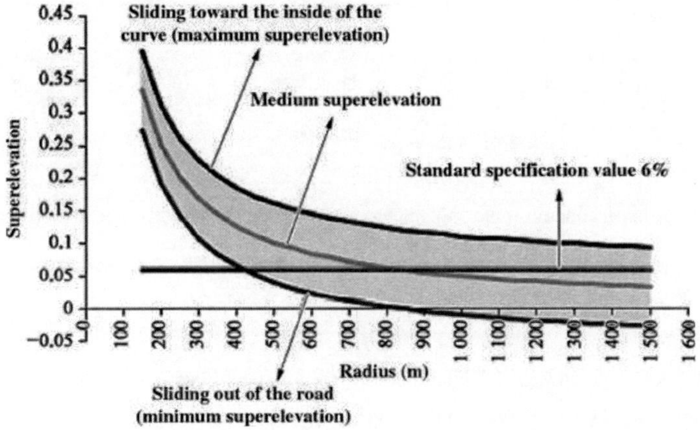

Fig. 6.22 Maximum and minimum superelevation under the most unfavorable conditions (80 km/h). Superelevation, sliding toward the inside of the curve (maximum superelevation), Medium superelevation, standard specification value 6%, Radius, Sliding out of the road (minimum superelevation)

Table 6.31 Ideal superelevation values under different curve radius at 80 km/h

Radius (m)	450	500	550	600	650	700	750	800	850	900	950
Recommended superelevation			–				6%	6%	6%	6%	5%
Radius (m)	1000	1050	1100	1150	1200	1250	1300	1350	1400	1450	1500
Recommended superelevation(%)	5	5	5	4	4	4	4	4	4	3	3

Table 6.32 Limit radius under different superelevation conditions at 80 km/h

Limit value (superelevation 6%)	Standard	270 m
	Proposal	430 m
Limit value (superelevation 4%)	Standard	300 m
	Proposal	510 m

respectively. Since the intermediate value between the maximum and the minimum superelevation value is the most ideal one, the superelevation value suggestions under different curve radii are shown in Table 6.33.

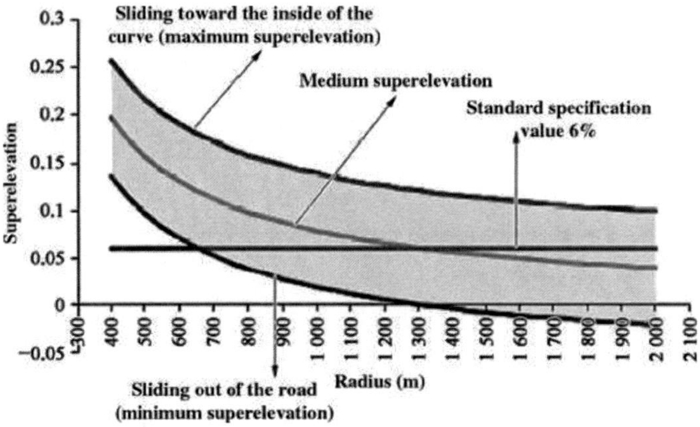

Fig. 6.23 Maximum and minimum superelevation under the most unfavorable conditions (100 km/h). Superelevation, sliding toward the inside of the curve (maximum superelevation), medium superelevation, standard specification value 6%, radius, sliding out of the road (minimum superelevation)

Table 6.33 Ideal superelevation values under different curve radius at 100 km/h

Radius (m)	1000	1050	1100	1150	1200	1250	1300	1350	1400	1450
Recommended superelevation	–	–	–	–	–	6%	6%	6%	6%	5%
Radius (m)	1500	1550	1600	1650	1700	1750	1800	1850	1900	1950
Recommended superelevation(%)	5	5	5	5	5	4	4	4	4	4

When the curve radius is less than 1,250 m, the ideal superelevation value is greater than 6%, so it is recommended that when the speed is 80 km/h, the radius can be greater than or equal to 1250 m if conditions permit. See Table 6.34 for the limit radius under different superelevation conditions at 100 km/h.

The ideal superelevation value at the speed of 120 km/h is shown in Fig. 6.24.

The shaded area in Fig. 6.24 represents the safety range of superelevation. Outside the shaded area, vehicles may slide out of the road or toward the inside of the curve. Under the most unfavorable condition of road icing, the value of transverse friction coefficient is the minimum. Under the condition of superelevation of 6% and 4%, it is recommended that the minimum radius of circular curve should be 960 m and 1,160 m

Table 6.34 Limit radius under different superelevation conditions at 100 km/h

Limit value (superelevation 6%)	Standard	440 m
	Proposal	660 m
Limit value (superelevation 4%)	Standard	500 m
	Proposal	800 m

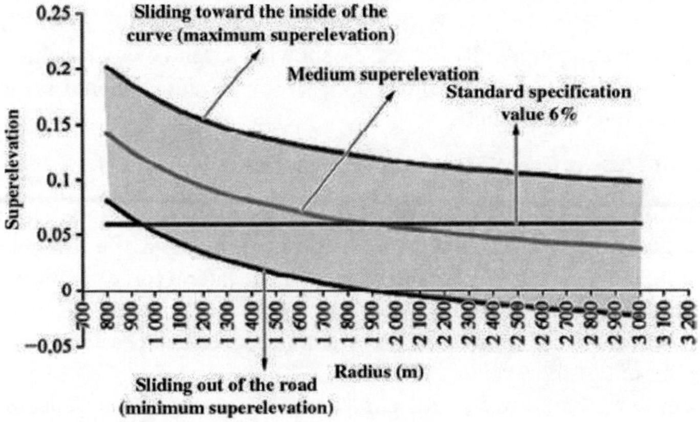

Fig. 6.24 Maximum and minimum superelevation under the most unfavorable conditions (120 km/h). Superelevation, sliding toward the inside of the curve (maximum superelevation), medium superelevation, standard specification value 6%, radius, sliding out of the road (minimum superelevation)

respectively. Since the intermediate value between the maximum and the minimum superelevation value is the most ideal one, the superelevation value suggestions under different curve radii are shown in Table 6.35.

When the curve radius is less than 1,800 m, the ideal superelevation value is greater than 6%, so it is recommended that when the speed is 80 km/h, the radius can be greater than or equal to 1,800 m if conditions permit. See Table 6.36 for the limit radius under different superelevation conditions at 120 km/h.

The specific selection of superelevation value is directly related to the value of curve radius. Superelevation affects the curve radius, and the latter also affects the former. It is one-sided to just discuss superelevation value regardless of curve radius. Therefore, the selection of reasonable superelevation values mentioned above is

Table 6.35 Ideal superelevation values under different curve radius at 120 km/h

Radius (m)	1300	1400	1500	1600	1700	1800	1900	2000	2100
Recommended superelevation	–	–	–	–	–	6%	6%	6%	5%
Radius (m)	2200	2300	2400	2500	2600	2700	2800	2900	3000
Recommended (%) superelevation	5	5	5	5	4	4	4	4	4

Table 6.36 Limit radius under different superelevation conditions at 120 km/h

Limit value (superelevation 6%)	Standard	710 m
	Proposal	960 m
Limit value (superelevation 4%)	Standard	810 m
	Proposal	1160 m

based on a specific radius value. With the increase of curve radius, the change range of ideal superelevation gradually decreases, showing a trend of approaching stability. Under the conditions of ice and snow, a larger curve radius should be selected to reduce the superelevation value.

(5) **Value of Crown Cross Slope under Ice and Snow Conditions**

When the radius of a circular curve is greater than a certain value, superelevation may not be arranged, but reverse superelevation equal to the crown of a straight section is allowed. From the perspective of driving comfort, the lateral force coefficient must be controlled to the minimum value, while from traffic safety, the vehicle must maintain its driving stability in the process of driving in the reverse superelevation section to avoid slipping out of the road.

The minimum radius of circular curve without superelevation specified in the *Technical Standard for Highway Engineering* of China shall be calculated according to the following method: when the crown transverse slope is less than or equal to 2%, the value range of the transverse force coefficient is 0.035 ~ 0.040; when that is greater than 2%, the value range of the transverse force coefficient is 0.040 ~ 0.050. Under the most unfavorable condition of road icing, the transverse friction coefficient of the road is 0.06, which is less than the transverse force coefficient used to calculate the minimum radius of circular curve without superelevation. Therefore, the minimum radius of circular curve without superelevation given in the current specification meets the safety requirements. However, considering the actual situation, the area has a high altitude, a long cold winter, a high proportion of trucks, and most of the vehicles are fully loaded, in order to avoid accidents caused by vehicles driving out of the road, it is recommended to take a smaller value for the crown cross slope.

6.1.6 Stopping Sight Distance Under Ice and Snow Conditions

When driving in the plateau area, on the one hand, the oxygen content is low, and the driver's operational agility is reduced; on the other hand, the plateau area has frozen snow all the year round, especially in winter, it is very likely that heavy snow will suddenly fall, leading to the reduction of road friction coefficient. The particularity of these two aspects will have a certain impact on the parking sight distance in plateau areas.

However, it can be seen from the above analysis that at an altitude of 4,533 m, the driver's perception time in the event of unpredictable events is 0.75 s, which is less than the current standard of 1.5 s, so the response time of 2.5 s used in the calculation of stopping sight distance in plateau environment can still meet the requirements. Therefore, next it will demonstrate the requirements for stopping sight distance on plateau roads from the perspective of friction coefficient reduction under ice and snow conditions.

(1) Relevant Requirements of the Standard for Stopping Sight Distance

According to the *Technical Standards for Highway Engineering* in China, the design of freeway must meet the requirements for stopping sight distance. Stopping sight distance refers to the shortest distance required by the driver from seeing the obstacles in front to reaching the safe stop of the obstacles when the vehicle is traveling at a certain speed. During the inspection of stopping sight distance, the driver's sight distance used for the stopping sight distance of passenger cars and heavy trucks is 1.2 m and 2.0 m respectively, and the top height of obstacles in front of the sight point is 0.10 m. For general highways, Class I highways and Class II and III highways with a high proportion of trucks, the truck's stopping sight distance should be used to inspect the relevant sections.

For the stopping sight distance in snow and ice sections, considering that the speed of driving in these sections will be greatly reduced, it may not be increased. However, for principal arterial highways, the stopping sight distance can be appropriately increased according to the minimum speed required by various regions to ensure safety.

Stopping sight distance is mainly composed of two parts: i. the distance driven by the driver's reaction time; ii. braking distance (the distance from the start of braking to the stop of braking). The safety distance of 5 ~ 10 m shall be increased. The calculation formula is as follows:

$$S_{停} = \frac{v}{3.6}t + \frac{(v/3.6)^2}{2gf_1} \tag{6.24}$$

where,

f_1 Longitudinal friction coefficient, depending on the speed and road conditions; in this book, the longitudinal friction coefficient of ice snow pavement is 0.18.

t Driver reaction time, taken as 2.5 s, which is composed of two parts: the sensing time of 1.5 s and the braking effective time of 1.0 s.

According to the above formula, the parking sight distance of the car in the wet state of the road surface in China's *Technical Standards for Highway Engineering* is shown in Table 6.37.

(2) **Longitudinal Friction Coefficient under Ice and Snow Conditions**

The Qinghai-Tibet Plateau is covered with frozen snow all the year round, and natural disasters occur frequently. After freeway construction completion, ice and snow will become an important factor affecting traffic safety. According to the hourly meteorological observation data of meteorological stations near the main highways

Table 6.37 Stopping sight distance in wet status (specification value)

Design speed (km/h)	running speed (km/h)	f_1	Calculated value (m)	Specified value
120	102	0.29	212.0	210
100	85	0.33	153.7	160
80	68	0.31	105.0	110
60	54	0.33	73.2	75
40	36	0.38	38.3	40
30	30	0.44	28.9	30
20	20	0.44	17.3	20

in Qinghai Province from October to April of the next year in the nine years from 2004 to 2012, the length of continuous freezing along the Qinghai-Tibet Highway from Kunlun Mountain to Tanggula Mountain is 20 ~ 30d/year. In high altitude locations and areas with frequent cold air activities, road icing starts early, ends late, and lasts a long time. After the completion of Qinghai-Tibet Freeway, ice and snow will become an important hidden dangers of traffic safety.

Under the conditions of ice and snow, the friction coefficient of pavement will be greatly reduced. Cheng Guozhu and others, by using the Pendulum Friction Tester and the electronic temperature and humidity meter, observed the urban road's friction coefficient of snow-covered pavement, rough ice pavement, smooth ice pavement and ice snow mixed pavement respectively, as well as atmospheric temperature and humidity of the ice snow road surface. See Table 6.38 for the observed data.

Du Xuesong et al., based on the climate characteristics of Northeast China, by using the British Pendulum Friction Tester, measured the friction coefficient of the rough, medium and fine pavement in winter dry and snow free, snow grains, snow powder and ice blocks status. See Table 6.39 for the specific measurement results.

Zheng Mulian and others, by using pendulum instrument and SAFEGATE friction test vehicle, measured the corresponding pendulum number and longitudinal friction coefficient, and finally collected 66 groups of valid data. Through regression analysis, the correlation equations of two measured values were established:

$$\mathrm{FB} = 95.93F_{\mathrm{P}} + 5.245,\ R = 0.9527$$

Table 6.38 Friction coefficient of different types of ice and snow pavements

S/N	Pavement status	Friction coefficient
1	Snow covered pavement	0.18 ~ 0.31
2	Rough ice pavement	0.12 ~ 0.20
3	Smooth ice pavement	0.07 ~ 0.15
4	Ice snow mixed pavement	0.06 ~ 0.17

Table 6.39 Average british pendulum number (bpn) corresponding to each measuring point

Filling	C	M	X	Average number
Status of snow free	70.0	64.7	67.5	66.5
Status of snow grains	41.0	40.3	40.0	40.3
Status of snow powder	47.0	34.0	35.5	36.7
Status of ice blocks	26.0	24.7	23.5	24.5

Note C, M and X represent the measuring point with relatively rough surface, moderate surface roughness, and dense surface respectively

where,

FB British Pendulum Number (BPN);
F_p Longitudinal friction coefficient.

According to the above formula, the value of longitudinal friction coefficient in Table 6.40 can be obtained by conversion.

Xie Jingfang and others studied the friction coefficient of freeway pavement under different meteorological conditions. The field test results show that on the same road section the friction coefficient is 0.72 in sunny days, 0.5 in rainy days, and less than 0.4 in snow-covered road. According to the test results, the classification standard of pavement friction meteorological indicator of Jilin Freeway is formulated. See Table 6.41 for the specific results.

Li Bachen believed that the main reasons for the formation of ice and snow pavement on the highway are the road snow were not be completed cleared and a small amount of snow brought by wind-blown. The ice snow covered pavement is scattered on the highway, and there is no large area of continuous ice and snow pavement. Therefore, when calculating the friction coefficient of ice and snow pavement, the friction coefficient cannot be selected according to the most unfavorable state under the ice and snow pavement. Taking 0.15 as the friction coefficient can represent the minimum value of friction coefficient of the road section.

Ye Ruimin takes 0.1 and 0.2 respectively to calculate the stopping sight distance under ice and snow conditions. When Liu Jing selecting the road friction coefficient to calculate the stopping sight distance, he takes 0.28 for the snow road surface and 0.18 for the icy road surface.

It can be seen from the above that different scholars have similar and different classifications of pavement conditions. According to the test results of Cheng Guozhu,

Table 6.40 Values of longitudinal friction coefficient under different conditions

S/N	Status	Mean value of longitudinal friction coefficient
1	Snow free	0.64
2	Snow grains	0.37
3	Snow powder	0.33
4	Ice blocks	0.21

Table 6.41 Relationship between friction indicator and skid resistance, friction coefficient and pavement condition

Friction indicator	Skid resistance	Actual friction coefficient	Corresponding pavement conditions
Level 1	Good	≥ 0.65	Normal temperature, dry, without impurities
	Normal	0.56 ~ 0.64	Wet, a small amount of water, low temperature
Level 2	Slightly poor	0.51 ~ 0.55	Water accumulation, low temperature
Level 3	Poor	0.41 ~ 0.50	Water accumulation, floating snow, frost
Level 4	Very bad	0.31 ~ 0.40	Accumulated snow
Level 5	Range	≤0.30	Icing

the friction coefficient of ice snow mixed pavement is 0.06 ~ 0.17, and that of rough ice pavement is 0.12 ~ 0.20; According to Zheng Mulian's research results, the BPN under the ice block state is 24.5, which is converted into the friction coefficient of 0.21. According to the classification standard of pavement friction meteorological indicator in Jilin Province, when the pavement is ice-covered, the actual friction coefficient is less than or equal to 0.30. Based on the above research results, the friction coefficient under ice and snow conditions is approximately 0.06 ~ 0.30, and the intermediate friction coefficient is 0.18. In this book, we take 0.18 for stopping sight distance calculation.

(3) Stopping Sight Distance on Highways under Plateau Ice and Snow Conditions

When calculating the highway stopping sight distance, if the grade *i* is considered, its formula is

$$S_{停} = \frac{v}{3.6}t + \frac{(v/3.6)^2}{2g(f_1 \pm i)} \tag{6.25}$$

where, "+ " is taken for the uphill section and "—" for the downhill section.

As the friction coefficient decreases under ice and snow conditions, the grade cannot be ignored and should also be considered. When the minimum design speed of freeways in China is 60 km/h, considering that the running speed of vehicles slows down under ice and snow conditions, only the design speeds of 40 km/h, 60 km/h, 80 km/h, and 120 km/h are selected to calculate stopping sight distance, the corresponding running speeds are 36 km/h, 54 km/h, 68 km/h, 85 km/h, and 102 km/h respectively, and the maximum grade is 6%. Therefore, the stopping sight distance required for different grades is calculated separately for 0 ~ 6%. See Table 6.42 for the final calculation results.

Table 6.42 Stopping sight distance corresponding to different design speeds and slopes (m)

Speed (km/h)	Specification value		0%	1%	2%	3%	4%	5%	6%
	Car	Trucks							
40	40	50	53	55	57	59	61	64	68
60	75	85	101	105	109	114	119	126	133
80	110	125	148	154	161	169	177	187	199
100	160	180	217	226	237	249	262	278	296
120	210	245	298	312	327	344	363	386	412

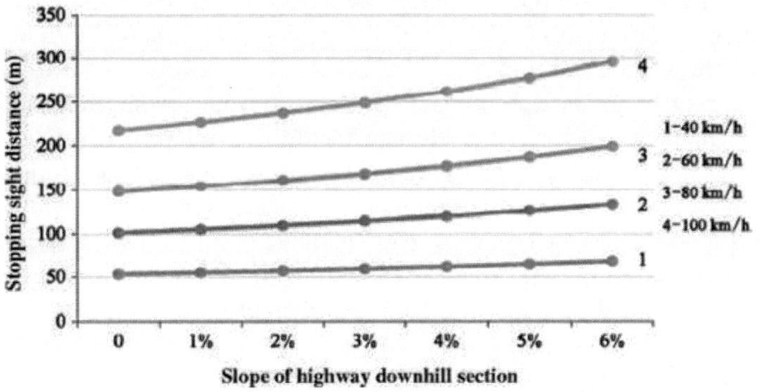

Fig. 6.25 Calculated stopping sight distance at different grades and design speeds. Stopping sight distance (m), Slope of highway downhill section

It can be calculated from Table 6.42 that the stopping sight distance at different grades and speeds under ice and snow conditions is as shown in Fig. 6.25.

It can be seen from the above analysis that:

i. Under ice and snow conditions, the calculated stopping sight distance is larger than the specified value in the current specification;
ii. At the same speed, the required stopping sight distance increases with the increase of downhill slope;
iii. At the same slope value in the downhill section, the greater the speed is, the greater the stopping sight distance is, and the greater the increase of sight distance is.

In Kunlun-Tanggula section of G109, the annual continuous icing time is 20 ~ 30 days, driving environment is harsh. From the perspective of ensuring the stopping sight distance requirements, smaller grades should be used as far as possible. In consideration of the frequent occurrence of malignant accidents on freeways in China under ice and snow conditions, Qinghai-Tibet Highway currently has a high proportion of trucks and is dominated by long-distance transportation, therefore, from the perspective of traffic safety, in order to reduce traffic accidents under ice and snow conditions, it is recommended to use the sight distance calculation value under such conditions for design inspection in high-altitude areas.

To sum up, in combination with the special geographical environment characteristics of the Qinghai-Tibet Plateau, in terms of highway horizontal alignment, the changes in the heart rate of drivers in the flat curve section and the curve slope combination section at different altitudes are studied respectively. From the perspective of driver comfort, the threshold value of the heart rate change rate is determined to determine the radius of the horizontal curve at different altitudes. At the same time, the reasonable values of superelevation and radius are calculated by the lateral friction coefficient under the condition of road icing. Use the difference between the actual superelevation and the critical superelevation, namely the net superelevation value, to evaluate the safety degree of superelevation value, balance between the maximum and the minimum superelevation, and finally calculate the reasonable value of superelevation and radius when the speed is 40 km/h, 60 km/h, and 80 km/h under ice and snow conditions.

In the aspect of grade design, first of all, according to the brake temperature calculation model considering the altitude factor, the test vehicle (six-axle articulated vehicle) and the vehicle are used to continuously descend the slope at a speed of 60 km/h. At the same time, the slope length related to the corresponding slope is determined according to the brake drum temperature of 200°C under the two braking modes of main brake braking and engine braking. According to the dynamic performance test results of the truck with the power weight ratio of 8.3 kW/t at different altitudes, the maximum grade and the corresponding slope length are determined by using the balance speed.

From the aspects of highway function, operation safety and permafrost subgrade scale effect, in this book it systematically analyzes the reasonable width value of each component element of freeway cross section in Qinghai-Tibet Plateau, and puts forward the recommended form of cross section suitable for the special environment of Qinghai-Tibet Plateau and the minimum width of each component element.

In terms of the value of parking sight distance, on the one hand, considering the long and cold winter in high-altitude areas, the heavy snowfall is likely to lead to the reduction of road friction coefficient; on the other hand, the low oxygen content in high-altitude areas and the increase of driver reaction time will lead to the increase of parking sight distance. Further considering the influence of grade, the stopping sight distance at the same speed under ice and snow conditions is calculated.

By studying the threshold value of the main technical indicator parameters of freeway in high altitude areas, it provides a strong support for establishing the main technical indicator system of freeway alignment in high altitude areas.

6.2 Reasonable Spacing of Service Facilities

According to the research results of fatigue characteristics of drivers in the plateau oxygen deficient environment in Chap. 4, the best time and driving distance for drivers in the plateau oxygen deficient area to rest and relieve fatigue on the way are proposed by comprehensively considering the driver's psychological and physiological needs, life support needs in the plateau area, emergency rescue, vehicle refueling and other needs; determine the reasonable distance between service facilities of Qinghai-Tibet Freeway and some specific requirements for service facility design.

6.2.1 Main Considerations for Designing Intervals of Service Facilities

The Qinghai-Tibet Freeway has a length of 1,110 km, featuring long distance, cold and oxygen deficient, few towns and villages along the line. Based on the statistical analysis of the traffic volume of the four inspection stations, namely, Naijigou Inspection Station, Wudaoliang Inspection Station, Anduo Inspection Station and Western Suburb Inspection Station, from January to September 2014, passenger cars and freight cars each account for half of the total traffic volume, and most of the freight cars are large trucks and extra-large trucks.

In China's *Technical Standards for Highway Engineering*, service facilities are divided into service areas, parking areas and passenger car stops. It is stipulated that the freeway should be provided with service areas with an average interval of 50 km. When the cities and towns along the line are sparsely distributed and water and electricity supply is difficult, the distance between service areas can be increased. The freeway service area shall be equipped with parking lots, gas stations, vehicle maintenance stations, public toilets, indoor and outdoor rest areas, catering, retail outlets and other facilities. Facilities such as personnel accommodation and vehicle water filling can be arranged according to the highway environment and needs.

The freeway shall be equipped with parking areas. One or more parking areas can be arranged between service areas. The distance between parking areas and service areas or parking areas should be 15 ~ 25 km. The parking area shall be equipped with parking lot, public toilet, outdoor rest area and other facilities.

In the *Opinions on the Implementation of Technical Standards for Freeway Construction in the Western Desert Gobi and Grassland Areas* it puts forward that for sections with small traffic volume and difficult water and power supply, the service area spacing can be appropriately increased, but the land area and building area of the service area should be increased accordingly. There is no quantitative criterion for the increase of service area spacing, building area and scale. The main factors to be considered for the service facility interval and the interval value are introduced as follows.

(1) **Refueling Needs**

It shall be ensured that vehicles can enter the service area to replenish fuel in time when necessary. Liu Yafei proposed an indicator of "vehicle driving distance under low fuel volume", and determined a recommended distance between service areas under the condition of meeting vehicle refueling demand through the indicator analysis. The "vehicle driving distance under low fuel condition" refers to the distance that the vehicle can travel when the running speed is kept near the operating speed after the low fuel warning light is on. Liu Yafei calculated the driving distance of three types of passenger cars, buses and freight cars under the condition of "low fuel consumption". See Table 6.43 for the calculation results.

It can be seen from Table 6.43 that under the condition of low fuel consumption, small cars have the shortest driving distance, with an average of 51.26 km, which is consistent with the service area spacing of 50 km under the current standard control in China.

(2) **Physiological Needs**

The arrangement of service areas must meet the physiological needs of passengers, such as toilets, catering, rest, etc. The fatigue time points of drivers at different altitudes have been discussed previously.

Figure 6.26 shows the investigation on the reasons why Chugoku Freeway in Japan used the parking area. A total of 1,154 users of the parking area or service area were counted, and nearly 86.9% of the respondents went there for the toilet.

According to the calculation of the International Health Organization, normal people usually urinate 4 ~ 6 times and defecate once in the daytime. According to the calculation of 8 h, they need to urinate once every 1.3 ~ 2 h. Based on the minimum speed limit of 60 km/h in China's freeway, the interval between service facilities is 60 ~ 120 km.

According to the field survey data along the Qinghai-Tibet Highway, about 81% of truck drivers hope to have a rest after driving for 3 ~ 4 h continuously; about 6.7% of drivers need to take a break and reorganize after driving for 1 ~ 2 h (Fig. 6.27). Based on the minimum speed limit of 60 km/h in China's freeway, the interval between service areas is 60 ~ 120 km.

In addition, the environment of the Qinghai-Tibet Plateau is special, with low temperature, low pressure and low oxygen content. The atmospheric oxygen content in Lhasa City, 3,658 m above sea level, is only about 60% of that in Beijing. When people in the plain area quickly enter the plateau above 3,000 m altitude, 50% ~ 75% of them have altitude sickness, and common symptoms include headache, insomnia, anorexia, fatigue, dyspnea, etc. In the process of arranging the service area, it is necessary to fully consider the situation that drivers and passengers may experience altitude sickness.

Table 6.43 Vehicle driving distance under low fuel volume (km)

Vehicle models	Passenger cars	Buses	Freight cars
Minimum value	35.04	73.96	92.51
85th percentile value	45.73	89.55	95.64
Average value	51.26	110.53	102.73
Maximum value	79.17	167.11	152.94

Note The "85th percentile value" refers to the 85th percentile value of the calculated driving distance of different vehicles but the same model from large to small in the case of low fuel consumption

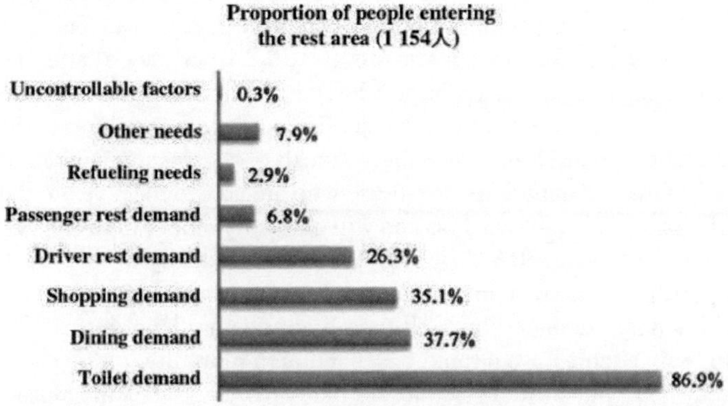

Fig. 6.26 Reasons for vehicles entering the parking area or service area. Proportion of people entering the rest area, uncontrollable factors, other needs, refueling needs, passenger rest demand, driver rest demand, shopping demand, dining demand, toilet demand

Fig. 6.27 Continuous driving time

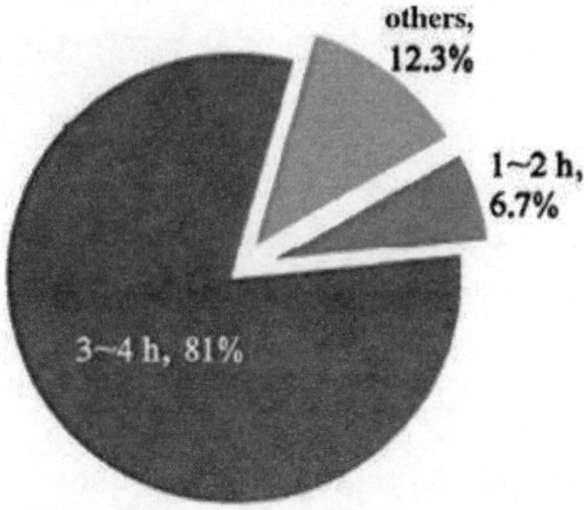

In the medical field, the area above 3000 m altitude is called plateau. Bao Zhengquan and others observed the changes of blood pressure, heart rate, blood oxygen saturation and ECG of the tested personnel at about 2,800 m, 4,200 m, 4,600 m, 5,100 m, 4,500 m and 3,600 m altitude respectively, and compared their data at 2,800 m altitude. It was found that blood pressure and heart rate increased with the altitude increase and decreased with the altitude decrease; blood oxygen saturation decreases with altitude increase, and increases with altitude decrease.

A survey was conducted by the Department of High-altitude Military Medicine of the Third Military Medical University on 113 young men to analyze the incidence of acute mountain sickness (AMS) after they took a plane from Chengdu to Xigaze, Tibet (3,850 m above sea level) in winter. The symptoms are headache, weakness, dizziness, insomnia and shortness of breath in turn. The research results show that young people in low altitude areas of the central and eastern regions are most likely to suffer from acute altitude sicknesses. The adaptability of people in different regions to altitude is different, and the overall incidence rate of the samples is 73.46%. Finally, the study suggested that in the future, the dispatched personnel at high altitude should try their best to ensure the rest time, and there should be a gradual adaptation time from low altitude to high altitude to reduce and mitigate the occurrence of altitude sickness.

The main impact of plateau on human body is the hypoxia and cold environment caused by its bad climate. Hypoxia can cause headache, chest tightness, shortness of breath, palpitations, nausea and vomiting, cyanosis of lips, insomnia, dreaminess, and blood pressure may also increase. For those who are new to the plateau, these symptoms are obvious in the first two days and will gradually reduce or disappear later; but there are also very few people whose symptoms may gradually worsen due to fatigue, cold, upper respiratory tract infection and other reasons, and develop into high-altitude pulmonary edema or high-altitude brain edema. High altitude pulmonary edema and brain edema have a rapid onset and high mortality.

Research shows that altitude sickness is likely to occur when the altitude is more than 2,500 m, various altitude sickness is likely to occur when the altitude is more than 3,500 m, and acute and severe altitude sickness with high mortality such as high-altitude pulmonary edema and high-altitude brain edema are likely to occur when the altitude is more than 4,500 m.

Yu Zhikang and others studied and showed that the risk of altitude sickness is the highest in winter and the lowest in summer. The risk indicator of altitude sickness increases gradually from 3,000 m above sea level (Fig. 6.28).

In addition, during the journey to Lhasa, the Qinghai-Tibet Railway started to supply oxygen at Golmud Station (2,829 m above sea level), which is divided into diffusion type oxygen supply and self-help type oxygen supply at the head of passenger bed. Through a series of safeguard measures of the Qinghai-Tibet Railway, the total incidence rate of brain and lung edema of construction personnel during the construction period of the Qinghai-Tibet Railway is 0.74% on average, and the incidence rate of brain and lung edema of passengers at high altitude during the operation period is 1/215 of those during the construction period.

In general, when reaching the altitude above 3,000 m, due to the reduction of oxygen content, there will be a certain degree of altitude sickness, and with the increase of altitude, the risk of altitude sickness increases rapidly. Once high-altitude

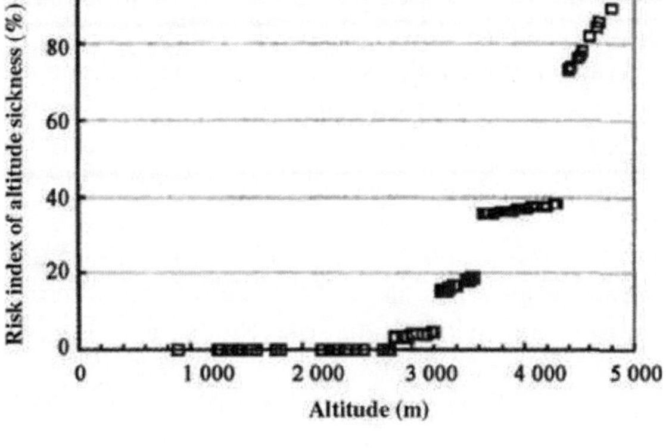

Fig. 6.28 Relationship between altitude and risk indicator of altitude sickness. Risk index of altitude sickness (%), Altitude (m)

pulmonary edema and brain edema occur, timely treatment is needed. It is suggested that people entering the plateau should try their best to ensure rest time and avoid fatigue driving, and there should be a gradual adaptation time from low altitude to high altitude to reduce and mitigate the occurrence of altitude sickness. Therefore, it is suggested that the service facilities arranged above 3,000 m altitude should strengthen the medical aid function, especially the oxygenation function.

(3) **Security Requirements**

The Regulations for the Implementation of the Road Traffic Safety Law of the People's Republic of China stipulates that driving a motor vehicle shall not commit the following acts: driving a motor vehicle continuously for more than 4 h without stopping for a rest, or stopping for a rest time of less than 20 min.

The research results of the University of California in the United States show that when the distance to the downstream of the parking area is more than 30 miles (48.3 km), the traffic accident rate related to fatigue will suddenly increase (Fig. 6.29).

Taylor et al. showed that the number of single vehicle accidents increased gradually 35 miles (56.3 km) downstream of the parking area. The specific results are shown in Figs. 6.30 and 6.31.

Through simulation and analysis at different altitudes, the time points when drivers enter into mild fatigue at different altitudes have been analyzed above.

According to the survey along the Qinghai-Tibet Highway, the expected running speed of passenger cars is mainly 100 km/h, and that of freight cars is mainly 70 ~ 80 km/h. The specific survey results are shown in Figs. 6.32 and 6.33.

(4) **Truck Demand**

The traffic volume statistics of four checkpoints on the Qinghai-Tibet Highway Naijigou Inspection Station, Wudaoliang Inspection Station, Anduo Inspection

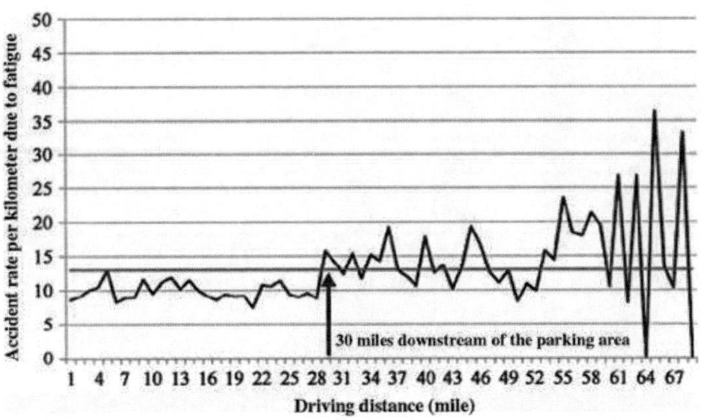

Fig. 6.29 Fatigue related traffic accident rate. Accident rate per kilometer due to fatigue, 30 miles downstream of the parking area, Driving distance (mile)

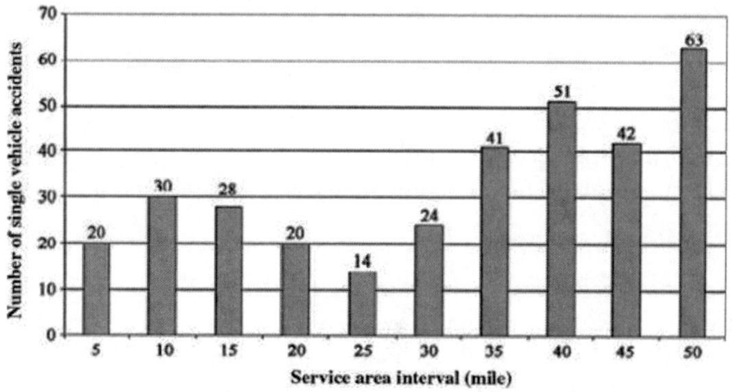

Fig. 6.30 Service area interval and number of single vehicle accidents. Number of single vehicle accidents, Service area interval (mile)

Fig. 6.31 Traffic accident of vehicle on plateau

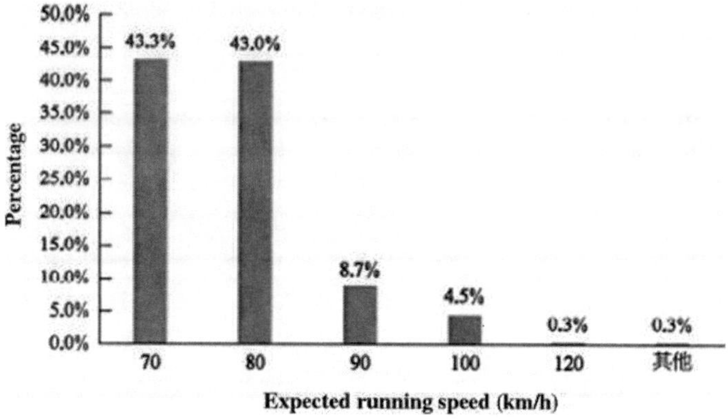

Fig. 6.32 Expected running speed of freight cars. Percentage, expected running speed (km/h)

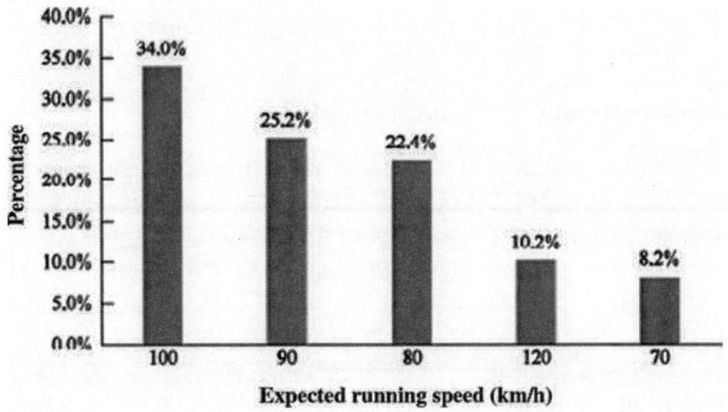

Fig. 6.33 Expected running speed of passenger cars. Percentage, expected running speed (km/h)

Fig. 6.34 Composition of average traffic volume. Minivans, medium trucks, passenger cars, large and extra-large trucks

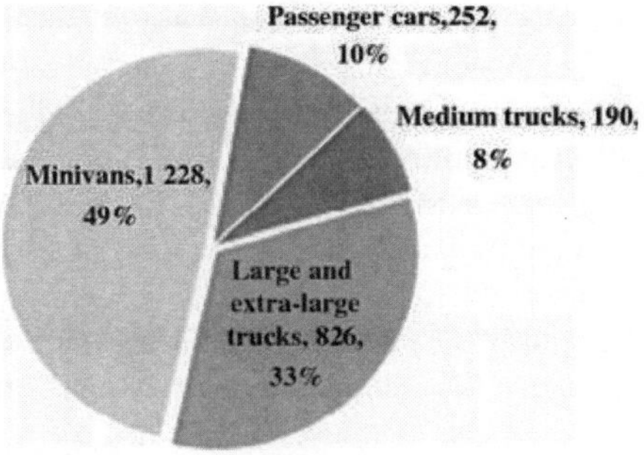

Station and Western Suburb Inspection Station from January to September 2014 are shown in Fig. 6.34.

It can be seen from Fig. 6.34 that passenger cars and freight cars each account for half of the total traffic volume, of which large and extra-large trucks account for 33%, medium trucks account for 8%, and minivans account for 10%. For the characteristics of greater number of large and extra-large trucks along the Qinghai-Tibet Highway, it must also be considered in the construction of the service area.

The demand of freight cars for service facilities is mainly driven by the demand for goods they transport. The goods transported by refrigerated vehicles that need strict freshness preservation will need to be iced in the service area due to their long journey. Fresh plants such as flowers, trees and bonsai need to be watered to maintain biological vitality during transportation, which can only be met through the corresponding configuration of the service area. Dangerous goods need more protection during transportation. When they stop in the service area, it is better to have a special parking area, which can provide relevant detection facilities and protection facilities to ensure the normal state of dangerous goods on the way; for long-distance transportation, the demand for such goods on the way has a higher dependence on the service area. The diagram of passenger and freight zoning parking in Germany is shown in Fig. 6.35.

(5) **Emergency Demand for Natural Disasters**

When natural disasters such as snowfall and road icing occur, the freeway is closed and passengers are stranded (Fig. 6.36). Therefore, as a shelter on the freeway, the service area needs to reserve a certain space for disaster emergency.

In 2008, a large-scale low temperature ice and snow disaster occurred in the south of China. Due to in disaster areas the lack of experience in the planning and design of service areas or the restriction of construction funds, there were some serious problems, such as insufficient parking spaces in parking areas and inadequate facilities and equipment, which have affected the service capacity and emergency response capacity of freeway to a certain extent.

The main disasters along the Qinghai-Tibet Railway are shown in Fig. 6.37. Since the alignment of the Qinghai-Tibet Highway is close to the Qinghai-Tibet Railway, Fig. 6.37 is also of reference significance.

From the total frequency along the railway, the total frequency of natural disasters at Tanggula Mountain Pass is the highest, with a total frequency of 86 times in 50 years, followed by Dulan, Huangyuan, Dangxiong and other regions, with a total frequency of more than 50 times. Almost every year, natural disasters occur. The combination types of natural disasters in Dulan and Dangxiong are relatively complex; the Qiangtang Plateau is prone to snow and wind disasters; the Qingnan Plateau is the section of the road south of the Kunlun Mountains and north of the Tanggula Mountains where snow and earthquake disasters are likely to occur.

Fig. 6.35 Passenger and freight cars parked in different areas (Germany). Parking area for trucks, Parking area for light trucks, Parking area for passenger cars

Fig. 6.36 Traffic congestion caused by heavy snow on freeway

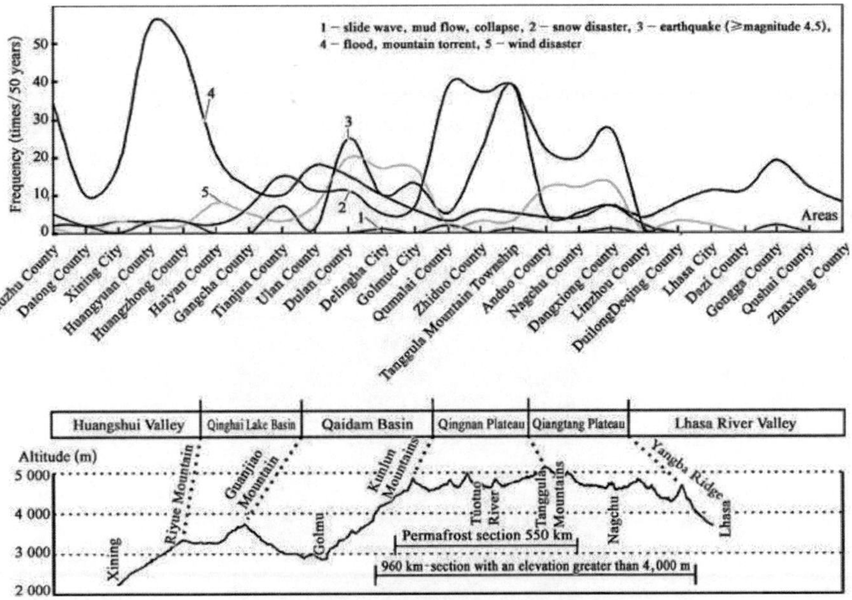

Fig. 6.37 Frequency distribution of natural disasters along Qinghai-Tibet Railway. Frequency (times/50 years), 1—slide wave, mud flow, collapse, 2—snow disaster, 3—earthquake (≥magnitude 4.5), 4—flood, mountain torrent, 5—wind disaster, Areas, Huzhu County, Datong County, Xining City, Huangyuan County, Huangzhong County, Haiyan County, Gangcha County, Tianjun County, Ulan County, Dulan County, Delingha City, Golmud City, Qumalai County, Zhiduo County, Tanggula Mountain Township, Anduo County, Naqu County, Dangxiong County, Linzhou County, Duilong Deqing County, Lhasa City, Dazi County, Gongga County, Qushui County, Zhabao County

Figure 6.38 shows the location of the Kunlun Mountain Pass M8.1 earthquake, which occurred on November 14, 2001. More than ten years have passed since

Fig. 6.38 Location of the M8.1 earthquake in West of Kunlun Pass. Surface fracture trace

then, and there are still traces left by the earthquake on the surface. The Wenchuan earthquake in 2008 was also 8.1 magnitude.

Figure 6.39 shows a large trench with a depth of more than 5 m and a width of nearly 4 m formed on the fracture zone after the earthquake. The M8.1 earthquake at the Kunlun Pass is the largest earthquake in China in the past 50 years. The surface rupture and other phenomena caused by the earthquake are also the earthquake ruins unique in China, rare in the world, and ever the most complete, spectacular and latest.

Fig. 6.39 Deep trench after Kunlun pass earthquake

(6) **Sightseeing Demand**

The scenery along the Qinghai-Tibet Highway is beautiful. In combination with the terrain conditions, viewing platforms are designed in some scenic places for tourists to rest and enjoy the scenery.

6.2.2 Classification and Functional Positioning of Service Facilities

In Ireland the service facilities are divided into two types, namely, full service area and rest area. The full service area includes shops, toilets, gas stations, parking lots and other comprehensive facilities. The rest area will be smaller, without gas stations, and only serves vehicles in one direction.

In Canada the service facilities are divided into large service areas, medium service areas and small service areas. The annual average daily traffic of the freeway where the large service areas are located is 7,500 vehicles, the annual average daily traffic of the freeway where the medium service areas are located is 2,000 ~ 7,500 vehicles, and the annual average daily traffic volume of the freeway where the small service areas are located is less than 2,000 vehicles. The interval of service facilities is arranged as 80 km or 1 h's drive.

The service facilities on freeway are divided into rest areas and service centers (Table 6.44).

In addition, for arterial highways, if the daily traffic of large trucks is more than 500 and that of light trucks is more than 30,00, truck parking areas should be designed. See Table 6.45 for specific categories.

Table 6.44 Classification and interval of service facilities

Items	Rest areas	Service centers
Interval of service areas	20 ~ 30 km (AADT > 40,000) 30 ~ 50 km (10,000 < AADT < 40,000) 50 ~ 60 km (AADT < 10,000)	30 ~ 50 km (AADT > 25,000)
Location	Serves two-way traffic	Serves two-way traffic

Table 6.45 Classification and spacing of truck service facilities

Classification	Spacing (km)	Parking space (pcs)
Truck parking area	100	20
Category II truck parking area	30 or 35	8
Small parking area	10 ~ 15	/
Light truck parking area	50	/

Freeway service facilities are an important part of freeway construction, and also an important supporting facility for providing traffic safety guarantee and comprehensive services for highway users and vehicles in the freeway operation stage. In order to meet the above requirements, service facilities are divided into three categories according to different functions, namely, the first category, the second category and the third category.

The third type of service facilities corresponds to the parking area in China's specifications: parking lots, public toilets, outdoor rest and other facilities should be arranged, mainly for road users to take a short rest, to achieve simple functions such as changing drivers, checking vehicles, and going to the bathroom.

The second type of service facilities corresponds to the service area in our country's specifications. Our country's specifications stipulate that the service area facilities in the freeway service area should be provided with parking lots, fuel stations, public toilets, outdoor rest rooms, etc. If possible, catering, commodity retail and other facilities can be arranged. According to the highway environment and needs, personnel accommodation, vehicle water filling, etc. can be arranged. Similar to the plain service area, it mainly provides road users with long time rest and reorganization to meet the needs of refueling, shopping, dining and accommodation. In addition, it must also provide oxygen supply and other functions.

The first type of service facilities is mainly to realize the life support function.

The total length of Golmud Lhasa section is about 1,100 km. Considering the harsh plateau environment, the deficiency of oxygen and long distance, it is difficult to carry out emergency rescue once an accident occurs. From the perspective of traffic safety, it is recommended to add oxygen stations on the basis of the second type of service facilities to meet the oxygen intake needs of some personnel; a medical rescue station shall be arranged to rescue the personnel with sudden altitude reaction in time; to arrange an emergency rescue center for traffic accidents to provide emergency rescue for vehicles involved in traffic accidents at the first time; to arrange a vehicle maintenance center to carry out simple maintenance on the faulty vehicle, such as changing the tire, so that it can drive normally.

The division of three types of service facilities is shown in Fig. 6.40. The first type of service facilities is a comprehensive service facility, which can not only provide catering, accommodation, refueling and other functions, but also achieve medical assistance, accident emergency rescue, vehicle maintenance and other functions. If conditions permit, rescue helicopters can be equipped to shorten rescue time and improve rescue efficiency. The second type mainly provide functions such as rest, catering, refueling and shopping. The third type is mainly for short rest and reorganization. The grade of service facilities has gradually increased from the third to the first type, and the facilities and functions are becoming more and more perfect.

Due to the long distance of the Qinghai-Tibet Freeway, the accommodation is recommended to provide, and its location is in places with complete facilities and low elevations, the standard can be classified into different levels and grades, and the price setting can also be handled flexibly. It is not recommended to arrange accommodation areas at high altitude.

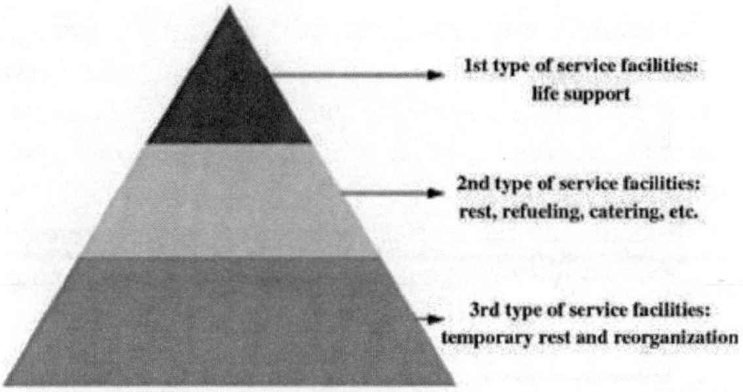

Fig. 6.40 Division of three types of service facilities. 1st type of service facilities: life support, 2nd type of service facilities: rest, refueling, catering, etc., 3rd type of service facilities: temporary rest and reorganization

In addition, the Qinghai-Tibet Plateau is covered with snow and ice all the year round, and natural disasters occur frequently. After the completion of the freeway, ice and snow will become an important factor affecting traffic safety. In the snow and freezing environment, the adhesion coefficient of the road surface decreases, and the emergency braking of the motor vehicle during driving is easy to cause sideslip. The braking distance on ice and snow roads is greatly extended, which is very easy to cause rear end collision and even chain collision accidents. It is more difficult for cars parked on the grade to start, which may cause skidding and sliding. Moreover, due to the perennial snow in high-altitude areas, the driver's vision is tired and it is difficult to identify the road conditions due to the reflection of snow light. Extreme weather on the plateau happens frequently, which is very easy to cause road traffic accidents. Affected by the rainy and snowy weather, at about 9:00 a.m. on November 21, 2016, at 65 km + 500 m in Taiyuan direction of Pingyang section of Beijing Kunming Freeway, multiple vehicles collided successively, causing 17 deaths, 37 injuries and 56 vehicle damage (Fig. 6.41).

Fig. 6.41 Scene of accident on Beijing Kunming freeway

Considering the road network around the freeway is single, the populated area along is sparse, and the total mileage is long, the service facilities of Qinghai-Tibet Freeway should not only have the above functions, but also bear the emergency rescue function under ice and snow conditions. Specific functions are described as follows:

Take the service area as the carrier to release relevant traffic information, including weather conditions, road traffic conditions and other information. Take the service area as the carrier to diverge the traffic flow under ice and snow conditions, so the service areas along the line should be provided with sufficient parking spaces and relevant supporting facilities. Take the service area as the carrier for emergency rescue of traffic accidents; if rescue personnel can rush to the scene of the highway traffic accident as soon as possible and carry out relevant work the first time of it, the number of casualties can be minimized and the impact time of the accident on traffic operation can be shortened. In high-altitude areas, the oxygen content is low, and the human tolerance is significantly reduced compared with that in plain areas, which puts forward stricter requirements on the emergency rescue time. Therefore, the service area should be equipped with a certain number of rescue personnel and corresponding equipment.

6.2.3 Second Class Service Facility Interval of Qinghai-Tibet Highway

(1) **Intervals of Service Facilities Worldwide**

Jiang Cailiang mentioned in *Analysis of Highway Service Area Construction Requirements in Ice and Snow Disaster Areas* that, for ice and snow or other disaster-prone areas, reasonable spacing of highway service areas can avoid vehicles and passengers being stranded on the road section. For disaster-prone areas, spacing for highway service areas should be reduced in addition to the usual consideration, such as economic development trend of the road through the region, the prospective traffic volume, traffic flow characteristics and landscape along route, it is recommended that the service area should be arranged in accordance with 20 ~ 30 km interval.

The U.S. Federal Highway Administration's *U.S. Highway Design Geometry Manual* specifies the standard spacing between integrated service areas as 40 km, and the maximum positioning is 100 km. The Japanese *Highway Design Guidelines* specifies the standard and maximum spacing between integrated service areas as 50 km and 100 km respectively.

The Australian Road Research Center classifies service areas into three categories: major rest areas, minor rest areas, and truck parking bays. The major service areas are mainly for long rest periods, with separated parking areas for heavy vehicles and light vehicles; the minor service areas are mainly for short rest periods, and according to the actual situation, separate parking areas can also be arranged; the truck parking bays are mainly designed for heavy vehicles, to carry out cargo inspection, record travel logs, etc. The interval between the main service areas is 100 km, that of for small service areas is 50 km, and for truck parking bays is 30 km.

(2) **Intervals of Second Class Service Facilities**

Considering the above factors, interval for the second class service facility is finally determined. As for the determination of calculation speed, on one hand, the minimum speed limit of the freeway is 60 km/h, and on the other hand, it is found that the expected maximum speed of trucks is 80 km/h. Therefore, the interval between service facilities is calculated by 60 km/h and 80 km/h respectively.

The Qinghai-Tibet highway is long, with few surrounding towns and villages, and no large populated area, the vehicles driving on Qinghai-Tibet highway are mainly in a hurry, especially large buses and cargo transport vehicles, from the perspective of traffic safety, priority function for second class service facilities should be resting, once the driver appears fatigue state, it must be corrected in time, if excessive fatigue, coupled with cold and upper respiratory tract infection, etc., symptoms may gradually aggravate, and even develop into plateau pulmonary edema or plateau cerebral edema. Therefore, it should be avoided that there is no service area for the driver to rest, and forced to cause tired driving.

As can be seen from Table 6.46, when the driver's fatigue degree reaches 85%, calculated by 60 km/h and 80 km/h, the calculated distances of test points A, B and C are 92–123 km, 84–113 km and 63-84 km respectively, by taking average value, then the calculated distances of test points A, B and C are 107 km, 99 km and 74 km respectively. Similarly, the calculated distances for 15% positions of fatigue degree at test points A, B and C are 83 km, 74 km and 48 km respectively. In conclusion, it is recommended that the interval value of the service area should be arranged as 100 km near the altitude of 3500 m. Considering that the risk indicator of altitude sickness increases rapidly when the altitude rises above 4000 m, it is recommended that the interval value of the service area should be arranged as 80 km when the altitude exceeds 4000 m.

6.2.4 Spacing of Third Class Service Facilities on Qinghai-Tibet Highway

The third class service facilities correspond to the parking area in our regulations. The parking area is mainly for vehicles and drivers to rest for a short time, and only provides simple facilities such as toilets and outdoor rest in the parking lot to realize simple functions such as driver replacement and vehicle inspection.

In Europe, the freeway service area can be divided into three categories: comprehensive service area, parking area and refueling area. The functions of the comprehensive service area are roughly the same, mainly for vehicle maintenance, cleaning, refueling/water filling, rest, shopping, catering and other functions for drivers and passengers, with the distance of 30 ~ 50 km. Parking area plays an important role in freeway service area, which is an important facility to ensure driving safety and

Table 6.46 Calculation for intervals of service areas at different altitudes

Factors considered	Test point A (3500 m)	Test point B (4200 m)	Test point C (4600 m)
Driving distance under low fuel consumption conditions	Average value of passenger cars is 51.26 km Average value of large buses is 110.53 km Average value of trucks is 102.73 km		
Urinate every 1.3 ~ 2 h	78 ~ 120 km (60 km/h) 104 ~ 160 km (80 km/h)		
Taking rests for adjustment after 1 ~ 2 h of continuous driving for 6.7% drivers	60 ~ 120 km (60 km/h) 80 ~ 160 km (80 km/h)		
Taking rests after 4 h of continuous driving	240 km (60 km/h) 320 km (80 km/h)		
Fatigue of non-local drivers is 15% (60 ~ 80 km/h)	71 ~ 95 km	63 ~ 84 km	41 ~ 55 km
Fatigue of non-local drivers is 85% (60 ~ 80 km/h)	92 ~ 123 km	85 ~ 113 km	63 ~ 84 km
Emergency needs for snow and ice disasters	It is recommended that sufficient disaster emergency space be reserved in the service area, and two service areas connected through underground passageways		
Landscape requirements	Depends on the actual situation		

reduce the accident rate. European countries have different spacing for parking areas, mostly at 10 km interval to ensure that drivers can use parking facilities within 10 min of driving, so as to facilitate parking at any time. The spacing of service areas in some European countries is shown in Table 6.47.

In Japan's *Study on the Planning and Design of Rest Facilities on Freeways (II)*, the satisfaction rate of the corresponding spacing is given. The study shows that the maximum spacing of 25 km can meet 91% of the needs, while the standard spacing of 15 km can meet 98% of the needs. When the maximum distance between the

Table 6.47 Intervals of service area for some European countries

Country	Type of facility	Interval (km)
Britain	Service area	16 ~ 17
Germany	Parking area	5 ~ 10
	Service area	50
France	Parking area (A class)	8 ~ 10
	Parking area (B class)	25 ~ 30
	Refueling facilities	40 ~ 50
	Service area	100
Hungary	Parking area	20 ~ 30
Holland	Refuelling facilities	20 ~ 30

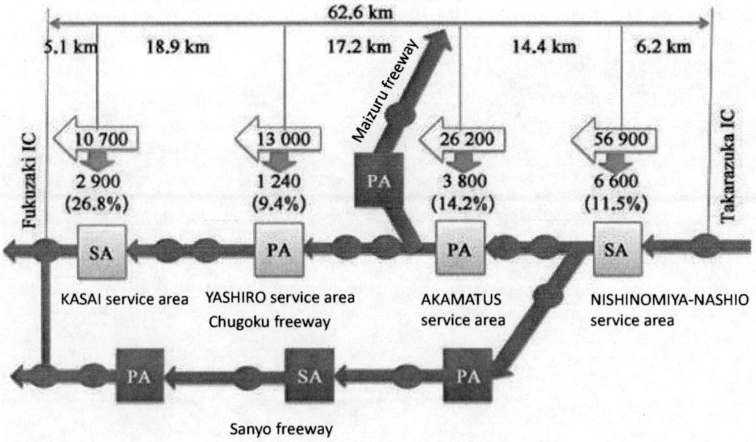

Fig. 6.42 Interval of service area and parking area

comprehensive service interval is 100 km, 61% of the needs can be met, and when the standard distance is 50 km, 89% of the needs can be met.

Figure 6.42 shows the spacing between parking area and service area of Chugoku freeway in Japan, and the average interval between parking areas is 16. 8 km.

Figure 6.43 shows that after the vehicle broke down in the direction of Riyueshan Section to Lhasa, because there was no parking area in this section, the vehicle stopped in the emergency lane for maintenance and occupied part of the lane. If there was a parking area nearby, the vehicle can enter the parking area for maintenance to improve its safety.

To sum up, the parking area is mainly for a short time rest and relieve fatigue, so outdoor benches, toilets and other facilities can be added, which should be simple and practical to save land and construction costs. Due to the plateau areas, a deficiency of oxygen and long distance of Qinghai-Tibet plateau, priority for most drivers is hurry on with their journey, and cost will be greatly increased if implement construction as per foreign construction standards, therefore, specific interval should be determined by considering the second class and the first class service facilities, it is suggested to arrange 1 ~ 2 parking area between service areas as per practical conditions, and the spacing of the parking area is 33 ~ 50 km near the altitude of 3500 m and 27 ~ 40 km above the altitude of 4000 m as per the intervals arranged above.

Finally, considering the demand of future traffic growth, a certain area can be reserved for current parking area for future upgrading and reconstruction, and it can be built into a service area. It is suggested to connect the parking areas with an underground channel for vehicle's turn-around and other functions.

Fig. 6.43 Faulty vehicle stopped in emergency lane

6.2.5 Interval of First Class Service Facilities on Qinghai-Tibet Highway

According to the results of the questionnaire survey along the Qinghai-Tibet highway, the vast majority of drivers will be physically unfit at Tanggula Pass. The Tanggula Mountain is 5231 m and is the highest point along the Qinghai-Tibet highway. The next places where the body may not adapt are: Wudaoliang (4665 m), Tuotuo River (4547 m), Anduo (4702 m), Kunlun Mountain (4676 m), Fenghuo Mountain (5010 m), Nagchu (4513 m). The survey results are shown in Figs. 6.44 and 6.45.

According to the survey results above, attention should be paid to places above 4500 m, which is also confirmed by the literature search results. When the altitude exceeds 4000 m, the risk indicator of altitude sickness increases rapidly, and this area is most prone to physical inadaptation. Medical aid stations should be built in the service area to provide timely assistance to drivers and passengers with altitude sickness.

The functions of the first class service facilities are more comprehensive than that of the service area, for example, to arrange oxygen stations for oxygen intake of some personnel. Medical aid stations will be arranged to provide timely aid to those suffering from sudden altitude sickness. The establishment of traffic accident emergency rescue center, in the first time to traffic accident vehicles emergency rescue; Arrange the vehicle maintenance center to carry out simple maintenance of the faulty vehicle, such as replacing the tire, so that it can drive normally; When conditions permit, equipped with a helicopter for emergency rescue, the first time to deal with the emergency situation, such as vehicle accidents.

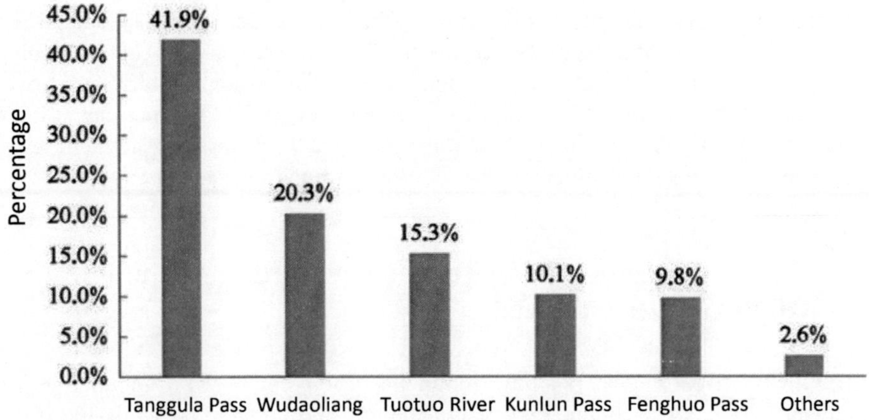

Fig. 6.44 Section prone to be physically unfit along Qinghai-Tibet Highway (Start from Golmud)

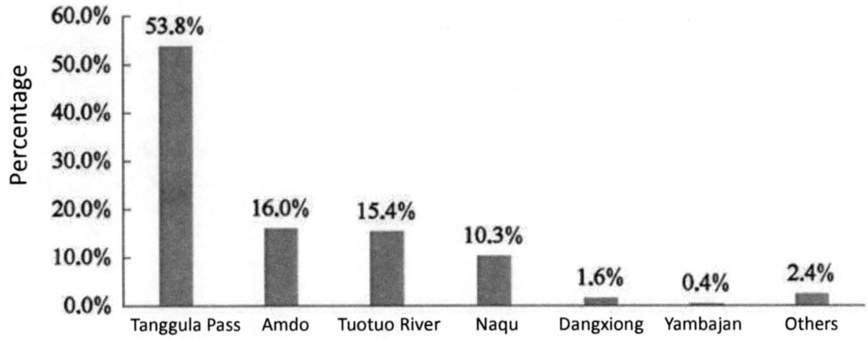

Fig. 6.45 Section prone to be physically unfit along Qinghai-Tibet Highway (Start from Lhasa)

The functions and equipment of the first class service facilities are the most complete. In practice, a first class service facility can be built at every two second class service facilities. That is, a first class service facility can be arranged every 300 km around altitude of 3500 m, and that of at intervals of 240 km above 4000 m. Location of first class service area should also take into account factors such as water and electricity supply.

According to research results above, the particularity of the plateau environment should be comprehensively considered in the arrangement of service facilities in engineering construction, and the distance between service facilities in the high altitude area should be arranged reasonably and the corresponding necessary configuration should be provided from the perspective of meeting the needs of vehicles and personnel and ensuring traffic safety. There are three classes of service facilities. The first class is mainly integrated service facilities for life support, arranging at every 300 km around altitude of 3500 m, and that of every 240 km above 4000 m. The second class service facilities correspond to the service area in our specifications. Around the altitude of 3500 m the service area is arranged every 100 km, regarding the risk indicator of altitude sickness increases rapidly when the altitude is over 4000 m, the interval of the service area is arranged every 80 km when the altitude exceeds 4000 m. It is recommended to arranged 1 ~ 2 third class service facilities in between of the second class service facilities according to the actual conditions.

6.3 Dynamic Segmentation Technology of High Altitude Highway Design Speed

The Qinghai-Tibet Plateau is a huge mountain system composed of mountains and highland terrace. The mountains are treacherous and undulating, while the highland terrace is flat and low undulating, which is mostly plain and hilly. These two significantly different terrain and geological conditions determine a large difference in the selection of technical standards in construction of highways in the Tibetan plateau area. The technical standards difference is reflected in the selection of design speed, which determines the values for main technical indicator parameters of the flat, vertical and horizontal indicators. Therefore, according to the topographic and geological conditions of the Qinghai-Tibet Plateau area, this section studies the design methods suitable for high altitude highway design speed segmentation technology and the speed transition between segments of different design speeds, taking into account the vehicle driving characteristics of the plateau area and other factors.

6.3.1 Design Speed Select by Function and Technical Level of Highway

To select the design speed by highway function and technical grade. According to the function of transportation, highway is divided into three categories such as arterials, collectors and local roads, among which, arterials are mainly subdivided into principal arterials and minor arterials.

(1) **Principal Arterials**

It connects large and medium-sized cities with a population over 200,000, transportation hubs, important foreign ports and military strategic locations; provides long-distance, high-capacity and high-speed transportation services between provinces and large and medium-sized cities.

(2) **Minor Arterials**

It connects cities with a population over 100,000 and regional economic centers; provides medium and long distance, higher capacity and higher speed transportation services within the region or province.

(3) **Major Collectors**

It connects counties (cities) with more than 50,000 people, major industrial and agricultural production bases, important economic development zones, scenic spots and commodity distribution areas; provides medium distance, medium capacity and medium speed traffic services; connects with arterials so that all counties (cities) are within a suitable distance of arterials.

(4) **Minor Collectors**

It connects counties (cities) with more than 10,000 people, large towns and other traffic generators; provides shorter distance, smaller capacity and lower speed traffic services; connects arterials, major collector highways and local roads, dispersing arterial traffic and gathering local traffic.

(5) **Local Roads**

The main function is to provide access to adjacent property for a variety of users, connecting with collector roads and serving local traffic and through traffic. l.

Highways with different functions correspond to different design speeds. Principal arterials provide higher travel speeds, generally over 80 km/h, and the design speeds of 80, 100 and 120 km/h can be applied. Minor arterials are important supplements to principal arterials, and the design speeds of 60, 80, 100 km/h and 120 km/h can be applied. The major collectors are connected to arterials, converging local traffic to arterials and diverging arterial traffic to local roads, and the design speeds of 30, 40, 60 km/h and 80 km/h can be applied. Minor collectors are mainly converging traffic from the local road, and the design speeds of 30, 40 and 60 km/h can be applied.

Different functions and technical levels correspond to different design speeds. Generally the corresponding design speeds of freeways are 120, 100 and 80 km/h.

6.3.2 Selecting Design Speed by Route Layout Factors Such as Altitude and Terrain Conditions

According to the research results in Sect. 7.1, the thresholds of the main technical indicator parameters of the highway and the altitude are correlated. When selecting the design speed, requirements of the technical indicators of the routes at different altitudes should be fully considered. On the basis of altitude factor, further consideration should be made on influence of terrain factor. The topography of the Qinghai-Tibet Plateau area is dominated by plains and high mountains, and the flatter terrain is surrounded by mountain ranges and the altitude rises in steps, forming a unique plateau terrace terrain.

In accordance with requirements of the current *Technical Standards for Highway Engineering*, the design speed of freeways should not be less than 100 km/h, subject to terrain and geological conditions, can be selected as 80 km/h.

(1) **Altitude Factor**

According to the research results on the main geometric indicators and parameters of high altitude freeways, the technical indicators of the route that are largely influenced by altitude factors include the radius of the circular curve of the plane, as well as the grade and slope length of the longitudinal section. As the altitude rises, the minimum value of the radius of the circle curve increases, and the grade to meet the requirements of vehicle dynamics decreases. Therefore, according to the initially determined route plan, the design speed that can be satisfied by the radius of the circular curve of the plane and the slope of the longitudinal section should be analyzed in combination with the altitude factor. The analysis criteria are shown in Table 6.48.

According to the minimum radius and maximum grade of the circle curve corresponding to the design speed under different altitude conditions shown in Table 6.48, the design speed of the preliminary proposal of the route can be determined under different altitude for the flat and longitudinal technical indicators, and then the design speed of the preliminary proposal of the route under different altitude conditions can be determined.

(2) **Terrain Factor**

After determining the design speed under different altitude interval conditions, the design speed should be further selected according to the terrain conditions in the altitude interval.

(1) **Highland Plain Micro-hill Terrain**

The plain terrain in this area is flat, without obvious undulations, and the natural slope of the ground is generally within 3° (Fig. 6.46); micro-hill terrain refers to low hills with little undulation, and the natural slope of the terrain is below 20°, and the relative height difference is below 200 m (Fig. 6.47). The alignment layout in this area is less affected by the terrain and geological conditions, and higher geometric design indicators can be used, and the design speed may not be less than 100 km/h when the design speed is selected, and 120 km/h of design speed can be used when the terrain is open.

Table 6.48 Design speed considering altitude factors corresponding to limit values of flat and longitudinal indicators

Design speed (km/h)	Min. radius of circular curve of the plane (m)		Max. grade (%)		
	Altitude < 4000 m	Altitude ≥ 4000 m	Altitude 3500 ~ 4500 m	Altitude 4500 ~ 5500 m	Altitude over 5500 m
120	650	650	2	1.5	1
100	400	400	2.5	2	1.5
80	300	350	3.5	3	2.5

Fig. 6.46 Plain micro-hill terrain

Fig. 6.47 Low hills

(2) **River Valley Plain Terrain**

The terrain is relatively gentle, the river valley is a few hundred meters to several thousand meters wide, is a wide shallow U-shaped river valley landform, the slope of the riverbed is mostly below 5°, the natural slope of the landform is below 20° (Fig. 6.48). The terrain along the river alignment is generally not restricted, the route grade is gentle or slightly undulating, the design speed can be selected as 100 km/h.

Fig. 6.48 River valley plain

For the section where terrain conditions are limited, the design speed can be selected as 80 km/h.

(3) **Mountain Canyon Terrain**

Mountain canyon terrain includes complex terrain changes such as ridges, steep slopes, cliffs, canyons, deep ravines, the terrain is steep, with serious cutting and narrow river valleys, the natural slope of the ground is mostly above 20° (Fig. 6.49). Most of the design indicators of the route flat, vertical and horizontal are restricted by the terrain, and the design speed is generally selected as 80 km/h.

6.3.3 Design Speed Selection by Considering Climatic Conditions

According to Sects. 6.1.5 and 6.1.6 of the book, for the area susceptible to snow and ice, the radius of the circle curve and the sight distance indicator should be increased appropriately. When the route passes through the area susceptible to snow and ice, design speed should be determined based on the radius of the curve and sight distance that can meet the extreme freezing requirements, meanwhile sufficient length should be ensured for the design speed segment length.

Fig. 6.49 Mountain canyon

6.3.4 Design Speed Selection by Considering Operating Speed of Typical Vehicles

By studying the influence of altitude on the power performance and braking performance of the vehicle, it is found that the climbing ability of the same slope under different altitudes varies greatly, and the speed that the vehicle can reach on the slope section of different slopes gradually becomes lower with the increase of altitude and the decrease of power performance.

The effect of altitude on operating speed of passenger cars is relatively small, and that of on trucks such as heavy trucks and articulated combination of vehicle is larger. In the research results of Chap. 3, for a heavy load truck with load capacity of 30t and power-to-weight ratio of 8.3 kW/t, it's vehicle performance is reduced by 10–15% below 3000 m, and the vehicle performance is reduced by nearly 35–40% from 3500 ~ 5500 m. According to the changing characteristics of the power performance of large trucks, the equilibrium speed of the corresponding slope at each altitude is calculated and the equilibrium curve is drawn, and the equivalent slope theory based on the altitude is used to calculate the equivalent slope and slope section offset values at different altitudes to obtain the grade correction amount in the large truck operating speed model, and the operating speed of the large truck is deduced.

According to the function of highway, terrain and geological conditions, the design speed segment of the corridor belt route can be determined in general, and then get the preliminary highway route location. Affected by high altitude environment and highway flat vertical and horizontal design indicator combination, the operating speed of trucks change at different altitudes and under different alignment

Table 6.49 Min. allowable speed in uphill direction at different design speed (Km/h)

Item	120	100	80	60
Min. allowable speed	60	55	50	40

conditions, especially being influenced largely by grade. Therefore, on the basis of the preliminary highway alignment, combined with the characteristics of the speed distribution of trucks, the design speed segmentation must be further revised.

From Table 6.49, it can be seen that when the design speed is 100 km/h, the minimum allowable speed for heavy trucks is 55 km/h; when the design speed is 80 km/h, the minimum allowable speed for heavy trucks is 50 km/h. By verifying the minimum allowable speed that each road section can meet, corresponding design speed that can be used along the route can be determined.

For section with high technical indicators overall, but low in truck operating speed locally, when the difference between design speed determined by minimum allowable speed and the overall design speed of the section is greater than 20 km/h, alignment should be optimized to control the variation in design speed differences.

6.3.5 *Speed Segmentation Selection Considering Road Network Nodes Along the Route*

Different design speed segmentation nodes should be arranged in interchanges, mainline toll stations. The arrangement of interchange and mainline toll station should be determined according to the distribution of major towns, residential areas and national and provincial arterial highways along the highway.

Considering the above factors, the technical flow chart of dynamic segmentation of freeway design speed in high altitude area as shown in Fig. 6.50.

6.3.6 *Speed Transition Design*

The maximum speed difference between sections with different design speeds can be 40 km/h, which will result in a huge difference in the speed change of sections with different design speeds. From the perspective of improving vehicle operation safety, the overall operating speed section of freeway shall be designed. This section proposes the design method of speed transition section with different design speed sections based on the coordination of operating speed.

(1) **Design Speed ≥ 100 km/h**

For freeways located in plateau, plain, hilly area and river valley plain with flat terrain and design speed greater than or equal to 100 km/h, because they are basically located in the area with flat terrain, geometric design indicators are high, and the

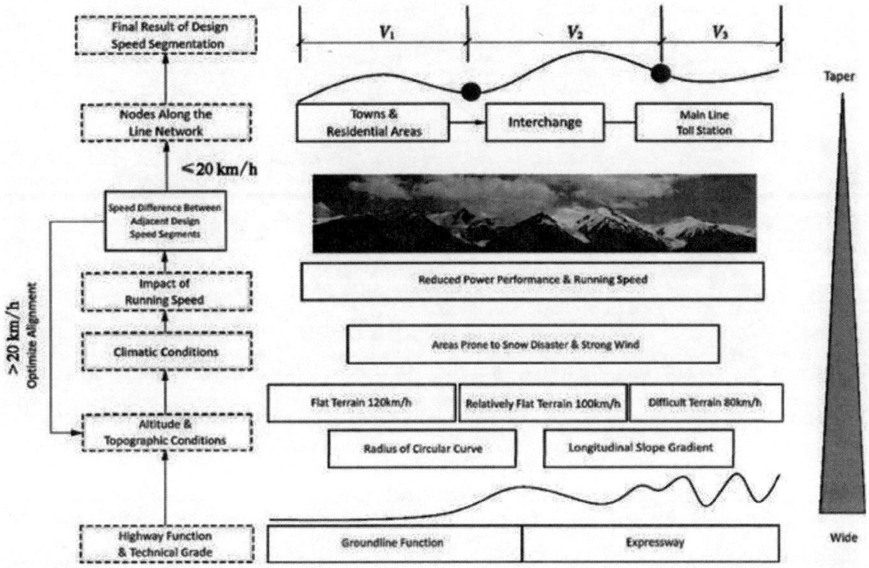

Fig. 6.50 Flow chart of dynamic segmentation technology for design speed of freeway in high altitude areas

operating speed is basically not affected by geometric alignment indicators, the consistency between operating speed and design speed is generally well. In the design of such highway projects, the general requirements of the current geometric design specifications for indicators are generally considered, because the geometric indicators of such highway projects are relatively high. As is shown in Fig. 6.51, Section 2 is designed as a speed transition section to achieve a smooth transition. At the same time, in the design process, it is necessary to check and calculate the superelevation and sight distance according to the actual operating speed, and properly strengthen the superelevation and guarantee sight distance.

(2) **Design Speed < 100 km/h**

For freeways located in high mountains, canyons, river valleys and plains with complex topographic and geological conditions, with a design speed of 60 ~ 90 km/h, the operating speed is basically greater than or equal to the design speed. In order to ensure a good connection and transition between the operating speed of different sections, it is necessary to consciously increase the speed transition section during the design, and ensure that the section has sufficient transition length. As is shown in Fig. 6.52, Section 2 between sections with a design speed of 80 and 100 km/h, is a deliberately added speed transition section to ensure continuous speed changes. During the design of the transition section, the route technical indicators that are suitable for the transition speed shall be adopted as far as possible to achieve the purpose of controlling the transition of the operating speed.

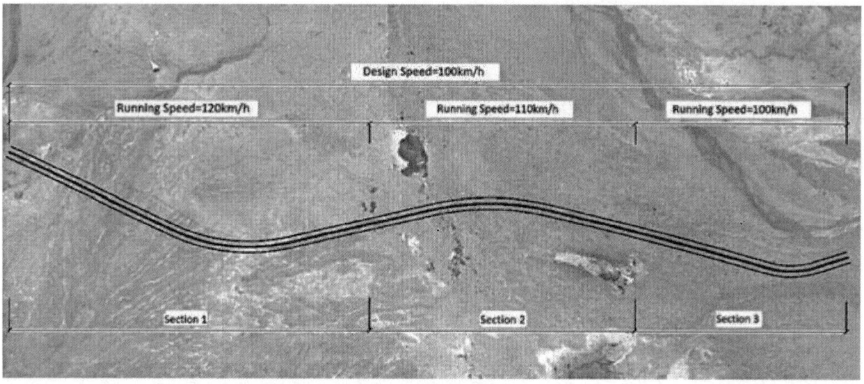

Fig. 6.51 Example of transition design for operating speed of freeway with design speed ≥ 100 km/h

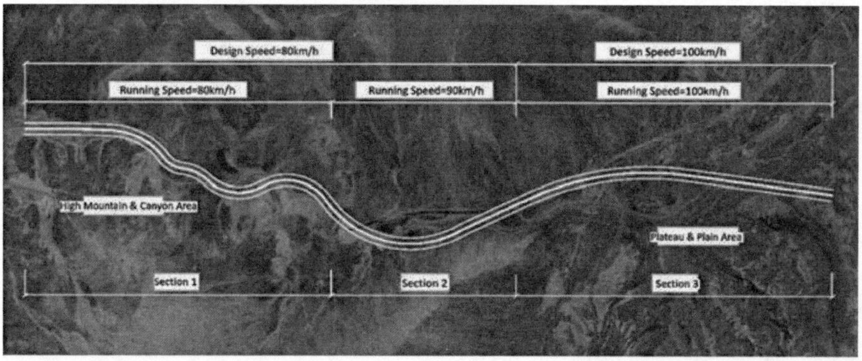

Fig. 6.52 Example of transition design of freeway operating speed with design speed less than 100 km/h

Since the operating speed is usually greater than the design speed, and the operating speed changes greatly, the designer needs to predict, analyze, and evaluate the operating speed changes during the design process of such projects, and adjust and optimize the geometric alignment indicators according to the operating speed evaluation conclusions. In addition, while paying attention to the change of operating speed between adjacent sections, the relationship between operating speed and design speed should also be well coordinated, and the subgrade superelevation transition, inspection and guarantee of driving sight distance, and timely supporting traffic safety measures should be arranged according to the change of operating speed. When necessary, check the traffic capacity and service level in the uphill direction, check the traffic safety in the downhill direction, demonstrate and add climbing lanes and escape lanes. This kind of highway project is the key point to realize the route safety optimization design by applying the operating speed.

Table 6.50 Variation range of operating speed and its relationship with design speed (km/h)

Type of highway	Design speed	Variation range of operating speed
Freeway	120	100 ~ 120
	100	80 ~ 120
	80	60 ~ 100

Note The operating speed is the reference value selected for the geometric design section indicator of the highway, not the design value

(3) **Operating Speed Variation Range**

Under different geometric alignment conditions, the general rule of operating speed mainly shows the following characteristics: acceleration on straight sections, deceleration on curves, deceleration on uphill sections of grades, and acceleration on downhill sections of grades. In order to ensure the continuous change of the operating speed and meet the requirements of indicators such as coordination, it is necessary to consciously carry out transition design for the operating speed of the whole line section by section according to the terrain and other comprehensive conditions of the area where the highway passes through in the route proposal design stage, consider the reasonable change interval of the operating speed, and select the geometric indicators and parameters corresponding to the speed. Table 6.50 shows the relationship between the reasonable operating speed variation range and the design speed.

6.4 Optimization Method of Route Safety Design

After determining the design standards for different sections of freeway, the geometric design indicators of the route shall be designed. In order to improve and ensure the safety of highway design, this section proposes the selection principle of route geometric alignment indicators and the safety optimization design process in combination with the special geographical environment of the plateau area.

6.4.1 Selection Principle of Alignment Indicators

The selection principle of alignment indicators is mainly based on the following factors:

(1) **Fully Analyze and Study the Traffic Composition, Vehicle Type Proportion and Typical Design Vehicle Types of Highway Projects**

The highway traffic composition, vehicle type composition and proportion have a great impact in the design of extreme gradient and slope length, long and large grade, superelevation of different lanes, emergency escape lane and climbing lane, as well as the inspection of parking sight distance and the arrangement of safety facilities, which should be fully analyzed and studied.

(2) **Be Careful When Determining Indicators**

When determining indicators, it is necessary to conduct actual survey, consult the opinions of relevant parties, get first-hand data, and then determine them by careful comparison according to the requirements of the design specification and in combination with current and future use requirements.

(3) **Fully Consider the Impact of Different Plateau Environments on People and Vehicles**

The terrain, landform, altitude, climate, and other natural conditions along the freeway, as well as the terrain and landform features along the freeway in the Qinghai-Tibet Plateau are complex and changeable, and the vehicle power, tire and braking performance models have also changed under different altitudes. Therefore, the selection of technical standards should take full account of the different natural conditions around.

(4) **Overall Balanced and Continuous Transition of Horizontal and Vertical Design Indicator Parameters**

For all types of highway projects, especially in the freeway design projects with design speed less than 100 km/h, the horizontal and vertical design indicators should not adopt the high limit indicators in a large range, and should pay attention to the balance and transition of the indicators.

(5) **In Plane Wiring, Curves Should be Used More to Increase the Proportion of Curves in the Total Route Mileage**

As the plateau area is generally flat, the selection of technical standards should also consider the use of geometric indicators of the route, and try to use higher indicators. It can create better operating conditions, shorten mileage, and reduce transportation costs. However, long straight lines should be avoided as far as possible, and the proportion of curves in the total mileage of the route should be increased to reduce the possibility of drivers' fatigue driving.

(6) **Pay Attention to the Coordination of operating speed**

The coordination of operating speed means that the operating speed difference or gradient of operating speed of adjacent sections should be less than a certain critical value, making the transition between different operating speeds as smooth as possible.

(7) **Superelevation Design**

The superelevation should be calculated and determined according to the operating speed of different vehicle types, and it can be considered to design the superelevation by lane in strict lane division driving sections. For sections with long freezing period, shady slopes, and other sections vulnerable to snow and freezing, the maximum superelevation shall be selected considering the impact of snow and freezing.

(8) **Pay Attention to Roadside Design**

Reflect the core of tolerance design. Within the roadside clean area, the fewer the obstacles, the better; appropriate protective measures shall be taken for the obstacles that cannot be removed within the roadside clear area to reduce the severity of the accident as much as possible.

6.4.2 Route Safety Optimization Design Process

For freeway projects with a design speed of 100 km/h or higher, due to the overall high geometric indicators, during the design process, it is necessary to check and calculate the superelevation and sight distance according to the actual operating speed, and properly strengthen the superelevation and guarantee sight distance.

For freeway projects with design speed less than 100 km/h, the optimization design method is introduced as follows.

(1) **Determine the Value Range of Plane Radius**

The route selection is designed according to the main control points in the highway corridor area, and the radius value range to be used for plane design is selected according to this speed range. The radius range is generally above the limit corresponding to the design speed and the general minimum critical value. The value of plane radius in plateau area should consider the physiological feelings of drivers, the impact of ice and snow conditions, etc., and avoid using limit values.

(2) **Highway Graphic Design**

The graphic design takes the design speed as the overall design control, and the change of operating speed is taken into consideration to guide the design parameters of specific curve units to adopt and combine, to carry out the initial plane alignment design. The design shall meet the general provisions and requirements of plane design and alignment design in the specification, and the three alignment elements of straight line, circular curve and spiral line shall meet the requirements of "general value" or "minimum value" of corresponding indicators of operating speed.

(3) **Highway Longitudinal Design**

The vertical section design takes the subgrade design elevation, the positions of bridges, long tunnels, interchanges, railway crossings, etc., along the line as the elevation control points for slope test design. In the design, attention shall be paid to the selection of maximum and minimum gradient, general slope length, limit slope length, vertical curve radius, etc., and the actual operating speed of the road section shall be considered according to the relationship between operating speed and slope length. In addition, the reduction of vehicle dynamic performance at different altitudes should be considered in plateau areas, so the value of maximum grade and slope length will be reduced to a certain extent compared with that in plain areas.

(4) **Combination Design of Horizontal and Vertical Alignment**

When designing horizontal alignment, vertical alignment must be considered; similarly, when designing the vertical alignment, it must also coordinate with the horizontal alignment. The principle of horizontal and vertical linear combination design is "mutual correspondence", and the horizontal curve is slightly longer than the vertical curve. According to foreign research data, the radius of vertical curve is generally 10 times the design value of horizontal curve radius. When the horizontal curve radius is less than 2000 m and the vertical curve radius is less than 15000 m, the alignment combination of horizontal and vertical curves is very important; with the increase of the radius of horizontal and vertical curves, their influence decreases gradually; when the radius of the horizontal curve is greater than 6000 m and the radius of the vertical curve is 25000 m, the influence on the alignment is insensitive. Therefore, the degree of correspondence and conformity must be mastered in line design according to the radius of horizontal and vertical curves. At the same time, the alignment design shall focus on the coordination with bridges, tunnels, and facilities along the alignment.

(5) **Inspection and Evaluation of operating speed**

According to the initially determined horizontal and vertical geometric alignment proposal, the operating speed of each section unit is calculated through the operating speed prediction model, and the balance and safety of the design indicators are evaluated and checked based on the coordination of the operating speed along the alignment, to test and correct the initial horizontal and vertical design.

(6) **Sight Distance Inspection, Superelevation Arrangement**

According to the adjusted horizontal and vertical alignment of the route and the calculation results of operating speed, the curve sight distance, superelevation, widening and other design indicators are finally determined. The Qinghai-Tibet Plateau is covered with snow and ice all the year round. Therefore, in the superelevation design, the snow and even ice on the pavement in winter should be considered, and a large superelevation value should not be used. This design process has been adopted in the inspection geometric design proposal of supporting projects and actual design projects.

(7) **Design of Facilities and Safety Facilities along the Alignment**

The layout of main alignment toll station, ramp toll station, service area, parking area and other facilities along the alignment, and the design of signs, markings and traffic safety facilities should be consistent with the distribution of traffic speed alignment. Improve the design of facilities and safety facilities along the alignment from the perspective of improving traffic safety. Considering the heavy snowfall and high wind level in winter on the Qinghai-Tibet Plateau, the wind and snow prevention

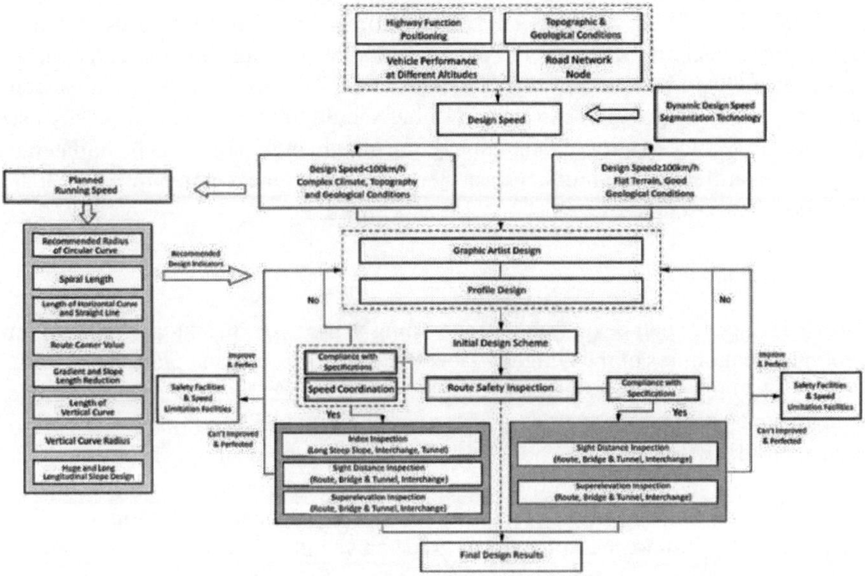

Fig. 6.53 Procedures for safety optimization design of freeway routes in Qinghai-Tibet Region

function should be considered when selecting the guardrail form, and low wind load signboards should be used as far as possible to reduce wind resistance.

Based on the above research contents, the process and steps of highway route safety optimization design are shown in Fig. 6.53.

6.5 Research on Technical Standards of Qinghai-Tibet Freeway

The selection methods and principles of technical standards for Qinghai-Tibet Freeway are proposed by combing and summarizing the relevant research results of highway operating speed, vehicle performance and driving characteristics in the high-altitude oxygen deficient area in Chaps. 2 to 7 above, and combining the special construction environment along the Qinghai-Tibet Freeway and the functional characteristics of Qinghai-Tibet Freeway.

(1) **General Principles for Selection of Technical Standards**

The selection of technical standards for Qinghai-Tibet Freeway shall reasonably apply the ***Technical Standards for Highway Engineering*** (Version 2014). When determining the design speed, design vehicles, cross section form and width and other indicators, the characteristics of the traffic environment on the Qinghai-Tibet Plateau shall be fully considered, and the factors affecting vehicle performance, traffic prevalence and driving characteristics, traffic composition characteristics, operation safety, project economy, etc., in the plateau cold, low pressure and oxygen deficient environment shall be comprehensively considered. The selection of technical standards should avoid going to extremes. Neither should limit indicators be easily used to affect the service quality of the highway, nor should high indicators be unilaterally pursued regardless of the number of projects, resulting in excessive investment and increased land occupation.

(2) **Design Speed**

The design speed shall be determined according to the two typical topographical and geological conditions of the Qinghai-Tibet Plateau mountains and plateau platforms. In principle, the design speed should be less than 100 km/h for sections with complex climatic and geological conditions such as strong wind sections, pass sections and permafrost distribution; for sections with flat terrain, good geological conditions and little impact on the route layout, the design speed above 100 km/h can be adopted.

Different altitudes have different effects on vehicle dynamic performance. With the increase of altitude, the influence of grade on vehicle performance increases, the dynamic performance of large trucks decreases significantly, and the speed difference between large and small trucks increases, which has a great impact on traffic operation safety. Therefore, for areas with good topographic and geological conditions, but the operating speed of trucks is greatly affected by the altitude, in order to reduce the impact of the large difference between large and passenger car speeds on driving safety, the selection standard of design speed can be appropriately reduced.

Segments with different design speeds shall be of sufficient length, and generally shall not be less than the general minimum distance between two adjacent interchanges of freeways in plateau areas. A speed transition section shall be arranged between sections with different design speeds, and the speed difference between the operating speed of the speed transition section and adjacent design speed sections shall not be greater than 10 km/h.

(3) **Design Vehicle**

The overall dimensions of design vehicles adopted for the design of Qinghai-Tibet Freeway shall meet the relevant provisions of the of ***Technical Standards for Highway Engineering*** (Version 2014).

The design vehicles shall be determined according to the analysis and prediction results of the traffic volume and its composition. The selected design vehicles shall be emphasized in the selection and formulation of grade design, cross section design, sight distance inspection, superelevation, speed control plan and other indicators.

For freeways with national defense requirements, the selection of design indicators should also meet the traffic requirements of major military vehicles in China. Military vehicles mainly include combat vehicles, traction vehicles, transport vehicles and special vehicles. Among them, the 60t semi-trailer tractor should be used as the design inspection vehicle due to the large difference between its overall dimensions, load capacity, power performance and other indicators and the design vehicle.

(4) **Cross Section Form and Width**

The freeway in Qinghai-Tibet Plateau can adopt two forms of subgrade cross section: integral type and separate type. For subgrade cross section crossing permafrost areas, separate type shall be adopted, integral type subgrade cross section can be adopted for non-permafrost areas, and separate type cross section can also be adopted for bridge and tunnel transition sections.

The width of each component element of the cross section shall meet the basic requirements of traffic operation safety. For the freeway mainly used for medium and small buses, the width of the inner lane can be 3.5 m.

(5) **Interchange Spacing**

China has stipulated the maximum spacing, average spacing and minimum spacing of interchanges in general areas. However, most sections of Qinghai-Tibet Freeway are in areas where there are few people and villages, and the road network is sparse. The value of interchange spacing should be determined flexibly and objectively according to the economic development situation, traffic conversion demand and other factors in the area.

The average spacing of interchange on Qinghai-Tibet Freeway should not be greater than 50 km; in case of exceeding, "U-Turn" facilities separated from the main alignment shall be arranged.

(6) **Traffic Engineering and Facilities along the Alignment**

(1) Traffic safety facilities

Traffic safety facilities shall provide systematic and complete instructions, directions, warnings, prohibitions, and other information for road users to ensure safe and comfortable driving.

Traffic safety facilities shall be equipped with signs, markings, sight guidance signs, barriers, protective nets, anti-glare panels, guardrails, anti-collision facilities, etc.

Traffic safety facilities such as wind barriers, snow (sand) barriers, falling nets, snow cover scales, etc. shall be arranged at sections where wind, snow, sand, falling rocks and other roads endanger the safety of the highway. At the same time, when designing traffic safety facilities such as signs and guardrails, the plan of reducing wind load, preventing snow accumulation, and facilitating maintenance shall also be adopted.

Emergency exits for first aid, firefighting, management, and other specific vehicles in emergency situations can be arranged at appropriate locations.

(2) Service Facilities

The service facilities of Qinghai-Tibet Freeway are divided into three categories. The distance between different service facilities is 4000 m above sea level as the boundary. Among them, the third type of service facilities are places where drivers can temporarily stop, rest and be convenient; the second type of service facilities shall be equipped with parking, refueling, vehicle maintenance, medical care, catering and other facilities; the first type of service facilities should be arranged in combination with the second type of service facilities, and should be equipped with high altitude response rescue, accident emergency rescue equipment, various drugs and professional medical personnel for treating common distresses at high altitude.

(3) Management Facilities

The management facilities shall provide clear, complete, clear, and accurate highway information for road users, provide scientific and advanced technical means for highway managers, and ensure the safety, comfort, and efficiency of highway operation.

The management facilities shall be equipped with management, monitoring, charging, communication, power distribution, lighting, and maintenance facilities, among which the monitoring facilities shall be of Class A.

Tibetan architecture is a unique architectural system that Tibetan people combine religious cultural traditions and customs, adapt to the terrain and climate characteristics, and reflect the national style. The architectural style of highway management and service facilities should consider such humanistic factors.

The particularly harsh meteorological conditions in Tibet place high requirements on the normal operation and function of electromechanical equipment. The influence of strong light shall be considered in the design and equipment selection of information acquisition and release terminal equipment. Meanwhile, drought, flood, frost, hail, strong wind, and heavy snow are the main disastrous weather in Tibet. When determining the technical parameter indicators of equipment, the impact of these disasters shall be considered, and the equipment with disaster resistance indicators shall be selected, while meeting a certain number of spare equipment, accessories, and spare parts, so that they can be quickly replaced during emergency rescue and repair.

6.6 Design Technology for Dynamic Speed Control (Speed Limit) of Freeway in High Altitude Area

According to the survey data, based on the analysis of highway operating speed characteristics in high altitude areas, considering the influence of altitude and the proportion of large cars on operating speed, the relationship model between various influencing factors in high altitude areas and 85%-bit speed is established by using statistical analysis and research methods.

6.6.1 Speed Limit Mode and Its Distribution

According to the site survey, at present, the method of combining interval speed limit with location speed limit is adopted for the Gonghe-Maduo section of G214 Qinghai Section and G109 Qinghai-Tibet Freeway. This section takes G109 Golmud-Lhasa Section as an example to analyze the advantages, disadvantages, and applicability of different combinations of speed restriction methods for highways in high altitude areas.

(1) **Section Speed Limit**

Section speed limit, that is, the public security traffic police arranged inspection station along some key road sections and use the "speed limit card" method to monitor the driver's speed.

According to the site survey data, the total length of G109 Golmud-Lhasa Section is 1215 km, and the speed is limited by receiving a speed limit card along the alignment. A total of 9 public security traffic police inspection station is arranged, including Golmud Nanshankou Inspection Station, Yanshiping Inspection Station, Anduo Inspection Station, Nagchu Inspection Station, Gulu Inspection Station, Wumatang Inspection Station, Dangxiong Inspection Station, Yangbajan Inspection Station and Yangda Inspection Station. The specific speed range and limit time are shown in Fig. 6.54.

It can be seen from Fig. 6.54 that the speed limit range is from Golmud Nanshankou to Yanshiping, with a total length of 475 km, a speed limit time of 7 h, and a speed limit standard of 50 and 70 km/h.

It can be seen from Table 6.51 that in the Golmud-Lhasa direction, considering the impact of urban density, altitude and roadside environment, the speed limit interval is shortened in turn, from 475 to 30 km, the longest speed limit section. In the speed limit section, except for special points, the speed limit value is kept at 70 km/h. The Dangxiong-Lhasa Lane is changed from the original two-way two-lane to two-way four-lane, and the speed limit value is correspondingly increased to 80 km/h.

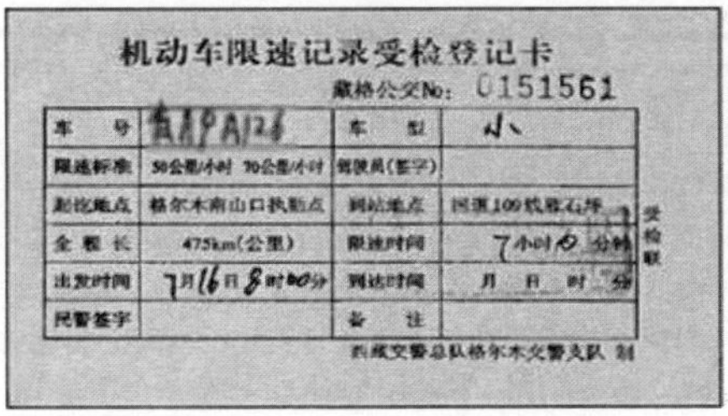

Vehicle Speed Limit Record Inspection Registration Card

Transportation of Tibet-Golmud Highway No. 0151561

License No.		Vehicle Type	Small
Speed Limit Standard	50km/h 70km/h	Driver's Signature	
Departure	Nanshan Station, Golmud	Destination	Yanshiping, G109
Total Length	475km	Speed Limit Time	7 Hour
Departure Time	08:00, 16 Jul.	Arrival Time	—
Police's Signature		Note	

Inspected sheet

Golmud Traffic Police Detachment of Tibet Traffic Police Corps

Fig. 6.54 Section speed limit sheet of G109

Table 6.51 Statistics of different speed limit sections and time of G109

Speed limit section	Distance (km)	Limit time (h)	Speed limit (km/h)
Golmud(K2772)—Yanshiping(K3214)	45	7	≤70
Yanshiping(K3214)—Anduo(K3435)	200	4	≤50
Anduo(K3435)—Nagchu(K3575)	126	2.5	≤70
Nagchu(K3575)—Gulu(K3652)	77	1.3	≤70
Gulu(K3652)—Dangxiong(K3739)	75	1.1	≤70
In which, Wumatang(K3697)—Dangxiong(K3739)	30	0.5	≤60
Dangxiong(K3739)—Yangbajan(K3819)	80	1.1	≤80
Yangbajan(K3819)—Maxiang(K3860)	41	0.5	≤80
Maxiang(K3860)—Lhasa(K3890)	30	0.5	≤60

(2) **Site Speed Limit**

Site speed limit, that is, speed limit in special sections, including sections with continuous long and steep grades, sharp turns, continuous curves, sections crossing densely populated villages and towns, county sections or urbanization sections, sections with frequent traffic accidents, and sections seriously affected by bad weather.

According to the site survey data, there are 50 speed limits at G109 Golmud Lhasa section. Table 6.52 shows the specific location characteristics and speed limits.

The specific arrangment mode of speed limit sign is shown in Fig. 6.55.

6.6.2 Analysis of Vehicle Overspeed

In order to study the result of speed limit on the Qinghai-Tibet Plateau, the following two indicators, overspeed rate and overspeed range, are selected as criteria. Overspeed rate refers to the ratio between the statistics of overspeed passenger cars per hour and its corresponding hourly traffic volume. Overspeed range refers to the extent to which the operating speed of the overspeed vehicle exceeds the speed limit, that is, the difference between the operating speed and the speed limit.

Firstly, the speed limit value is compared with the 85th percentile speed. As shown in Fig. 6.56, when the speed limit values are 40, 50 and 60 km/h, the 85th percentile speed at all sections are greater than the speed limit values, and the highest overspeed range is 57.3 km/h; when the speed limit value is 70 km/h, the 85th percentile speed of 84% section is greater than the speed limit value, and the highest overspeed range reaches 30.18 km/h; when the speed limit is 80 km/h, the 85th percentile speed of 56% sections is greater than the speed limit value, and the highest overspeed range is 18.7 km/h. It can be seen that the 85th percentile speed of most sections is greater than the corresponding speed limit. With the increase of the speed limit, the proportion of

Table 6.52 Current combination mode of golmud-Lhasa directional speed limit and site speed limit of G109

Section speed limit	Site speed limit	Mileage (km)	Speed limit (km/h)	Altitude (m)	Research location
Golmud Nanshankou Inspection Station → Yanshiping Inspection Station 475 km, 7 h	Golmud Nanshankou Inspection Station	K2772	70	3100	K2795 K2852 K3077 K3088 K3091 K3096 K3168 7 Points, 22 Sections
	Nachitai Town	K2825	40	3500	
		K2828	40	3565	
	Sharp Curve	K2862	40	4089	
	Xitaitan Town	K2865 + 500	40	4100	
		K2868 + 500	40	4100	
	Permafrost Hot Rod	K2948	60	4420	
	Wudaoliang Town	K3004	40	4665	
		K3006	40		
	Sharp Curve	K3126	40	4595	
	Tuotuo River	K3154	40	4533	
	Yanshiping	K3214	40	4900	
Yanshiping Inspection Station → Anduo Inspection Station 200 km, 4 h	Long Straight Alignment	K3275	80	4800	K3329 K3334 K3356 K3422 Total 4 Points, 12 Sections
		K3358	70	4970	
		K3375	70	5000	
		K3387	70	4914	
	Permafrost Hot Rod, Sharp Curve	K3392	30	4858	
	Down Steep Slope to Sharp Turn	K3402	30	4800	
	Continuous Downhill	K3404	30	4778	
	Sharp Curve, Frequent Accidents	K3416	30	4700	
	Sharp Turn, Downhill	K3417	40	4710	
	Sharp Curve, Frequent Accidents	K3419	30	4696	

(continued)

Table 6.52 (continued)

Section speed limit	Site speed limit	Mileage (km)	Speed limit (km/h)	Altitude (m)	Research location
	Sharp Curve, Frequent Accidents	K3420	30	4680	
	Sharp Curve, Frequent Accidents	K3421	30	4680	
Anduo Inspection Station → Nagchu Inspection Station 126 km, 150 min	Sharp Curve, Radar Speed Measurement	K3438	30	4745	None
	Long Downhill	K3493 + 700	70	4798	
	Curve	K3494 + 300	30	4796	
	Long Straight Alignment	K3501	70	4770	
	Curve	K3507	30	4779	
	Reverse Curve	K3512	40	4745	
	Long Straight Alignment	K3527	70	4590	
Nagchu Inspection Station → Gulu Inspection Station 77 km, 80 min	Sharp Turn Downhill	K3593	40	4572	K3621 Total 1 Point, 3 Sections
	Curve with Large Radius	K3606	70	4667	
	Sharp Turn	K3618	40	4700	
	Long Downhill	K3629	70	4862	
	Sangxiong Town	K3633 + 500	30	4900	
	Gulu Town	K3657	40	4905	
Gulu Inspection Station → Dangxiong Inspection Station 75 km, 68 min	Long Straight Alignment	K3666	70	4725	None
	Longrenxiang Town	K3709 + 600	30	4710	
	Sharp Curve, Frequent Accidents	K3715	30	4735	
	Sharp Curve, Frequent Accidents	K3722	30	5100	

(continued)

Table 6.52 (continued)

Section speed limit	Site speed limit	Mileage (km)	Speed limit (km/h)	Altitude (m)	Research location
	Sharp Curve, Frequent Accidents	K3723	30	5115	
	Dangxiong Town	K3728	30	4283	
	Dangxiong Town	K3732	30	4288	
	Sharp Turn Ahead	K3735 + 700	30	4399	
	Sharp Turn Ahead	K3735 + 900	30	4398	
	Sharp Curve, Frequent Accidents	K3739	30	4300	

Fig. 6.55 Example of speed limit in towns of G109

the 85th percentile speed exceeding the speed limit as well as its extent of overspeed decrease. Some sections affected by the terrain and traffic volume, the 85th percentile speed is lower than the speed limit.

(1) **Overspeed Rate Analysis**

Survey points K157 of G214 and K3168 and K3761 of G109 were selected for analysis. Both speed limit for passenger cars and trucks at K157 is 60 km/h. Both

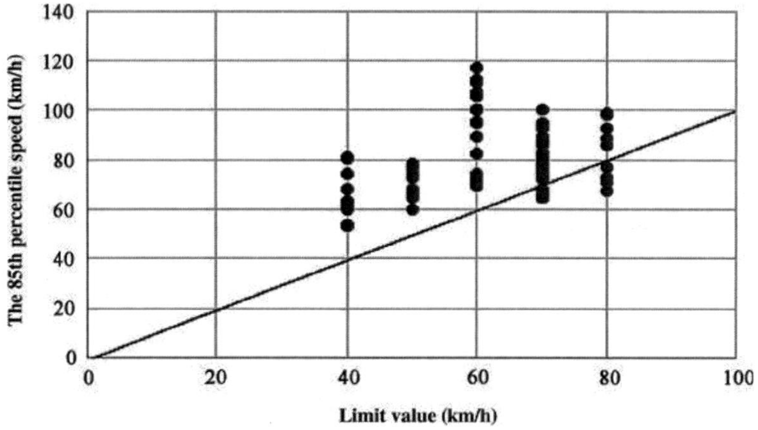

Fig. 6.56 Variation between 85% speed of the vehicle and the limit value

speed limit for passenger car and trucks at K3168 and K3422 is 70 km/h. The specific analysis is as follows.

As shown in Figs. 6.57 and 6.58, a total 1121 passenger cars passed through the K157 during the survey period, among which 1033 overspeed vehicles and the overspeed rate fluctuated around 90%. The traffic volume of trucks totaled 551, among which 397 overspeed overspeeds, resulting the highest overspeed rate of 72% and the lowest above 40%.

As shown in Figs. 6.59 and 6.60, the K3168 cross-section is a curve. During the survey time, a total of 961 passenger cars passed among 830 over limit speed. The overspeed rate was high from 7:00 to 16:00 in the daytime, fluctuating around 90%. At night, due to poor visibility, the lower speed resulted the overspeed rate fluctuated

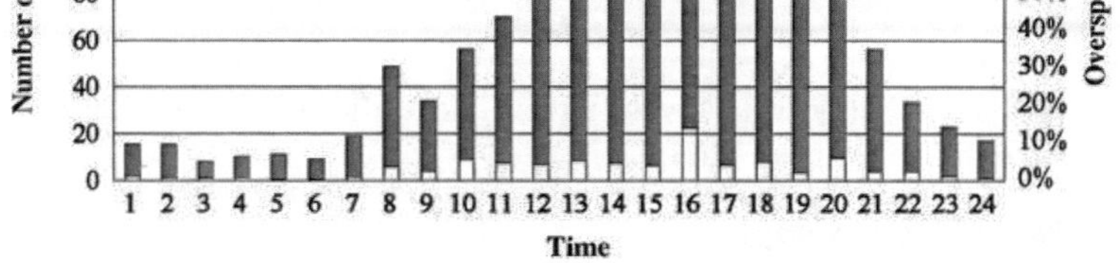

Fig. 6.57 Over-speed Passenger cars overspeed at K157 at different time

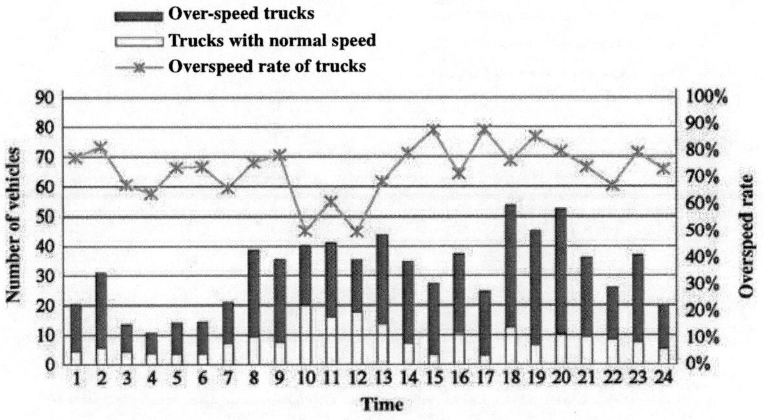

Fig. 6.58 Over-speed Trucks at K157 at different time

around 70%. Totally 604 trucks passed among 188 overspeed, which resulted the overspeed rate fluctuated around 30%.

As shown in Figs. 6.61 and 6.62, during the survey there were 3,376 passenger cars passed at K3176, among 2,151 overspeed. In the daytime the traffic volume reached 514 vehicles/h, affected by the traffic volume, the speed lowered and overspeed rate decreased, and the overall overspeed rate of passenger cars fluctuated around 70%. There were 1,059 trucks in total, among 342 overspeed, and the overspeed rate fluctuated around 30%.

(2) **Overspeed Range Analysis**

As shown in Fig. 6.63, at K157 the overspeed range of passenger cars was concentrated at 25~55km/h, accounting for 54.15% of the overspeed passenger cars, among

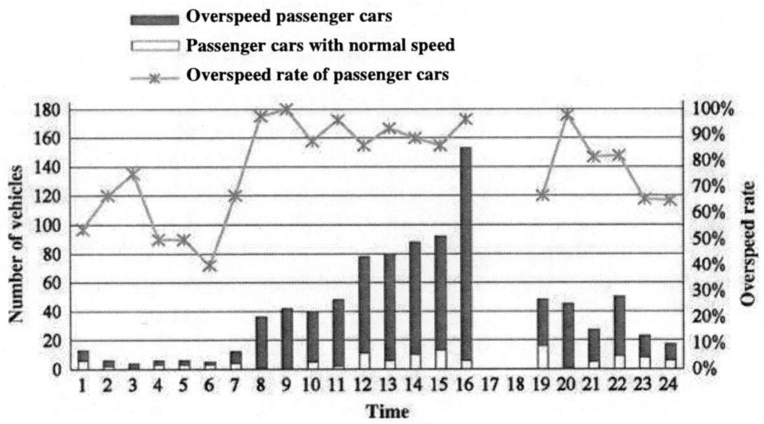

Fig. 6.59 Passenger cars overspeed at K3168 at different time

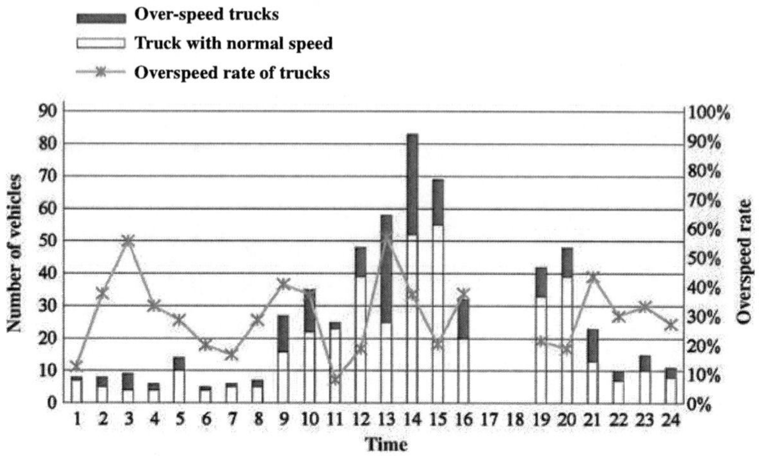

Fig. 6.60 Overs-peed Trucks at K3168 at different time

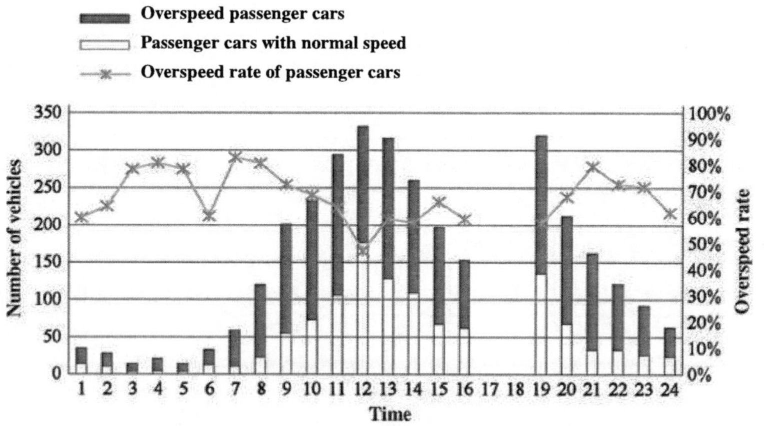

Fig. 6.61 Overspeed Passenger Cars at K3761 at different time

19.58% exceeded 30~35km/h than the speed limit, accounting the highest proportion. Overspeed range of trucks was concentrated at 0~20km/h, accounting for 68.68% of the overspeed trucks, among 18.63% exceeded 5~10km/h than speed limit, accounting the highest proportion.

As shown in Fig. 6.64, at K3168 the overspeed range of passenger cars was concentrated at 10~30km/h, accounting for 78.07% of the overspeed passenger cars, among 19.87% exceeded 20~25km/h than the speed limit, that's the highest proportion. Overspeed range of trucks was concentrated at 0~15km/h, accounting for 93.09% of the overspeed trucks, among 56.91% exceeded 0~5km/h, the highest proportion.

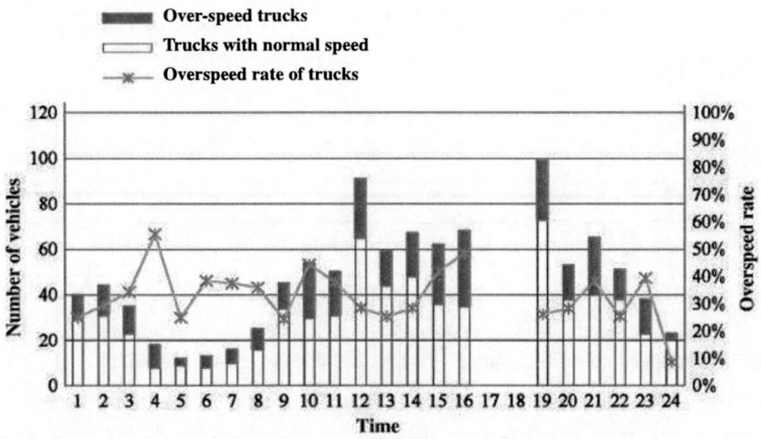

Fig. 6.62 Overspeed Trucks overspeed at K3168 at different time

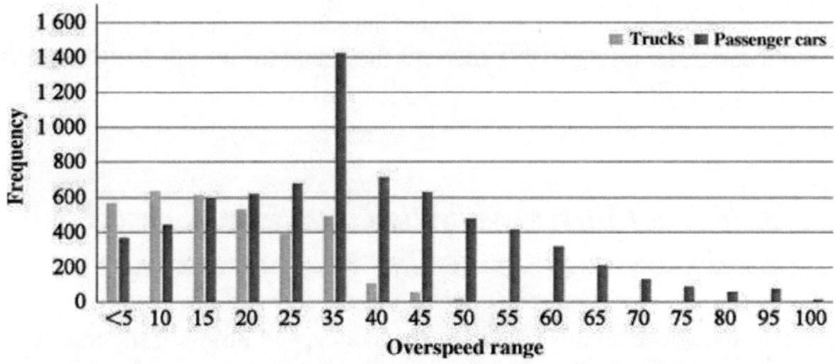

Fig. 6.63 Over-speed range distribution of Trucks and passenger cars at K157

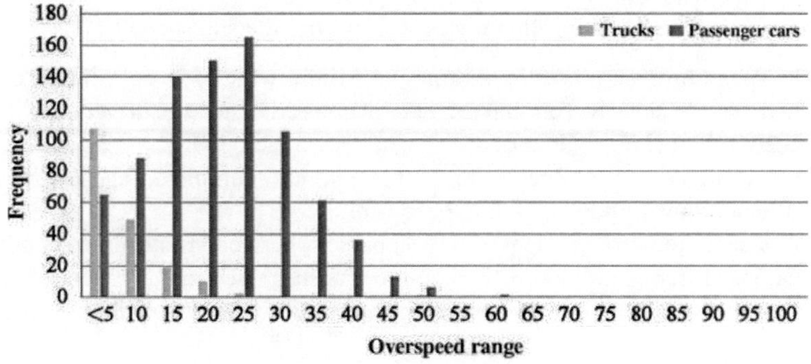

Fig. 6.64 Over-speed range distribution of trucks and passenger cars at K3168

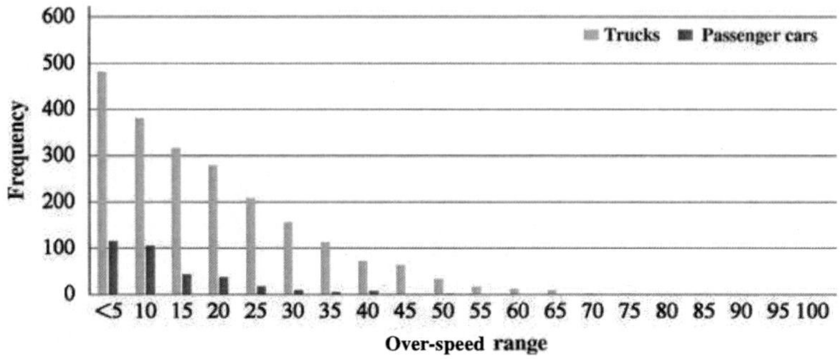

Fig. 6.65 Over-speed range distribution of trucks and passenger cars at K3761

As shown in Fig. 6.65, at K3761 the overspeed range of passenger cars was concentrated at 0~20km/h, accounting for 67.78% of the overspeed passenger cars, among 22.41% with overspeed range of 0~5km/h accounted the largest proportion. Overspeed range of trucks was concentrated at 0~20km/h, accounting for 88.30% of overspeed trucks, among 33.91% exceeded the speed limit of 0~5km/h, accounting the highest proportion.

6.6.3 Speed Limit Effect Evaluation of Existing Speed Limit Facilities

According to the above analysis, neither the local speed limit nor the regional speed limit (in the form of receiving a speed limit card) has achieved the corresponding speed limit effect, and the phenomenon of over-speed is very serious. The following are the problems of the two speed limiting methods.

At present, there are the following problems with the site speed limit:

i. The speed limit sign is adopted for speed limit, which has poor binding force. There is no penalty for the driver's overspeed, and the driver has a large "freedom" to choose the speed;
ii. The arrangement of speed limit signs is not complete, and there is no corresponding ending signs for speed limit;
iii. The speed limit arrangement is unreasonable. The speed limit values of adjacent sections differ greatly, and there is no transition sections. It is not easy for vehicles to adjust the running speed according to the restrictions of the limit values.

The regional speed limit method of obtaining speed limit cards at checkpoints along the way has the following problems:

i. The speed limit mode of receiving "speed limit card" has weak binding force, and there is no punishment measure for the driver who exceeding the speed limit. The driver has a greater "freedom" to choose the speed;
ii. The biggest problem is that drivers will drive too fast to save the queuing time at the checkpoint. Since this section is a two-lane road, roadside parking increases the driving risks;
iii. There is no uniform standard for the speed limit distance of this section, which is generally determined according to the experience of law enforcement personnel;
iv. The length of the speed limit zone is too long. Some sections reach more than 400 km, and drivers will have meals and rest within the speed limit zone, to level down the average speed below the limit value. Thus, the actual running speed of vehicles is not constrained.

6.6.4 Speed Limit Decision and Arrangement Technology Under Special Environment

Due to the special environmental characteristics of the Qinghai-Tibet Plateau, such as flat and wide area, high altitude distribution, thin air, etc., the measured traffic data in the study reflects that the vehicle operation is almost not affected by the restrictive measures. In the research process, based on the 85th percentile speed of trucks and passenger cars on highways in Qinghai-Tibet Plateau, a decision model of speed limit value was established, and the control standard and arranging basis of vehicle operating speed under different altitudes, alignment features and terrain conditions in high-altitude areas with a deficiency of oxygen were proposed.

The operating speed of vehicles in the Qinghai-Tibet Plateau is affected by altitude, grade, traffic volume and the proportion of large and passenger cars. Through the multivariate linear regression of the operating speed under the free flow state, the correlation between the influencing factors and the operating speed and its correlation function are as follows:

$$\left.\begin{aligned} S_{\mathrm{L}} &= V \\ V &= f(h, P, T) \end{aligned}\right\} \tag{6.26}$$

where,

S_L Limiting speed (km/h)
V Operating speed (km/h), i.e. 85% speed;
P Passengere car proportion (%);
T Equivalent traffic volume (pcu);
h Altitude (m).

A regression model with good linearity has been established through multiple revisions. See Table 6.53 for the summary of the model.

(1) **Judge the Advantages and Disadvantages of Multiple Linear Regression Model**

The criteria for measuring the superiority of multiple linear regression models are usually negative correlation coefficient R, determination coefficient R^2, and corrected determination coefficient $R_{adj}{}^2$. Table 6.54 shows that the larger R is, the closer the linear relationship is. The larger R^2 is, the better the model is, $0 \leq R^2 \leq 1$. When the added variables in the model are not statistically significant, the $R_{adj}{}^2$ value will decrease. Compared with the model established in the previous section, $R_{adj}{}^2$ increases, indicating that the variables are statistically significant. The R^2 of this model is 0.657, which is relatively good.

(2) **Hypothesis Testing of Regression Models**

F test results of all independent variables in the model with regression coefficients equal to 0. As shown in Table 6.55, F = 12.478, P < 0.05, indicating that the regression

Table 6.53 Regression coefficient

Model	Unstandardized coefficients		Standardized coefficients	Test value t of regression coefficient	Significance
	Partial regression coefficent	Standard error	Standardized regression coefficient		
Constant	−2462.101	1156.043		−2.130	0.037
Equivalent traffic volume T (pcu)	−0.030	0.010	−0.349	−3.018	0.003
ln P (P = portion of Passenger car)	22.847	10.660	0.255	2.143	0.035
Altitude h	−0.184	0.109	−12.992	−1.697	0.024
h^2	9.998E-6	0.000	5.162	1.261	0.011
ln h	379.508	176.165	7.925	2.154	0.034

Table 6.54 Summary of models

Multiple correlation coefficient R	Determination coefficient	Adjusted determination coefficient	Change Statistics				
			Determination coefficient change	Change of F	Degree of freedom 1	Degree of freedom 2	Significant F change
0.811	0.657	0.634	0.457	12.478	5	74	0.000

coefficient of at least one independent variable is not 0, and the regression model established is statistically significant.

(3) **Residual Analysis and Test of Model Trial Conditions**

(1) Linear Relationship Between Independent Variable and Dependent Variable

The linear relationship between independent variable and dependent variable is determined by nonstandard residual. It can be seen from Fig. 6.66 that the points are almost evenly distributed on both sides of the horizontal line 0, and there is no obvious positive or negative trend, indicating that the linear relationship between the variables introduced in the current model and the operating speed is correct.

(2) Normal Distribution of Residuals

According to the statistical significance, the residual should follow the normal distribution. By drawing the standardized residual histogram, as shown in Fig. 6.67, the residual distribution does not conform to the normal distribution.

(3) Homogeneity of Residual Variance

As the same as the discriminant standard when the preliminary model was established, the standardized residual was used as the indicator of the homogeneity test of variance. As shown in Fig. 6.68, no matter how the standardized predictive value of *V* (operating speed) changes, the fluctuation range of the standardized residual is basically stable, indicating that the residual variance is homogeneous.

(4) **Decision Model of Speed Limit**

The running speed of vehicles in high altitude areas is almost not affected by the current speed limit value and speed limit mode. Therefore, the multiple linear regression analysis on the operating speed of each influencing factor is carried out first. The results show that R^2 is 0.235, its linear correlation is small, and the standardized residual of operating speed (dependent variable) fluctuates greatly on both sides and downward in the middle, indicating that the residual variance is uneven. In view of the above results, variable transformation was carried out for each influencing factor (independent variable), and multiple linear regression was carried out again. The results showed that R^2 was 0.657, and the standardized residuals of operating speed

Table 6.55 Regression model hypothesis test results

Model	Sum of squares	Degree of freedom	Mean square	F Inspection	Significance
Regression analysis	7358.540	5	1471.708	12.578	0.000
Residual	8727.658	74	117.941		
Total	16,086.199	79			

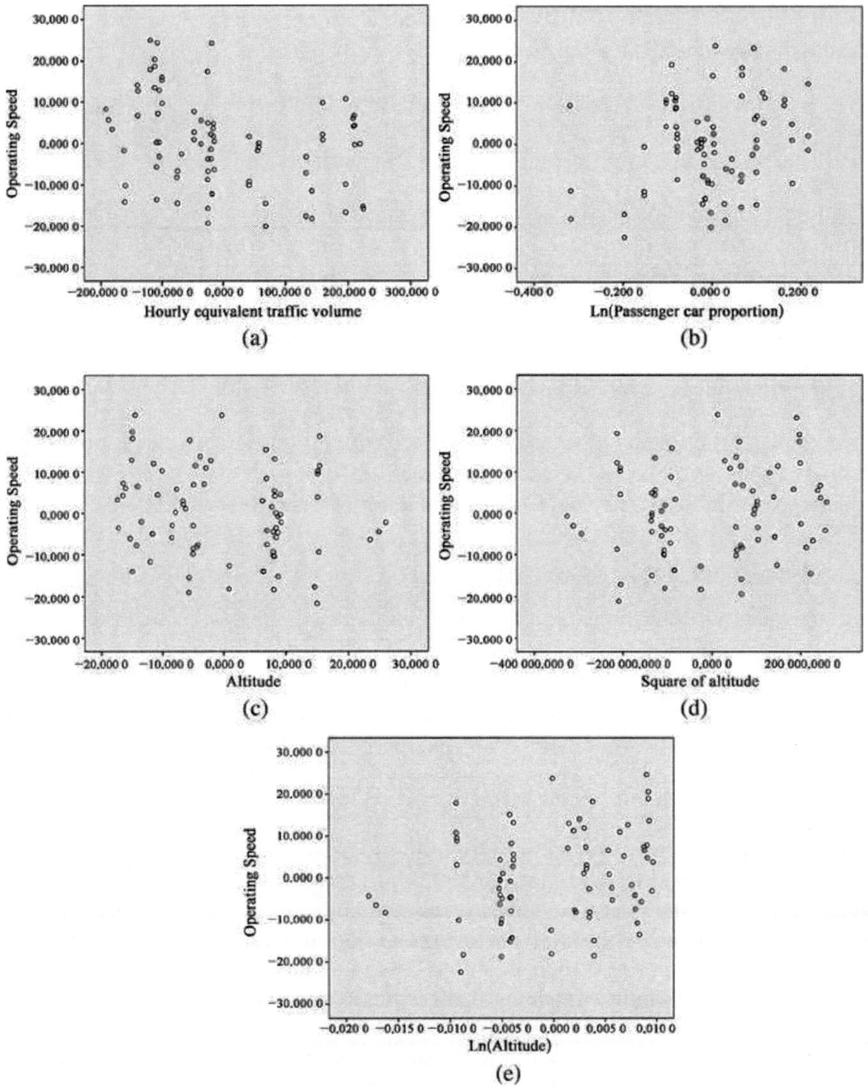

Fig. 6.66 Residual distribution diagram of independent variable

(dependent variable) were almost evenly distributed on both sides of the horizontal line of 0, with no obvious positive or negative trend, indicating that there was a linear relationship between variables introduced in the current model and operating speed. The final decision model of speed limit in high altitude areas is as follows:

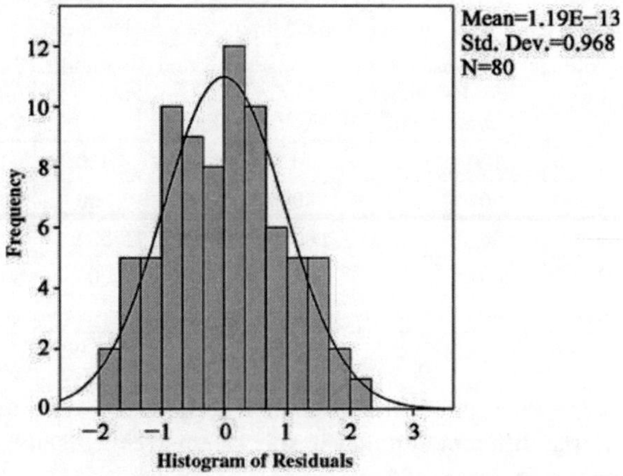

Fig. 6.67 Histogram of residuals

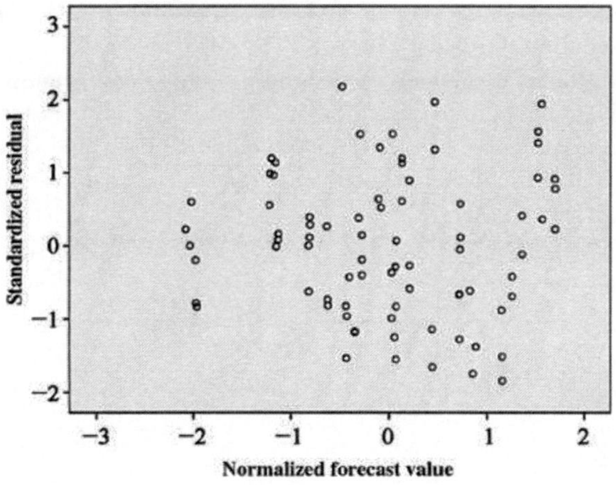

Fig. 6.68 Scatter plot of normalized predicted values and normalized residuals

$$\left.\begin{array}{l} S_{\mathrm{L}} = V \\ V = -0.03T + 22.844 \ln P + 9.998 \times 10^{-6} h^2 + 379.492 \ln h - 0.184h - 2462.003 \end{array}\right\} \tag{6.27}$$

where,

S_L Limited speed (km/h);
V Operating speed (km/h), i.e. the 85th percentile speed;
P Passenger car proportion (%);
T Equivalent traffic volume (pcu);
h Altitude (m).

Considering the altitude, the performance of large and passenger cars, the proportion of vehicle models and the characteristics of road grades, and combining the

Table 6.56 Recommended speed limits for vehicles at different altitudes (for the same lane)

Altitude (m)	Suggested Limit for Passenger Cars (km/h)	Suggested Limit for Large Vehicles (km/h)	Altitude (m)	Suggested Limit for Passenger Cars (km/h)	Suggested Limit for Large Vehicles (km/h)
2 000	100	80	4 500	80	60
2 500	90	80	5 000	60	50
3 000	90	80	5 500	60	50
3 500	80	60	6 000	60	50
4 000	80	60			

above relevant conclusions, the recommended speed limits for trucks and passenger cars at different altitudes in special environments in high altitude areas are given, as shown in Table 6.56.

The values in the table are the recommended vehicle speed limits for long straight sections in the special environment of high altitude areas. When ensuring the speed continuity ($\Delta V \leq 20$ km/h) of adjacent sections at different altitudes, the speed difference between trucks and passenger cars should also be controlled within the safe range to avoid traffic accidents caused by excessive difference. In addition, at the flat curve sections, curved slope sections and uphill and downhill sections, the values shall be reduced according to the roadside environment and vehicle types proportion.

Chapter 7
Application and Demonstration Relying on Project

7.1 Application Project Summary

Hua-Da Highway, located in the Golog Tibetan Autonomous Prefecture in Qinghai Province, is a part of Provincial Highway 209 (Delingha Gahai—Jiuzhi Hongtu Pass Highway), a part of the "three vertical and four horizontal and ten link" skeleton highway in Qinghai Province, and is also an important part of China's western regional economic corridor (Korla-Chengdu) planned from Xinjiang to Sichuan via Qinghai. The route is 155,831 km long, with two-way four-lane technical standards, and different design speeds of 80 km/h and 60 km/h for available sections, see details in Table 7.1. The subgrade widths are 19 m (for integral one) and 10 m (for separate one).

The 2003 version of *Technical Standards of Highway Engineering* was applied in the design of the example project. See Table 7.2 for the major technical indicators.

7.2 Adaptability Analysis of Design Speed Variation

7.2.1 Functions, Positions and Design Conditions

Hua-Da Highway serves as the province's principal arterial, the design standards of freeway will be properly for its functions and positions.

The safety design section is routed along the river valley on the northern slope of the Anyemaqen Mountains, and the corridor zone mainly contains three types of terrain, namely mountain canyons, hills and river valley plains. The mountain canyons spreads along the route Lower Dawu Township—Xueshan Township—Dongqingou Township on both sides; the hills are mainly located in the section from the starting point to Lower Dawu Town; and the river valley plains are mainly located in the Dongqingou Township and the section from Dawu Town to the end point.

J. Liu, *Technical Indicators and Safety Design of Freeway in High Altitude Area*,
https://doi.org/10.1007/978-981-99-0620-8_7

Table 7.1 List of design speeds for different sections of the project

S/N	Chainage	Length (km)	Design speed (km/h)
HD1	K0 + 000 ~ K44 + 000	44	80
HD2	K44 + 000 ~ K121 + 000	77.1	60
HD3	K121 + 000 ~ K133 + 000	11.9	80
HD4	K133 + 000 ~ K145 + 500	12.5	60
HD5	K145 + 500 ~ K155 + 700	10.2	80

Table 7.2 Major technical indicators

Indicators	Unit	Values	
Design speed	Km/h	80	60
Numbers of lanes	Nos	4	4
Subgrade width (intergral)	m	19	19
Subgrade width (separate)	m	2*10	2*10
Minimum radius of horizontal curves (general/limit)	m	400/250	200/125
Minimum radius of horizontal curves without superelevation	m	2500	1500
Minimum length of transitions	m	70	50
Maximum grade	%	5	5
Minimum length of grade	m	200	150
Minimum radius of crest (general/limit)	m	4500/3000	2000/1400
Minimum radius of sag (general/limit)	m	3000/2000	1500/1000
Minimum length of vertical curves (general/limit)	m	170/70	120/50
Flood frequency for bridges and culverts	Times/year	1/300 for super large bridges and 1/100 for large bridges	
Design loads for bridges and culverts	–	Highway Class I	

The adverse geological phenomena along the route mainly include collapse, landslide, debris flow, etc. The collapse is mainly distributed in mountain canyons; landslides are mainly distributed in Yangkao Gorge (K87 + 100 ~ K87 + 580) and the left bank of Duoxima Bridge (K138 + 030); debris flows always happen in section K123 + 880 ~ K147 + 557.

Fig. 7.1 Hills on the plateau

Sections with various design speeds:

Section K0 + 000 ~ K44 + 000 (Huashixia to Dawu): The terrain conditions are mainly hills (Fig. 7.1). The highway is arranged along the mountain discontinuous basin, with an altitude of about 4,000 m and a relative elevation difference of 100–200 m. A relatively high technical standard, compared to the whole line, should be adopted for this section.

Section K44 + 000 ~ K121 + 100: It starts at Dawu, passes through Xueshan Township and the pass of Chana Kado Gorge. The main topographic features of this section are mountain canyons (Fig. 7.2). On some sections, such as Yangkao Gorge, there are landslides, mudslides and other adverse geological conditions. In this road section, the terrain is steep, cut severely and with narrow river valleys, which are difficult for road alignment arranging. According to the approval of the Qinghai Provincial Department of Transportation, on the sections with special adverse conditions, the design speed may be 60 km/h.

K121 + 100 ~ K133 + 000 section: It is located in Dongqinggou Township. The terrain is mainly of river valley plains (Fig. 7.3) with flat landform, where a design speed of 80 km/h is desirable.

Fig. 7.2 Mountain canyons

Fig. 7.3 River valley plains

K133 + 000 ~ K145 + 500 section: It is located between Dongqinggou Township and Dawu Town. The geological conditions of this section are relatively complex. Landslides, debris flows and other unfavorable geological conditions are mainly distributed alongside. Affected by the topographic and geological conditions, the geometric design indicators of this section may only meet the technical standard requirements of the design speed of 60 km/h.

K145 + 500 ~ K155 + 700 section: It is located in Dawu Town. The terrain of this section is mainly river valley plain with flat landform. The technical standard of this section can meet the requirements of the design speed of 80 km/h.

7.2.2 Vehicle Operating Conditions

i. Through studying the influence of altitude on the power performance and braking performance of vehicles, it can be found that the climbing ability of the same gradient at different altitudes varies greatly, and the speed that vehicles can reach on the gradient sections of different gradients gradually becomes lower with the increase of altitude and the decrease of power performance.
ii. The effect of altitude is less on passenger cars and more on trucks such as laden wagons and articulated trains. With a load capacity of 30t and a power-to-weight ratio of 8.3 kW/t, the performance of vehicles is reduced by 10% to 15% with an altitude below 3,000 m, and by nearly 35% to 40% with an altitude between 3,500 m and 5,500 m. The longitudinal grade reduction of the road section dominated by trucks are shown in Table 7.3.

Table 7.3 Longitudinal grade reduction on plateau

Altitude (m)	3500–4500	4500–5500	Above 5500
Grade reduction (%)	1.0	1.5	2.0

The altitude of Hua-Da Highway is between 3,700 m– 4,600 m, affected by high altitude, the operating speeds of trucks such as heavy trucks and articulated trains are low and will be changed more frequently. If the design speed is 80 km/h, the minimum permissible speed of heavy trucks is 50 km/h; if the design speed is 60 km/h, the minimum permissible speed of heavy trucks is 40 km/h. The minimum permissible speed of each road section shall be checked, to determine the corresponding design speed that can be used along the route.

7.2.3 Road Network Node Conditions

Transition sections for shifting different design speeds shall be arranged at interchanges or mainline toll stations. As an important node of the road, the interchange can be arranged as the dividing point for different design speed sections. See Table 7.4 for the location of interchange arrangement alongside.

With comprehensive consideration about the road section's functions, terrain types and altitudes, vehicle dynamics changes and other conditions, the adaptability of sectional design speeds is appreciated on the road. However, the transition nodes were not arranged at interchanges, which will affect the speed control plan. When developing the speed control plan, special attention should be paid to justifying the high speed-limit between the design speed transition nodes and the interchanges.

7.3 Evaluation of Operating Speed Coordination

(1) Evaluation Method

When measuring the operating speed of the freeway, the operating speed V_{85} is measured both in the direction with ascending order of chainage (forward direction) and the direction with descending order of chainage (reverse direction). First, start

Table 7.4 List of interchanges alongside

S/N	Chainage	Interchange name	Intervals (km)	Remarks
1	K1 + 112.5	Huashixia Interchange		Huashixia Town, G214
2	K39 + 786.866	Dawu Interchange	38.67	Dawu Township, S209
3	K83 + 343	Xueshan Interchange	43.56	Xueshan Township, S209
4	K102 + 065.614	Anyemaqen Interchange	18.72	S209
5	K131 + 293	Dongqingou Interchange	29.23	Dongqingou Township, S209
6	K154 + 377	Dawu West Interchange	23.08	Dawu Town, S209

from the initial operating speed V_0 of the road section, and then calculate the operating speed V_{85} according to the type of the divided road section, by straight section, longitudinal slope section, curve section, curved slope section, tunnel section, interchange section, etc.

i. The project is measured according to the uniform acceleration or deceleration movement on the straight road section. Expected speed on straight road section: when the design speed is 80 km/h, it is 110 km/h for passenger cars and 80 km/h for trucks; when the design speed is 60 km/h, it is 90 km/h for passenger cars and 75 km/h for trucks. The acceleration of passenger cars in this section is 0.15–0.5 m/s^2, and that of trucks is 0.2–0.25 m/s^2.
ii. The dynamic performance of vehicles is greatly affected by the altitude. For longitudinal slope sections, the impact of longitudinal slope reduction on the operating speed shall be considered, and the operating speed of longitudinal slope sections shall be calculated by using the reduced longitudinal slope. According to relevant research results, the calculation formula considering longitudinal slope reduction is as follows:

$$\Delta i = -2.061 \times 10^{-3} + 4.004 \times 10^{-6} H - 1.770 \times 10^{-10} H^2$$

where,

H altitude (m);
Δi longitudinal grade reduction.

iii. For curve segment: the "speed prediction model on horizontal curve" is used to calculate the midpoint and exit speed of the curve.
iv. For the curved slope section, the longitudinal slope reduced by considering the altitude factor shall be adopted, and the speed of the center point and exit point of the curved slope curve shall be calculated by using the "prediction model of the operating speed under the curved slope combination alignment" [For the above calculation model, see the corresponding sections of the *Specifications for Safety Evaluation of Highway Projects* (JTGB05-2015)].
v. For tunnel sections and interchange sections, the operating speed prediction model for tunnel sections in the *Code for Safety Evaluation of Highway Projects* (JTGB05-2015) is adopted for calculation of operating speed.

(2) **Evaluation Criteria**

The operating speed coordination evaluation includes the operating speed coordination evaluation of adjacent sections and the operating speed and design speed coordination evaluation of the same section. In the design, the operating speed shall be calculated and evaluated according to the two typical models of trucks and passenger cars.

(1) Evaluation on Operating Speed Coordination of Adjacent Sections

It adopts $|\Delta V_{85}|$, the absolute value of the operating speed difference, and $|\Delta I_V|$, the absolute value of operating speed gradient, between adjacent sections to evaluate. If $|\Delta V_{85}| < 10$ km/h and $|\Delta I_V| \leq 10$(km/h·m), it is considered that the operating speeds of adjacent sections are well coordinated; If 10 km/h $\leq |\Delta V_{85}| < 20$ km/h and $|\Delta I_V| \leq 10$(km/h·m), it is considered that the operating speeds of adjacent sections are relatively well coordinated, and the horizontal and longitudinal design of adjacent sections should be optimized when the adjacent sections are decelerating; If $|\Delta V_{85}| \geq 20$ km/h or $|\Delta I_V| > 10$(km/h·m), it is considered that the operating speeds of adjacent sections are poorly coordinated, and the horizontal and longitudinal design of adjacent sections shall be adjusted if are decelerating.

(2) Evaluation on the Coordination between Operating Speed and Design Speed of the Same Road Section

The difference between the operating speed and the design speed of the same section is used for this evaluation. When the difference between the operating speed and the design speed is less than 20 km/h, it is considered that the two speeds are well coordinated and consistent, and the relevant technical indicators are well matched with the actual operating speed; on the contrary, the operating speed of the same road section is poorly coordinated with the design speed, that is, the design consistency is poor. The horizontal and vertical alignment should be optimized, or the safety check should be carried out for the relevant technical indicators of the road route.

7.4 Checking of Horizontal and Vertical Geometric Indicators Based on Operating Speed

7.4.1 Checking of Horizontal Geometric Indicators

(1) **Radius of Circular Curve**

If the radius of circular curve in the project is less than the general minimum value corresponding to the operating speed, the horizontal alignment shall be optimized as far as possible according to the terrain conditions. If the conditions are limited, the traffic safety facilities design shall use speed control management facilities to remind drivers to control the speed. The inspection conclusions of the demonstration section are listed in Table 7.5.

(2) **Horizontal Curve Length**

It is necessary to check the horizontal curve length of the road sections along the line. When the general value of the horizontal curve does not meet the operating

Table 7.5 Sections with unsatisfying horizontal curves radii

Chainage	Direction	Design speed (km/h)	Radii of horizontal curves (m)	Operating speed (km/h)	Lateral force coefficient	Superelevation (%)	Required radii (m)
K27 + 003.53 ~ K27 + 612.566	Ascending	80	850	106	0.05	4	>990
K28 + 075.843 ~ K28 + 921.448	Ascending	80	450	100	0.05	6	>715
K50 + 685.031 ~ K51 + 072.919	Ascending	60	800	106	0.05	4	>990
K66 + 907.535 ~ K67 + 117.957	Ascending	60	740	107	0.05	4	>1000
K67 + 397.957 ~ K67 + 728.009	Ascending	60	720	106	0.05	4	>990
K68 + 005.009 ~ K68 + 192.304	Ascending	60	748	105	0.05	4	>970
K84 + 480.537 ~ K84 + 840.237	Ascending	60	400	98	0.05	6	>690
K85 + 565.239 ~ K85 + 943.289	Ascending	60	800	108	0.05	4	>1020
K88 + 389.925 ~ K88 + 765.876	Ascending	60	638	102	0.05	5	>820
K90 + 483.452 ~ K91 + 025.681	Ascending	60	600	104	0.05	5	>860
K93 + 856.012 ~ K94 + 120.211	Ascending	60	750	106	0.05	4	>990
K99 + 454.941 ~ K99 + 744.737	Ascending	60	600	104	0.05	3	>1070

(continued)

speed requirements, the curve length should be optimized as far as possible; The curve length shall be lengthened for the section that only meets the limit value requirements.

Table 7.5 (continued)

Chainage	Direction	Design speed (km/h)	Radii of horizontal curves (m)	Operating speed (km/h)	Lateral force coefficient	Superelevation (%)	Required radii (m)
K100 + 011.55 ~ K100 + 147.44	Ascending	60	450	98	0.05	4	>840
K101 + 236.58 ~ K101 + 491.06	Ascending	60	500	90	0.05	4	>710
K111 + 823.43 ~ K112 + 058.56	Ascending	60	730	106	0.05	3	>1100

7.4.2 Checking of Vertical Geometric Indicators

(1) **Maximum Grade**

The longitudinal design shall check the arrangement of the maximum grade corresponding to the operating speed of the road section along the line. The longitudinal slope shall be less than the gradient requirements that can be supported by the operating speed. For the longitudinal slope that does not meet the operating speed requirements, the gradient shall be optimized as far as possible. Table 7.6 is the list of inspection conclusions of the demonstration road section.

(2) **Maximum Grade Length**

Checking whether the design of the slope length of the road section along the line meets the requirements of the maximum slope length corresponding to the operating speed, and optimize the longitudinal slope length as far as possible for the longitudinal

Table 7.6 Sections with unsatisfying grades

S/N	Chainage	Grades (%)	Operating Speed (km/h)
1	K51 + 110 ~ K51 + 810	4.65	100
2	K100 + 800 ~ K101 + 485	5	95
3	K103 + 066.425 ~ K103 + 670	5	100
4	K104 + 005 ~ K104 + 410	4.5	100
5	K110 + 955 ~ K111 + 500	5	110
6	K111 + 825 ~ K112 + 561.807	5	100

Table 7.7 Sections with unsatisfying grade length

S/N	Chainage	Grade (%)	Length (m)	Operating Speed (km/h)
1	K78 + 270 ~ K79 + 010	5	740	110
2	K97 + 800 ~ K98 + 620	4	820	110
3	K100 + 800 ~ K101 + 485	5	685	100
4	K103 + 066.425 ~ K103 + 670	5	603.575	100
5	K109 + 220 ~ K110 + 210	4	990	100
6	K111 + 825 ~ K112 + 561.807	5	736	100
7	K112 + 840 ~ K113 + 600	5	760	110

Table 7.8 Vertical curves in specifications

S/N	Chainage		Radii (m)	Operating speed (km/h)
1	Crest	K75 + 151.962	8000	110
2		K112 + 840	8000	110
3		K137 + 375	9000	100
4		K141 + 249.128	4500	110
5		K150 + 530	8000	110

slope that does not meet the operating speed. Table 7.7 is the list of inspection conclusions of the demonstration road section.

For sections where the longitudinal grade and grade length cannot be optimized due to limited conditions, lane management and speed control measures shall be taken to remind vehicles to control their speed and to remind trucks and passenger cars to drive in separate lanes.

(3) **Vertical Curve Radius**

Checking whether the vertical curve radius of the road section along the line meets the minimum radius requirements of the vertical curve corresponding to the operating speed. When the general value of the vertical curve does not meet the operating speed requirements, the length of the vertical curve should be optimized as far as possible; the length of vertical curve shall be lengthened for sections that only meet the limit value requirements. Table 7.8 is the list of inspection conclusions of the demonstration road section.

7.4.3 Sight Distance Checking

The driving sight distance of highways at all levels shall meet the requirements of sight distance, and the conventional inspection methods mainly include the calculation method of maximum transverse clear distance and graphical method (that is, drawing sight distance envelope diagram). The former can check whether the plane sight distance at a certain position on the curve meets the requirements, and the latter can more accurately determine the range affecting the sight distance on the horizontal curve (or vertical curve). These two methods are simple and practical, but both are two-dimensional inspection methods, which cannot accurately calculate the three-dimensional space sight distance under the actual driving state.

Spatial sight distance refers to the farthest distance that the driver can actually see the object on the carriageway in the most unfavorable lane (possibly due to the influence of curves, vertical curves, roadside facilities, vegetation and obstacles) in the real three-dimensional environment of the highway area according to the requirements of the view point height and object point height. It can also be understood as the maximum intervisibility distance between the drivers' line of sight and the object point on the roadway without occlusion under the influence of highway ancillary structures.

Spatial sight distance detection is to find out the area where the sight distance does not meet the requirements by calculating the spatial distance that the existing highway itself can provide intervisibility and comparing it with the stopping sight distance required by the design and the stopping sight distance required by the running sight distance. Figures 7.4 and 7.5 are the spatial line of sight block diagram and the spatial rendering.

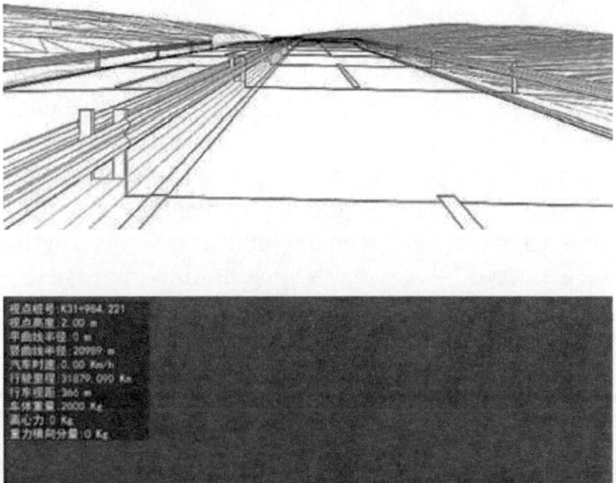

Fig. 7.4 Drawings of sight distance

Fig. 7.5 Rendering of sight distance

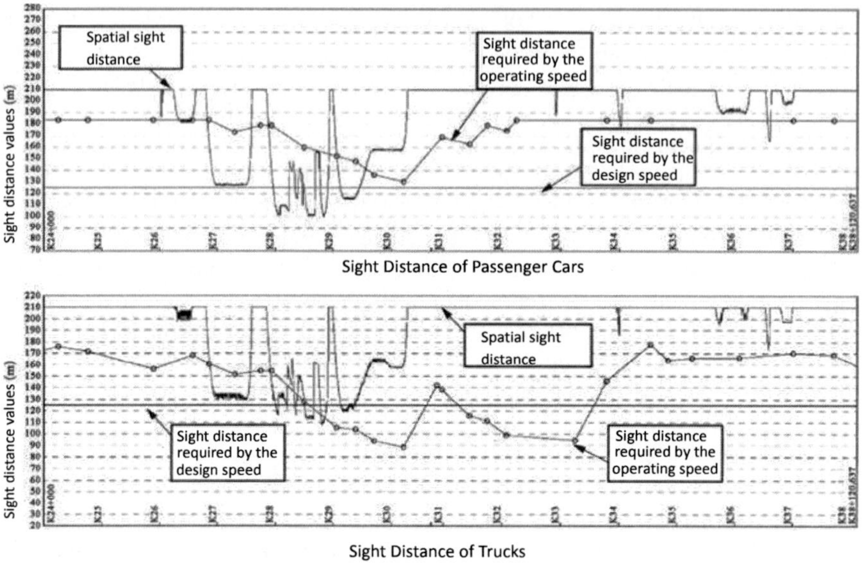

Fig. 7.6 Spatial sight distance comparison diagram

By comparing the technical indicators of the sight distance analysis chart (Fig. 7.6), it can be seen that the insufficient left turn sight distance is due to the insufficient radius of the left deflection curve, and the left guardrail blocks the sight of vehicles running in the inner lane; The insufficient sight distance of right turn is due to the insufficient radius of right deflection curve, and the right guardrail blocks the sight of vehicles running in the external traffic lane. For sections where the sight distance does not meet the requirements of operating speed/design speed, the horizontal and vertical alignment indicators shall be optimized or the sight distance shall be increased by moving the guardrail to increase the horizontal clear distance, and speed control and guidance facilities shall be arranged to remind drivers to control the speed.

7.5 Speed Limit Implementation Plan

The demonstration project in this chapter selects K27 + 000 ~ K42 + 000 section of Hua-Da Highway, with a total length of 15 km, to test the design technology of dynamic speed control (speed limit) of freeway under special environment. The road alignment is shown in Fig. 7.7.

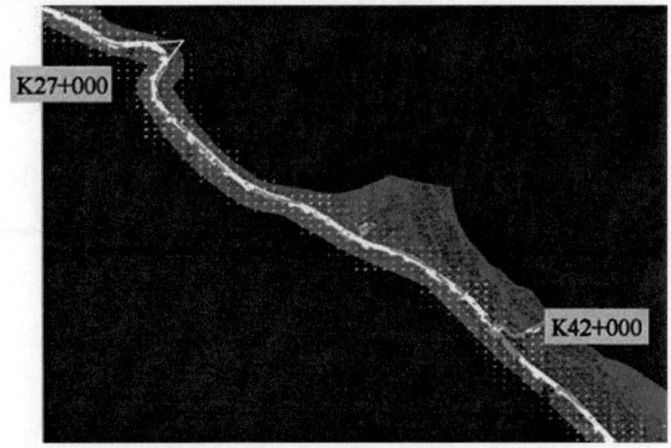

Fig. 7.7 Alignment of the demonstration section on hua-da highway

7.5.1 Main Contents and Technical Indicators of Demonstration Project

Based on the theoretical analysis of the comprehensive decision-making model of speed restriction in Chap. 6 of this book, the speed restriction standard and speed restriction implementation plan after reconstruction and expansion of Hua-Da Highway are formulated by using the relevant conclusions of the design technology of dynamic speed control (speed restriction) of freeway under the special environment mentioned above and combining with the actual design and construction of Hua-Da Highway.

Generally speaking, for freeways that have not been put into operation, the design speed should be taken as the speed limit basis, and the speed limit should be implemented for the whole line at the design speed. After a period of operation, collect the traffic volume, traffic accidents, operating speed, driver satisfaction and other data of the highway, adopt the process of formulating the highway speed limit plan in the operation stage, and then optimize and adjust the speed limit plan and speed control facilities. The general flow of speed limit for freeway not in operation is shown in Fig. 7.8.

7.5.2 Application of Speed Limit Comprehensive Decision-Making Model

According to the conclusion of the comprehensive decision-making model of speed limit in Chap. 6 of this book, the calculated speed limit value and recommended speed limit value are obtained by trial calculation of formula (6–27) model (the specific calculation process is omitted), as shown in Table 7.9.

In specific application, attention shall be paid to: for the sections with multiple curves on the upper and lower slopes controlled by the alignment, the speed limit value shall be calculated considering the constraints of the design speed; in consideration of the attenuation of dynamic performance and braking performance of trucks

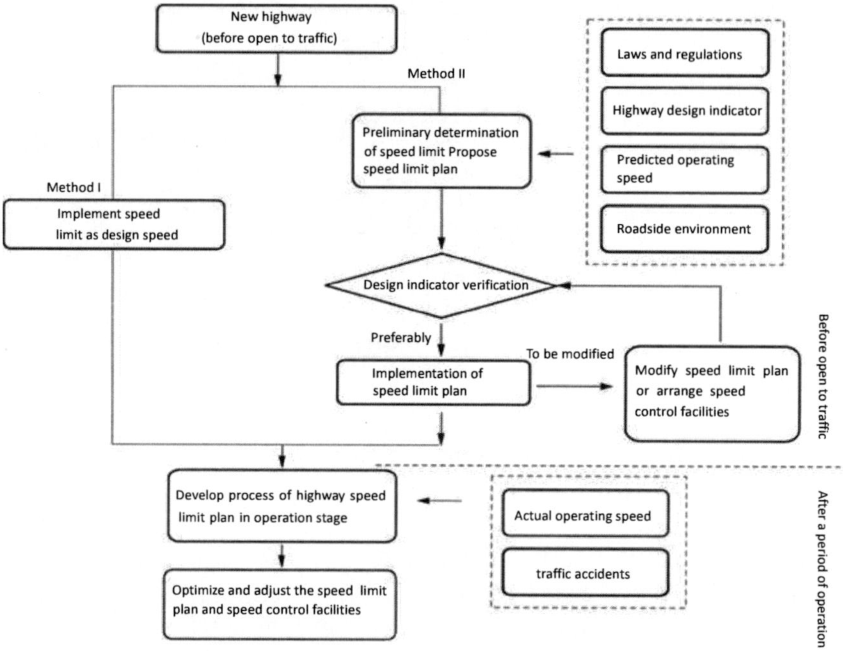

Fig. 7.8 General process of speed limit for a new highway before opening to traffic

on plateau, the speed limit value of trucks on uphill and downhill sections shall be reduced.

7.5.3 *Specific Implementation Plan*

Taking the relevant conclusions of the book on freeway speed control technology under special circumstances into account, and in combination with the characteristics of freeway driving under the conditions of high altitude, low pressure and oxygen deficiency on the plateau, the drivers' psychological and physiological characteristics, as well as the vehicle performance under the oxygen deficient conditions, and by analyzing the alignment, altitude and environmental characteristics of the demonstration section of Hua-Da Highway, and using the results of the research model of comprehensive speed limit decision in this chapter of the book, The following recommended speed limit plan is formulated from the aspects of vehicle operation safety and efficiency dynamic balance.

(1) **Speed Limit Plans Categorized by Vehicle Type**

i. Method and principle. Speed limit by vehicle type is a relatively new speed limit method proposed from the perspective of improving safety and operating efficiency, that is, the speed limit values of passenger cars and trucks

Table 7.9 Speed limit calculation of the demonstration section on hua-da highway

Chainage (km)	Altitude (m)	Design speed (km/h)	Calculated value for passenger cars (km/h)	Proposed value for passenger cars (km/h)	Calculated value for trucks (km/h)	Proposed value for trucks (km/h)
K26 + 570	3985.214	80	93.033	80	75.705	60
K27 + 635	3972.221	80	93.109	80	75.827	60
K28 + 105	3963.244	80	93.160	80	75.911	60
K28 + 655	3968.139	80	93.132	80	75.865	60
K29 + 185	3982.502	80	93.049	80	75.730	60
K29 + 455	3988.631	80	93.013	80	75.672	80
K30 + 205	4016.006	80	92.841	80	75.404	80
K30 + 980	4025.030	80	92.781	80	75.313	80
K31 + 450	4042.251	80	92.663	80	75.136	80
K31 + 750	4049.751	80	92.610	80	75.058	80
K32 + 085	4061.476	80	92.524	80	74.934	80
K33 + 250	4088.504	80	92.318	80	74.640	80
K33 + 790	4075.976	80	92.415	80	74.777	80
K34 + 540	4045.976	80	92.636	80	75.097	80
K34 + 845	4038.351	80	92.690	80	75.177	80
K35 + 260	4035.031	80	92.713	80	75.211	80
K36 + 075	4010.582	80	92.876	80	75.458	80

are considered separately, and trucks are prompted to drive on the right side through traffic signs.

ii. Features. The comprehensive consideration of the differences in operating characteristics and braking characteristics between passenger cars and trucks, as well as the differences in speed requirements of passenger car and truck drivers, is a speed limit method conducive to safety and traffic efficiency. At the same time, it is also conducive for drivers to identify and comply with the speed limit, reducing the difficulty of law enforcement.

(2) **Speed Limit Plans Categorized by Road Sections**

i. Method and principle. Based on the speed limit plan of the existing road section, combined with the measured data and driver demand, and referring to the speed limit value provided by the existing road traffic safety law, the speed limit plan is proposed from the perspective of paying more attention to traffic efficiency and giving full play to the road capacity.
ii. Features. Considering the characteristics of the road section and the speed demand of drivers, it is conducive to social harmony and improving road traffic efficiency, and can improve social satisfaction within a certain range.
iii. Speed limit value. According to the model calculation speed and relevant analysis and research, the speed limit of passenger car is 80 km/h, that of truck is 60 km/h, and that of special sections is 40–80 km/h.

(3) **Speed Limit Plans for Special Road Sections**

i. Speed limit plan for flat curve line section. For the road section of the demonstration project where the horizontal curve radius is less than the general value corresponding to the design speed but greater than the limit value, warning signs, sharp turn warning signs and road vibration strips shall be arranged 50 m in advance to control the speed.
ii. Speed limit plan for long and steep uphill and downhill sections. For long and steep uphill and downhill sections, a uniform speed limit of 60 km/h shall be considered.

(4) **Overall Speed Limit Plans**

To sum up, see Table 7.10 for the speed limit values and methods of specific sections.

Table 7.10 Speed limit plans

Speed limit for the entire section	Limit value (km/h)	Speed limit for partial section	Limit value (km/h)
K27 + 000 ~ K42 + 000	80 km/h for Passenger car; 60 km/h for Trucks; Trucks use the right lane	K29 + 455 ~ K30 + 205, K30 + 979.500 ~ K31 + 450	60 km/h for long and steep upward section
		K33 + 790 ~ K34 + 540, K37 + 015 ~ K37 + 715	60 km/h for long and steep downward section

7.5.4 Arrangement of Speed Limit Signs

Arrange the speed limit signs according to the speed limit plans:

i. Add speed limit signs by vehicle type near the end of the acceleration lane at the entrance of the interchange;
ii. Gantry type or attached type speed limit signs shall be added in general sections of freeway;
iii. Add the sign of "speed measurement area ahead" at the section easy to overspeed;
iv. Add speed limit signs for different vehicles on long and steep downhill sections;
v. Add unified speed limit signs on long and steep uphill sections.

7.5.5 Other Speed Management Suggested Plans

(1) **Speed Management with Adverse Weather Conditions**

Hua-Da Highway is located in the plateau area with an average altitude of more than 4,000 m. It is characterized by low pressure and oxygen deficiency, harsh natural environment, frequent rain, fog, snow and other adverse weather conditions, resulting in wet pavement, unclear vision and certain potential safety hazards. How to control and manage the speed in bad weather conditions such as rain, fog and snow is an important and practical problem. It is suggested to use the existing variable speed limit information board to display information in time under bad weather conditions such as rain, fog and snow, mainly displaying speed limit information.

(2) **Improve Supporting Safety Facilities**

The arrangement of traffic signs, markings, warning lights, etc. is an important part of freeway speed management. Improving the supporting safety facilities can prompt drivers timely and effectively, so that they can make timely and accurate judgments on the road conditions. It plays an important role in ensuring the safety of highway driving, especially improving the driving safety after speed limit.

The above speed limit and speed control plan has put forward relevant layout suggestions on traffic signs, markings, etc., and added safety facilities such as sharp turn signs, distance confirmation signs, signs for trucks to pull to the right, visual deceleration markings, yellow flashing lights, etc., which can meet the requirements for safe driving of vehicles after the road section is opened for operation. It is suggested that Hua-Da Highway implement the speed limit plan gradually by stages, and promote it after verifying the effectiveness:

i. Strengthen the management of key speed restricted sections. Patrol vehicles can be used to patrol on the road or radar speed detection and electronic police equipment can be installed;
ii. Strengthen the public's understanding of the Road Traffic Safety Law, let the public understand the relationship between speed and safety, understand the

determination of speed limit, abide by the speed limit regulations, and fasten the seat belt as required;

iii. Strengthen management, prohibit low-speed vehicles, motorcycles, non-motor vehicles and pedestrians from entering the freeway, and strictly manage overloading.